STUDY GUIDE

CHARLES M. SEIGER

ESSENTIALS *of* ANATOMY & PHYSIOLOGY

SECOND EDITION

MARTINI ◆ BARTHOLOMEW

PRENTICE HALL, Upper Saddle River, NJ 07458

Editor in Chief: Paul F. Corey
Supplement Editor: Damian Hill
Special Projects Manager: Barbara A. Murray
Production Editor: James Buckley
Supplement Cover Manager: Paul Gourhan
Supplement Cover Designer: Liz Nemeth
Manufacturing Manager: Trudy Pisciotti

Upper Saddle River, NJ 07458

Printed in the United States of America

10 9 8 7 6 5 4 3 2 1

ISBN 0-13-082193-4

Prentice-Hall International (UK) Limited, London
Prentice-Hall of Australia Pty. Limited, Sydney
Prentice-Hall Canada, Inc., Toronto
Prentice-Hall Hispanoamericana, S.A., Mexico
Prentice-Hall of India Private Limited, New Delhi
Prentice-Hall (Singapore) Pte. Ltd.
Prentice-Hall of Japan, Inc., Tokyo
Editora Prentice-Hall do Brazil, Ltda., Rio de Janeiro

Contents

Preface

This revised Study Guide has evolved from many years of teaching and writing experience. In preparing it, I have tried to incorporate my own classroom experience, evaluate student input about the ways they learn, draw on my discussions with colleagues about the dynamics of teaching and learning, and pay close attention to the comments of users and reviewers of previous editions and other publications. The result is a resource that I hope, when used in conjunction with *Essentials of Anatomy and Physiology, 2nd Edition*, by Frederic H. Martini and Edwin F. Bartholomew, will not only excite students about this course, but also provide valuable reinforcement of difficult concepts.

The sequence of topics within the Study Guide parallels that of the text and incorporates *Objective Based Questions* along with *Chapter Comprehensive Exercises.* The objective based questions are keyed directly to chapter objectives and include Multiple Choice, Completion, True–False Questions, and Labeling Exercises. They are designed to offer a review of facts and terms used in Anatomy & Physiology. The chapter comprehensive exercises include Word Elimination, Matching, Concept Mapping, Crossword Puzzles, and Short Answer Questions, all geared to a more provocative thought-processing procedure which encourages and develops the student's ability to apply information and think critically. Significant changes have been made in many exercises as a result of reviewer feedback.

I would like to acknowledge the students and colleagues who have had an impact on this work through their careful and thoughtful reviews. Special thanks goes to my editors, Byron Smith and James Buckley, for their support and ideas to complete this time-intensive project.

Any errors or omissions found by the reader are attributable to the author, rather than to the reviewers. Readers with comments, suggestions, relevant reprints, or corrections, should contact me at the address below.

Charles M. Seiger
c/o Prentice Hall
One Lake Street
Upper Saddle River, New Jersey 07458

CHAPTER

1

An Introduction to Anatomy and Physiology

Overview

For centuries, the study of the human body has aroused the curiosity of the human mind, just as it does today. A few years ago our knowledge of the human body was limited to what our senses could perceive; however, modern technology has changed the way we view and understand the physical and chemical makeup of living organisms. Even today, for many of you the study of anatomy and physiology may be an "adventure into the unknown." Are you curious to know the what, why, and how of your body? The knowledge of the body structures described by anatomists, and understanding the physical and chemical processes developed by physiologists, may provide the framework and foundation for your future career, your own personal health, or perhaps some necessary clinical applications.

Chapter 1 is an introduction to Anatomy & Physiology, citing the basic functions of living things, defining the various specialties of Anatomy & Physiology, and examining the major levels of organization in living organisms. The organ systems and their components are identified and the significance of homeostasis is explained. The chapter concludes with the use of anatomical terms to describe body sections, body regions, and major body cavities.

Review of Chapter Objectives

1. Describe the basic functions of living organisms.
2. Define anatomy and physiology and describe the various specialties of each discipline.
3. Identify the major levels of organization in living organisms from the simplest to the most complex.
4. Identify the organ systems of the human body, their functions, and the major components of each system.
5. Explain the significance of homeostasis.
6. Describe how positive and negative feedback are involved in homeostatic regulation.
7. Use anatomical terms to describe body sections, body regions, and relative positions.
8. Identify the major body cavities and their subdivisions.

Part I: Objective Based Questions

Objective 1 Describe the basic functions of living organisms.

_____ 1. The function which refers to all the chemical operations occurring in the body is

a. responsiveness
b. metabolism
c. reproduction
d. growth

_____ 2. The capacity to respond to an immediate change in an organism's environment is

a. movement
b. differentiation
c. adaptability
d. metabolism

_____ 3. Differentiation of cells during development is directly related to the function of

a. growth
b. metabolism
c. reproduction
d. adaptability

_____ 4. The term used to refer to the absorption, transport, and use of oxygen by cells is

a. excretion
b. breathing
c. metabolism
d. respiration

Objective 2 Define anatomy and physiology and describe the various specialties of each discipline.

_____ 1. The word structures relates to the study of _____________, while the word function relates to the science of _____________.

a. physiology; anatomy
b. cytology; histology
c. anatomy; physiology
d. biology; chemistry

_____ 2. The microscopic anatomy that analyzes the internal structure of individual cells is

a. physiology
b. cytology
c. microbiology
d. histology

_____ 3. The specialized science that involves the study and examination of tissues is

a. histology
b. cytology
c. cell biology
d. systemic anatomy

Objective 3 Identify the major levels of organization in living organisms from the simplest to the most complex.

_____ 1. The highest level of organization relative to living things is

a. tissue level
b. organ-system level
c. organismal level
d. cellular level

_____ 2. In the levels of organization, the level preceding the tissue level is the

a. cellular level
b. molecular level
c. organ level
d. systems level

_____ 3. The smallest stable units of matter are the

a. electrons
b. protons
c. neutrons
d. atoms

_____ 4. An example of the organ system level of organization is the

a. lungs
b. cardiovascular
c. heart
d. kidneys

Objective 4 Identify the organ-systems of the human body, their functions, and the major components of each system.

_____ 1. The system responsible for directing long-term changes in the activities of other organ systems is the ______________________ system.

a. nervous
b. lymphatic
c. reproductive
d. endocrine

_____ 2. The system responsible for defense against infection and disease is the

a. nervous
b. endocrine
c. lymphatic
d. integumentary

_____ 3. The organ system responsible for protection from environmental hazards and involved in temperature control is the

a. endocrine
b. respiratory
c. integumentary
d. cardiovascular

Labeling Exercise

The following figures represent the various body organ systems. Identify and name each organ system by filling in the blank below the drawing. Label the organs listed below in each system. Use these instructions for each one of the organ systems drawings.

Figure 1-1 Organ Systems

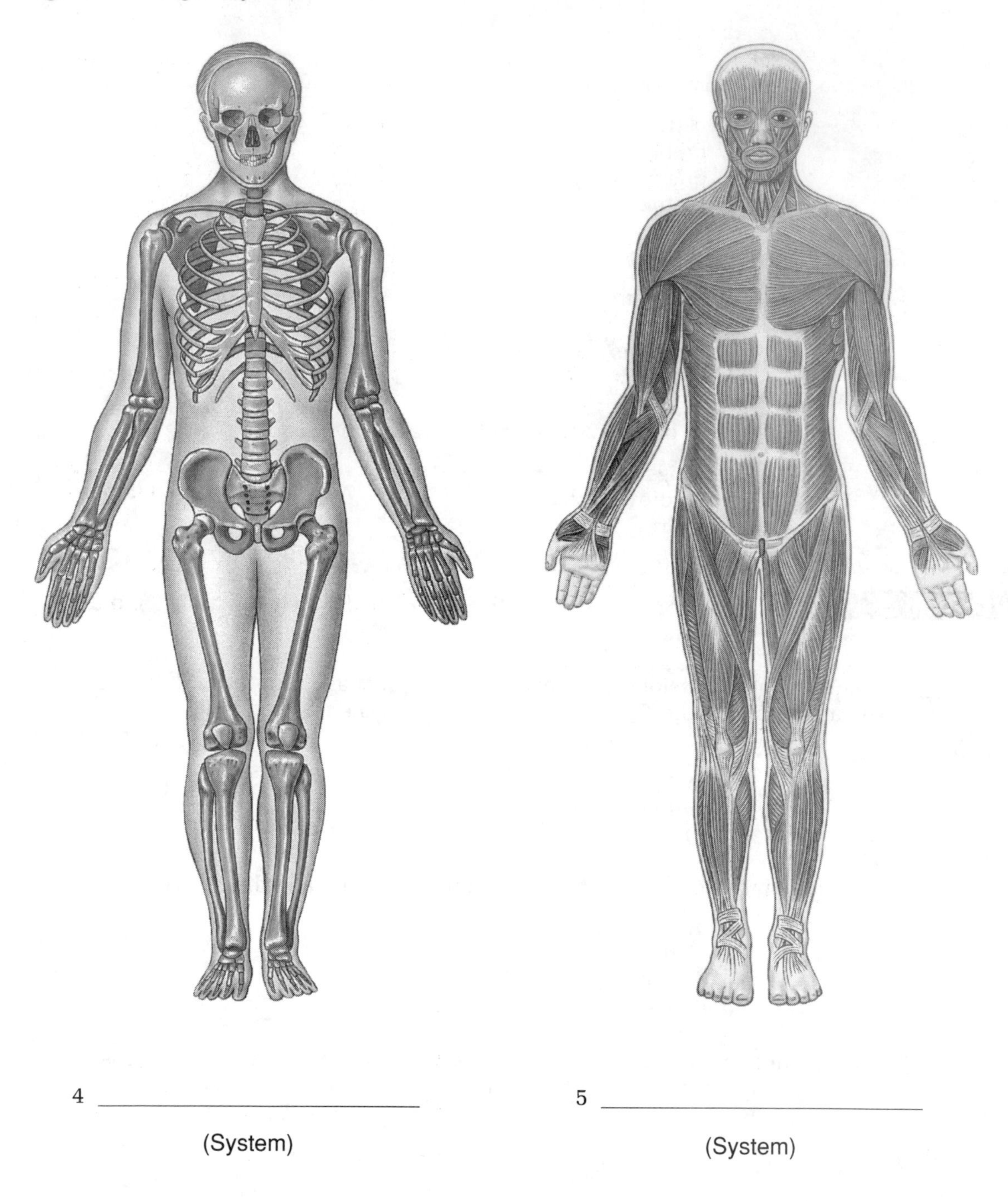

Figure 1-1 Organ Systems, continued

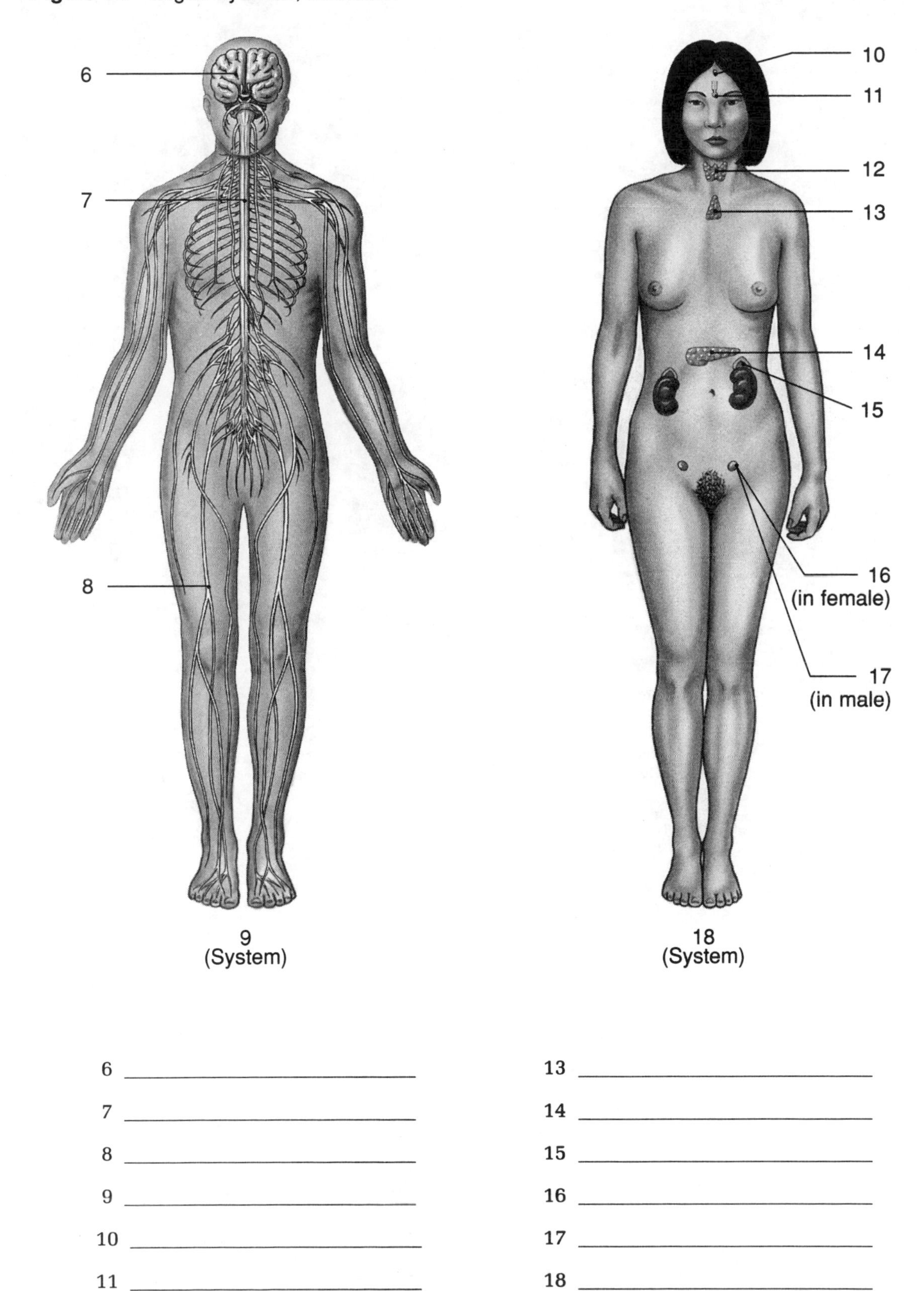

6 ____________________

7 ____________________

8 ____________________

9 ____________________

10 ____________________

11 ____________________

12 ____________________

13 ____________________

14 ____________________

15 ____________________

16 ____________________

17 ____________________

18 ____________________

Figure 1-1 Organ Systems, continued

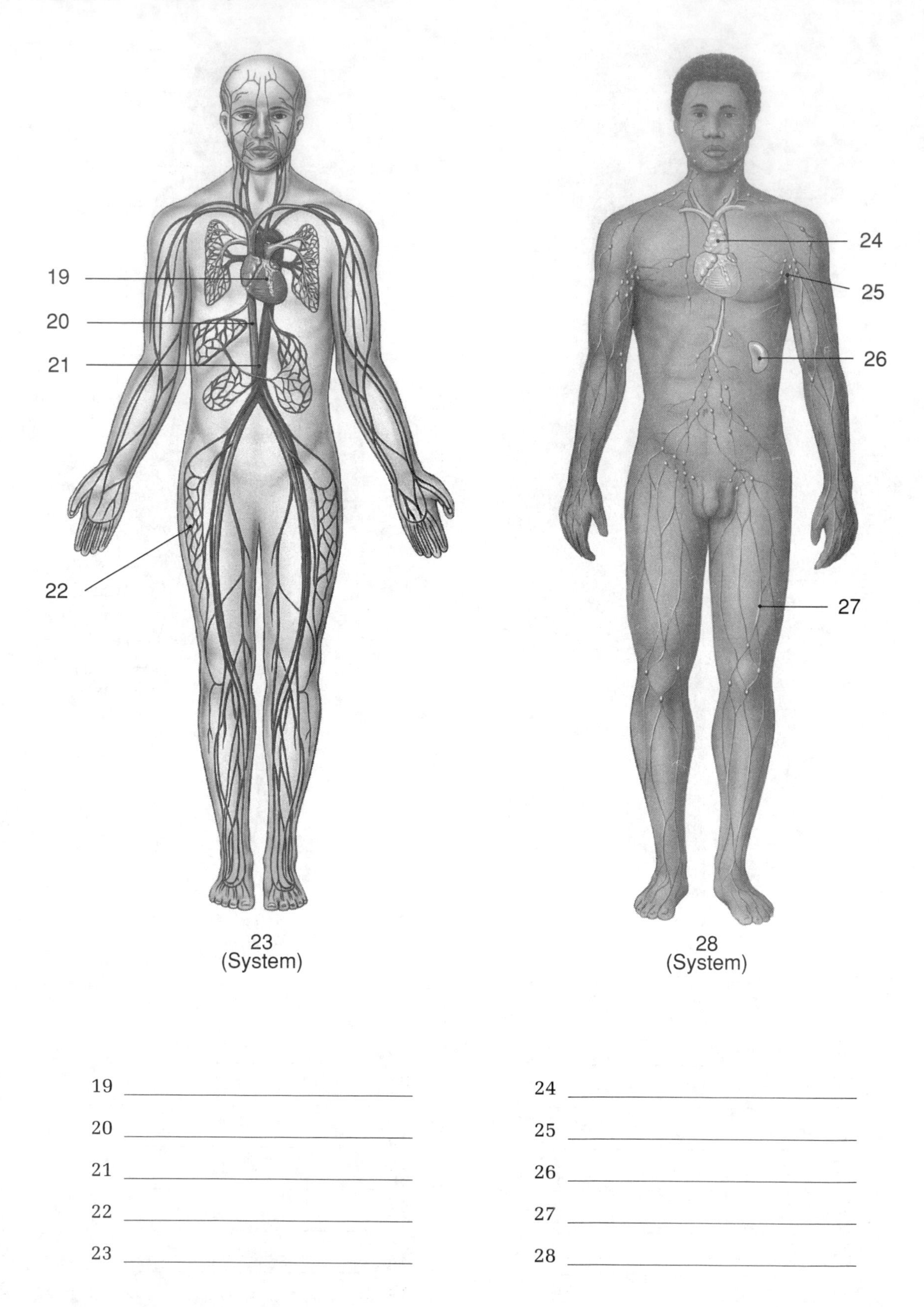

19 ______________________________

20 ______________________________

21 ______________________________

22 ______________________________

23 ______________________________

24 ______________________________

25 ______________________________

26 ______________________________

27 ______________________________

28 ______________________________

Figure 1-1 Organ Systems, continued

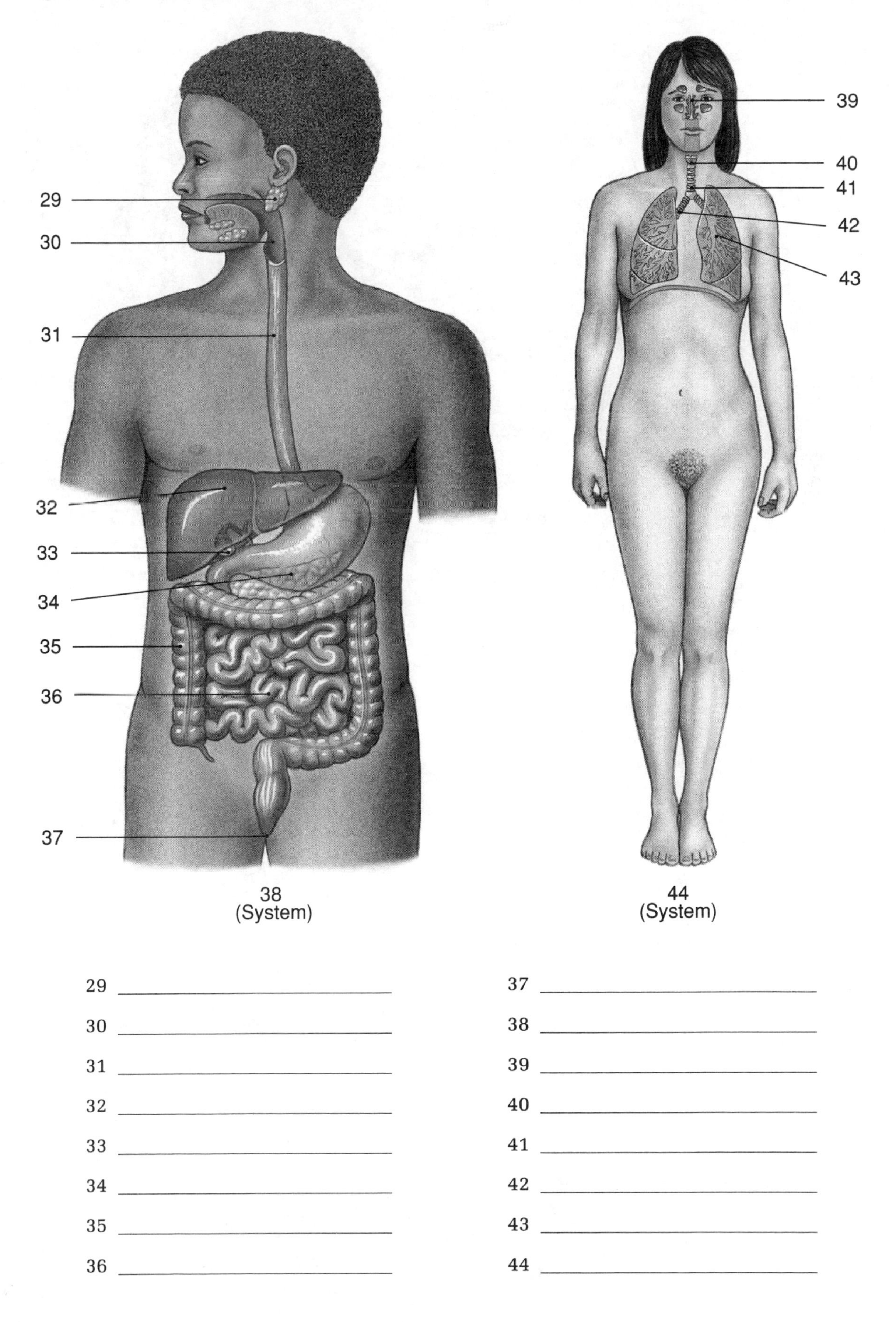

29 ______________________

30 ______________________

31 ______________________

32 ______________________

33 ______________________

34 ______________________

35 ______________________

36 ______________________

37 ______________________

38 ______________________

39 ______________________

40 ______________________

41 ______________________

42 ______________________

43 ______________________

44 ______________________

Figure 1-1 Organ Systems, continued

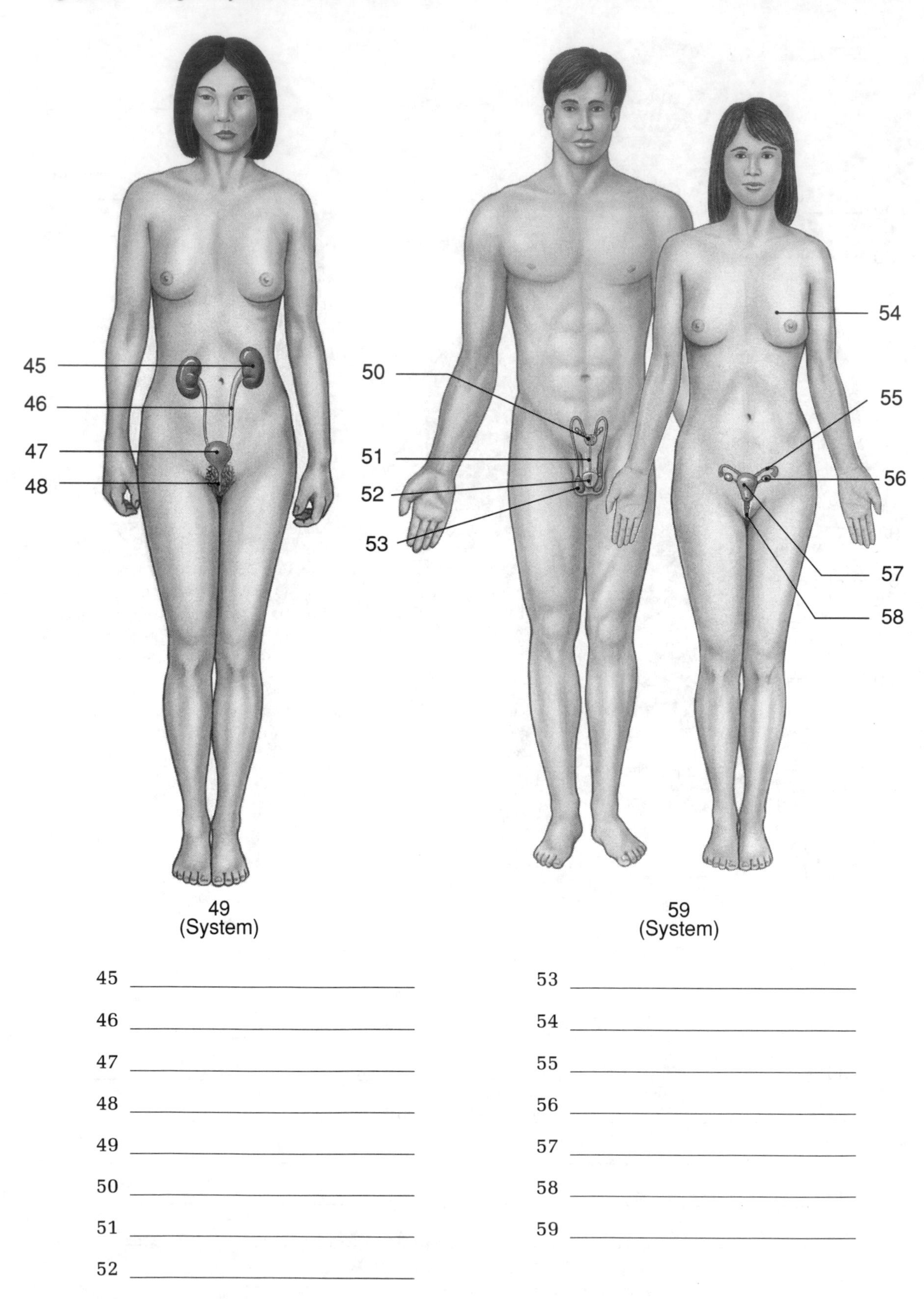

45 ______________________

46 ______________________

47 ______________________

48 ______________________

49 ______________________

50 ______________________

51 ______________________

52 ______________________

53 ______________________

54 ______________________

55 ______________________

56 ______________________

57 ______________________

58 ______________________

59 ______________________

Objective 5 Explain the significance of homeostasis.

_____ 1. The existence of a stable internal environment in the body is called

a. homeopathy
b. homeostatis
c. metabolism
d. feedback

_____ 2. Homeostatic regulation usually involves a receptor, a _______________ and an effector.

a. stimulus
b. response
c. control center
d. feedback

_____ 3. The mechanisms that control homeostasis are activated in response to

a. a stimulus
b. the control center
c. feedback mechanism
d. all of the above

_____ 4. Which one of the following statements is true?

a. When our organs begin to malfunction, the mechanisms that control homeostasis will begin to fail. We then become ill.
b. When the mechanisms that control homeostasis begin to fail, our organs malfunction. We will then become sick.
c. When we get sick, our organs begin to malfunction. The mechanisms that control homeostasis will then fail.
d. When we get sick, the mechanisms that control homeostasis begin to fail. Our organs will then begin to malfunction.

Objective 6 Describe how positive and negative feedback are involved in homeostatic regulation.

_____ 1. The regulatory mechanism that occurs when a variation outside of normal limits triggers an automatic response that corrects the situation is

a. negative feedback
b. positive feedback
c. effector regulation
d. receptor regulation

_____ 2. When the initial stimulus produces a response that reinforces the stimulus, the mechanism is called

a. thermoregulation
b. positive feedback
c. set-point integration
d. negative feedback

_____ 3. An example of a positive feedback mechanism in the body is

a. dilation and constriction of blood vessels
b. fluctuations of body temperature
c. blood clotting
d. response of the pupil of the eye to light

_____ 4. When body temperature rises, a center in the brain initiates physiological changes to decrease the body temperature. This is an example of

a. positive feedback
b. set-point integration
c. receptor regulation
d. negative feedback

_____ 5. Increasingly forceful labor contractions that lead to childbirth are an example of

a. negative feedback
b. positive feedback
c. effector regulation
d. receptor regulation

6. The system which responds rapidly, resulting in short-term changes to maintain homeostatis, is the __________________ system.

7. The system which produces effects that last for days or longer to maintain homeostatis is the __________________ system.

Objective 7 Use anatomical terms to describe body sections, body regions, and relative positions.

_____ 1. A person lying in the anatomical position is said to be

a. prone
b. recumbent
c. inclined
d. supine

_____ 2. The plane that divides the body into left and right sections is

a. sagittal
b. coronal
c. transverse
d. frontal

_____ 3. When viewing a cross-section of body tissue, the cut is along a __________________ plane.

a. mid-sagittal
b. coronal
c. transverse
d. frontal

_____ 4. The anatomical term that refers to the back of the human body is

a. ventral
b. posterior
c. coronal
d. cephalic

5. The plane that divides the body into anterior and posterior sections is the __________________.

6. The groin is anatomically referred to as the ______________________ region.

7. The buttocks is anatomically referred to as the ______________________ region.

8. The elbow is ______________________ to the shoulder.

9. Label the abdominopelvic quadrants and regions in the drawing below. Use the following terms to complete the exercise.

Figure 1-2 Abdominopelvic Quadrants and Regions

Umbilical region	**R. Hypochondrial Region**	**Hypogastric region**
Lumbar region	**L. Hypochondrial region**	**R. Lumbar region**
Iliac region	**Epigastric region**	**L. Iliac region**

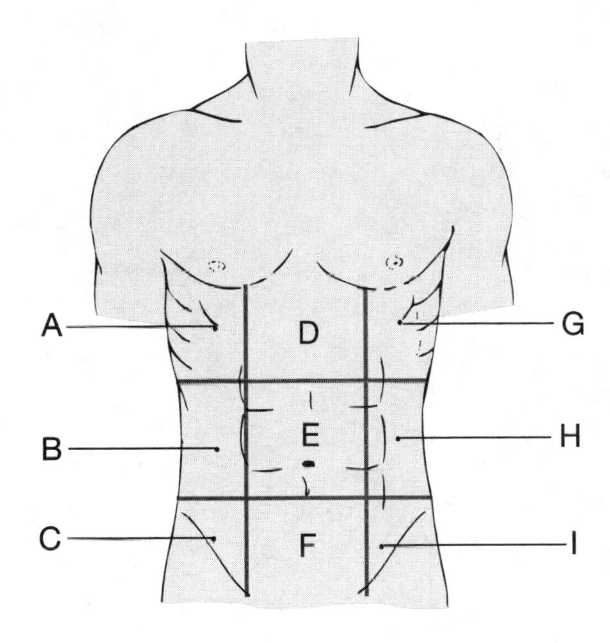

A ______________________

B ______________________

C ______________________

D ______________________

E ______________________

F ______________________

G ______________________

H ______________________

I ______________________

10. Label the body regions on the drawing below. Select the correct term from the following list. Place your answers in the spaces provided.

ANTERIOR

Orbital	Femoral	Umbilical
Patellar	Thorax	Palmar
Buccal	Pubic	Brachial
Axillary	Cubital	Abdominal

POSTERIOR

Calf	Deltoid	Gluteal
Lumbar	Popliteal	Occipital
Scapular		

Figure 1-3 Body Regions

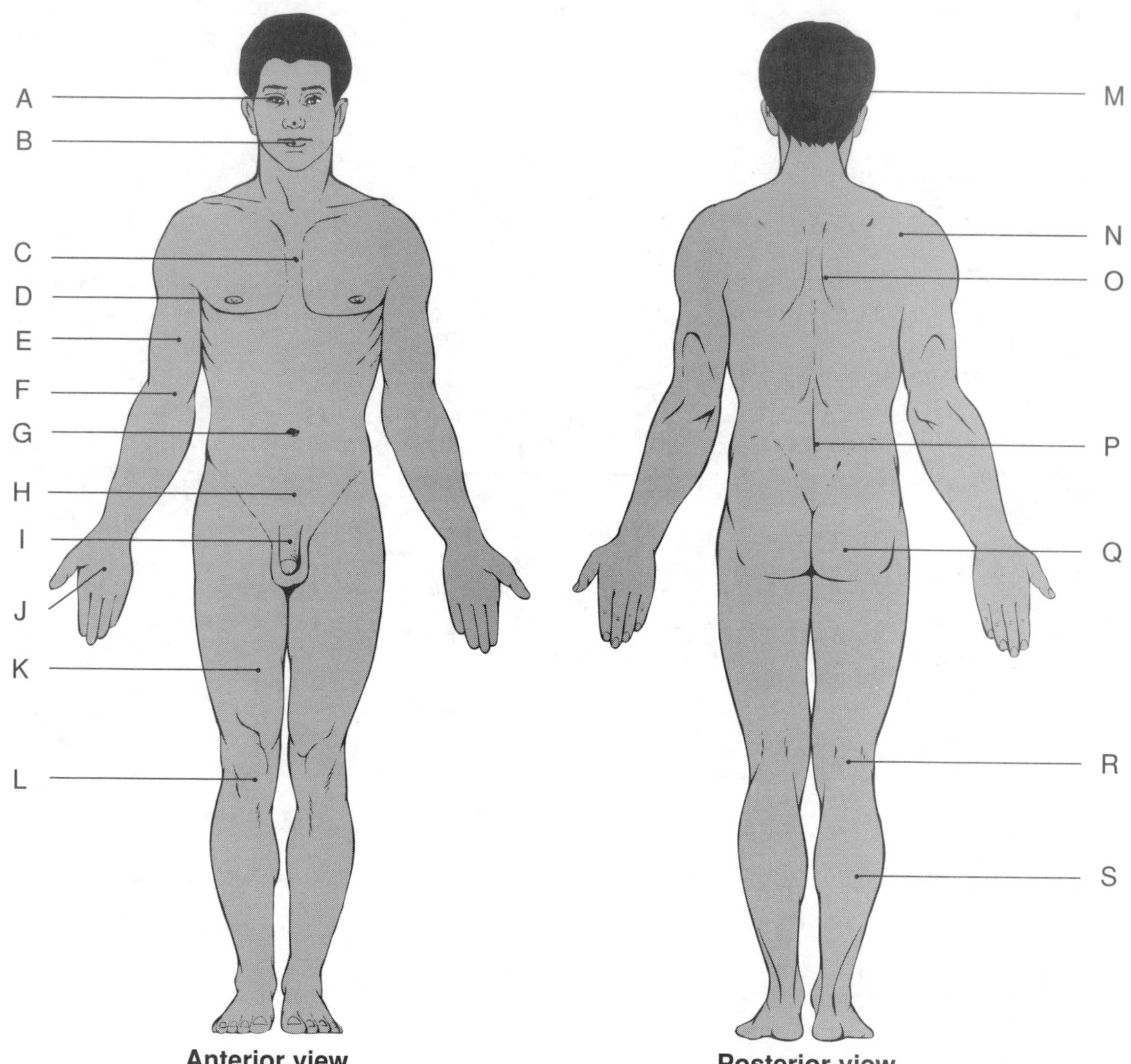

A ______________________ H ______________________ N ______________________

B ______________________ I ______________________ O ______________________

C ______________________ J ______________________ P ______________________

D ______________________ K ______________________ Q ______________________

E ______________________ L ______________________ R ______________________

F ______________________ M ______________________ S ______________________

G ______________________

Objective 8 Identify the major body cavities and their subdivisions.

_____ 1. The dorsal body cavity is subdivided into the

a. abdominal; pelvic
b. pleural; pericardial
c. cranial; spinal
d. a and b only

_____ 2. The ventral body cavity is divided into a superior ____________________ cavity and an inferior ____________________ cavity.

a. abdominal; pelvic
b. thoracic; abdominopelvic
c. pericardial; peritoneal
d. visceral; parietal

_____ 3. The serous membrane lining the abdominopelvic cavity is the

a. peritoneum
b. mediastinum
c. parietal pericardium
d. visceral pleura

_____ 4. The organs in the ventral body cavity make up the

a. thorax
b. viscera
c. abdomen
d. pelvis

5. The double sheets of serous membrane lining the peritoneal cavity are called ____________________.

6. The muscle that separates the abdominopelvic cavity from the pleural cavity is the ________________.

7. Label the body cavities in the drawing below. Use the following terms to complete the exercise.

Spinal cavity
Pleural cavity
Ventral body cavity
Abdominal cavity
Diaphragm
Dorsal body cavity
Pericardial cavity
Cranial cavity
Pelvic cavity
Abdominopelvic cavity

Figure 1-4 Body Cavities

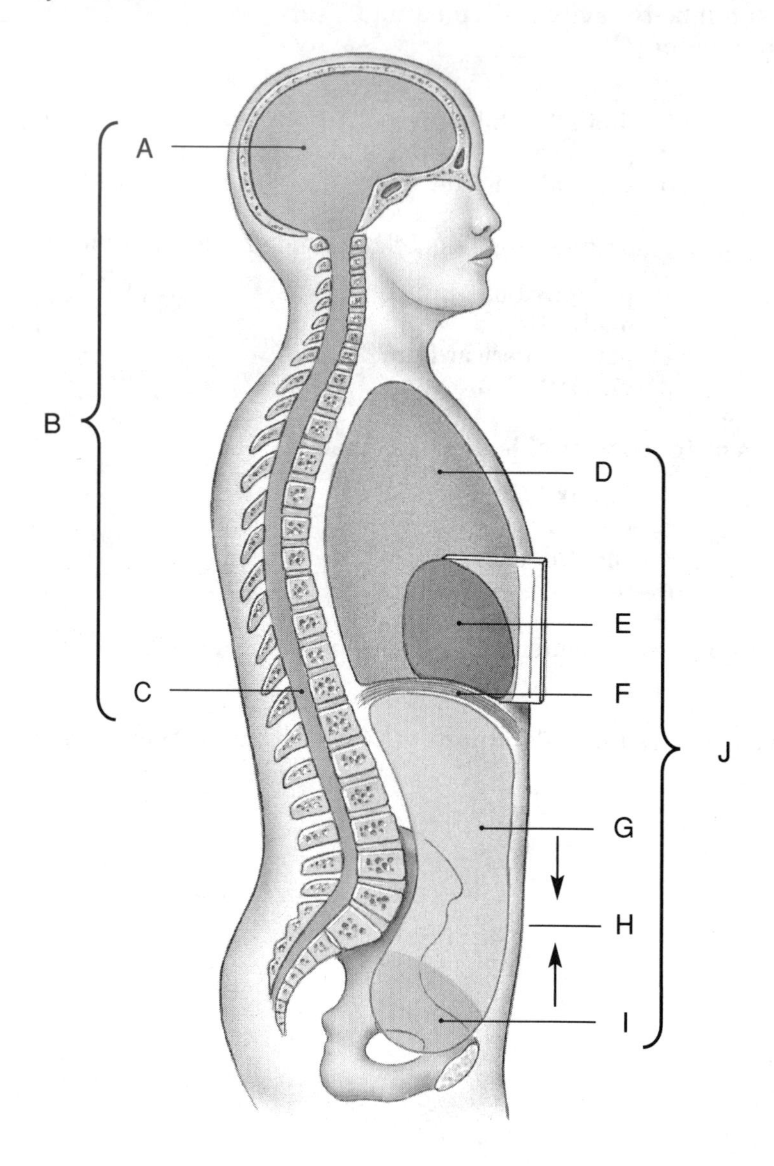

A ______________________

B ______________________

C ______________________

D ______________________

E ______________________

F ______________________

G ______________________

H ______________________

I ______________________

J ______________________

Part II: Chapter Comprehensive Exercises

A. Word Elimination

Circle the term that does not belong in each of the following groupings.

1. responsiveness growth reproduction biology movement
2. anatomy physiology cytology histology organism
3. cell digestion tissue organ organism
4. organism endocrine skeletal nervous reproductive
5. receptor effector temperature stimulus response
6. cephalic supine cervical axillary brachial
7. transverse frontal sagittal coronal prone
8. anterior posterior coronal cephalic proximal
9. caudal pelvic pericardial thoracic abdominal
10. pericardium mediastinum pleura peritoneum mesentery

B. Matching

Match the terms in Column "B" with the terms in Column "A." Use letters for answers in the spaces provided.

COLUMN A	COLUMN B
_____1. responsiveness	A. human being
_____2. renal physiology	B. labor contractions
_____3. organism	C. adjustments in physiological systems
_____4. muscular system	D. forms outer wall of body cavity
_____5. homeostatic regulation	E. head region
_____6. positive feedback	F. covers surfaces of internal organs
_____7. lying face down	G. irritability
_____8. cephalic	H. locomotion, support, produces heat
_____9. parietal	I. prone
_____10. visceral	J. kidney function

C. Concept Map

The concept map summarizes and organizes the basic information on body cavities in Chapter 1. Using the following terms, fill in the circled, numbered, blank spaces to complete the concept map. Follow the numbers which comply with the organization of the concept map.

Pelvic Cavity	**Spinal Cord**	**Cranial Cavity**
Heart	**Abdominopelvic Cavity**	**Two Pleural Cavities**

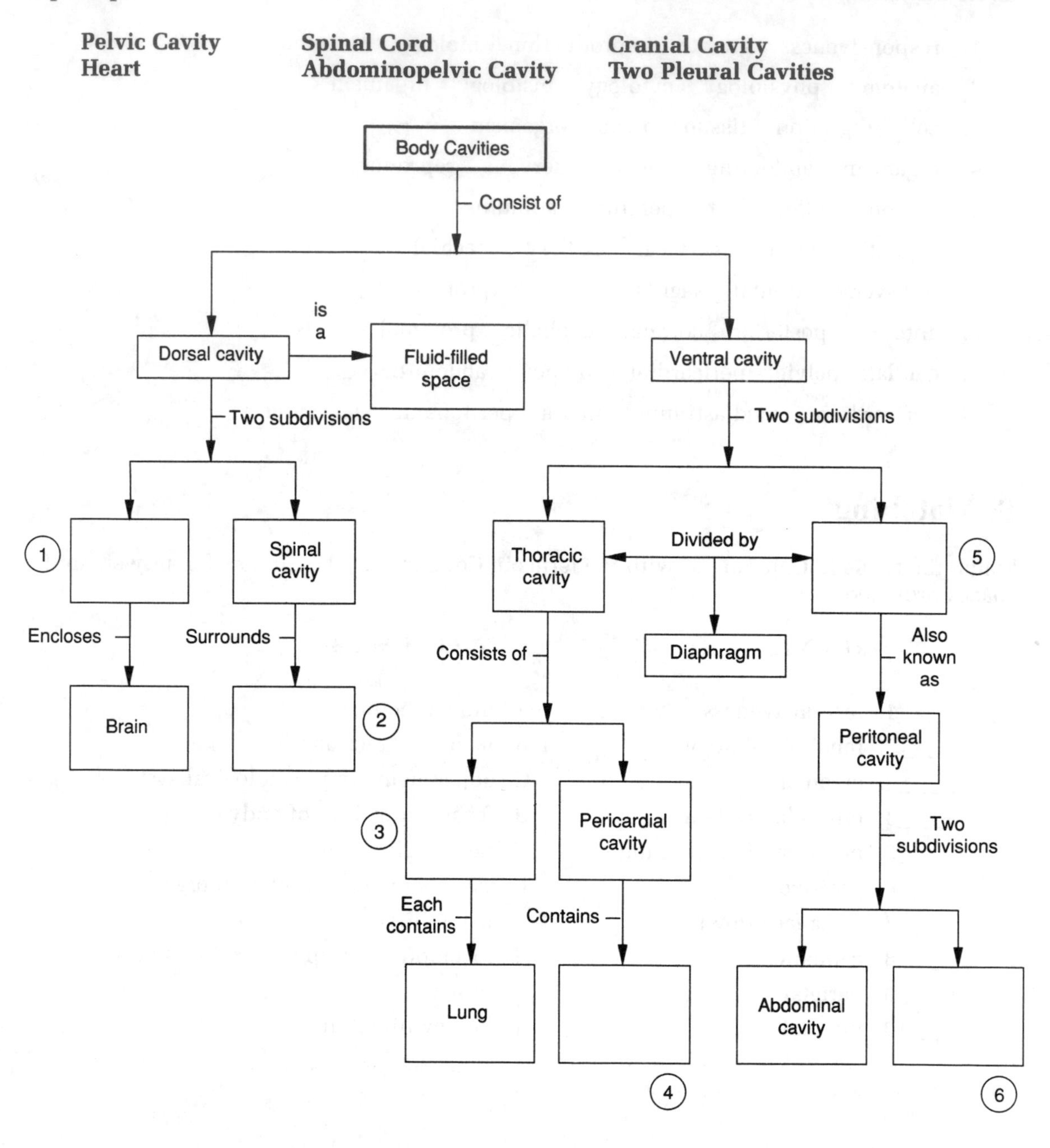

D. Crossword Puzzle

The following crossword puzzle reviews the material in Chapter 1. To complete the puzzle you must know the answers to the clues given, and you must be able to spell the terms correctly.

ACROSS

4. When the body is healthy, it is said to be in _____.
7. A lot of the scientific words we use are derived from the ____ language.
8. The body region typically referred to as the elbow is anatomically called the _____ region.
9. The visceral _____ membrane surrounds the lung.
10. The muscle that separates the thoracic cavity from the abdominal cavity is the _____.
11. The wrist bones are collectively called _____ (singular).

DOWN

1. The abdominal region located superior to the umbilical is the _____.
2. The study of how the body works.
3. The homeostatic mechanism that is constantly fluctuating is called _____ feedback.
5. A group of organized cells working together.
6. The depression located posterior to the knee is known as the _____ region.

E. Short Answer Questions

Briefly answer the questions in the spaces provided below.

1. Despite obvious differences, all living things perform the same basic functions. List five (5) functions that are active processes in living organisms.

2. List the *levels of organization* in a complex living thing by using arrows and listing in correct sequence, beginning with the molecular level, to the most complex level.

3. What is the major differences between *negative feedback* and *positive feedback*?

4. Describe the position of the body when it is in *anatomical position*.

5. References are made by the anatomist to the front, back, head and tail of the body. What anatomical direction reference terms are used to describe each of these directions? (Your answer should be in the order listed above).

6. What is the difference between a *sagittal section* and a *transverse section*?

7. What are the two (2) essential functions of body cavities in the human body?

CHAPTER

2

The Chemical Level of Organization

Overview

Most students of anatomy and physiology are surprised to learn that the human body is made up of atoms and that the interactions of these atoms control the physiological processes within the body. To fully comprehend the functioning of the body as a whole, the basic concepts of inorganic and organic chemistry are studied and mastery of chemical principles is necessary. Atomic structure and formation of molecules provide the basic framework for understanding how simple components interact to form more complex ordering and structuring found in all living things, whether plant or animal.

Even though over 100 chemical elements are known, only four (carbon, hydrogen, oxygen, and nitrogen) make up over 96% of all living matter. These four elements comprise the structure of the most abundant complex biological molecules such as proteins, fats, carbohydrates and nucleic acids, which provide the building blocks and functional processes in the human body.

The exercises in this chapter focus on many of the important organic and inorganic chemical principles. The tests and activities are set up to reinforce the student's understanding of basic chemical concepts and increase their ability to apply these principles to the functioning of the body as a whole.

Review of Chapter Objectives

1. Describe an atom and an element.
2. Describe the different ways in which atoms combine to form molecules and compounds.
3. Use chemical notation to symbolize chemical reactions.
4. Distinguish among the major types of chemical reactions that are important for studying physiology.
5. Describe the pH scale and the role of buffers in body fluids.
6. Distinguish between organic and inorganic compounds.
7. Explain how the chemical properties of water make life possible.
8. Describe the physiological roles of inorganic compounds.
9. Discuss the structure and functions of carbohydrates, lipids, proteins, nucleic acids, and high-energy compounds.
10. Describe the role of enzymes in metabolism.

Part I: Objective Based Questions

Objective 1 Describe an atom and an element.

_____ 1. The smallest chemical units of matter of which no chemical change can alter their identity are:

a. electrons
b. mesons
c. protons
d. atoms

_____ 2. The three subatomic particles that are stable constituents of atomic structures are

a. carbon, hydrogen, oxygen
b. protons, neutrons, electrons
c. atoms, molecules, compounds
d. cells, tissues, organs

_____ 3. The protons of an atom are found only

a. outside the nucleus
b. in the nucleus and outside the nucleus
c. in the nucleus
d. in orbitals

_____ 4. Isotopes of an element differ in the number of

a. neutrons in the nucleus
b. protons in the nucleus
c. electrons in the nucleus
d. orbital electrons

_____ 5. The atomic numbers represents the number of

a. protons and neutrons
b. protons in an atom
c. protons in an ion
d. neutrons in an atom

_____ 6. The mass number of an atom indicates the number of

a. protons and neutrons in the nucleus
b. protons in the nucleus
c. protons, neutrons, and electrons
d. protons and electrons

_____ 7. The chemical behavior of an atom is determined by

a. the mass of the atom
b. the number of neutrons
c. the number and arrangement of electrons
d. the number and location of protons

8. A chemical ____________________ is a substance that consists entirely of atoms with the same atomic number.

9. The center of an atom is called the ____________________.

10. The surrounding areas of the center of an atom represent ____________________ levels.

11. Draw an atom of *carbon* showing the number of protons, neutrons, and electrons it contains.

12. The atomic number of this element (atom) is ______________________.

13. The atomic weight of the element is ____________________.

14. How many electrons are needed to fill its outer shell?__________

15. Is this element inert or chemically active?__________________

16. The symbol for the carbon atom is ______________________.

Objective 2 Describe the different ways in which atoms combine to form molecules and compounds.

_____ 1. When a chemical reaction occurs, the chemical structures that contain more than one atom are called

a. ions
b. isotopes
c. molecules
d. buffers

_____ 2. Ions with a positive charge are called

a. cations
b. anions
c. radicals
d. isotopes

_____ 3. Unequal sharing of a pair of electrons between two atoms forms a

a. double covalent bond
b. ionic bond
c. hydrogen bond
d. polar covalent bond

_____ 4. In living cells, the weakest bond between two or more atoms is the

a. polar bond
b. ionic bond
c. hydrogen bond
d. metallic bond

_____ 5. In a molecule of nitrogen, three pairs of electrons are shared by two nitrogen atoms. The type of bond that is formed is a

a. triple covalent bond
b. triple polar bond
c. polar covalent bond
d. nitrogen molecule

_____ 6. Which one of the following molecules is drawn correctly to show the proper covalent bonding?

a. O–O
b. N ≡ N
c. H ≡ H
d. O–C–O

_____ 7. The formation of cations and anions illustrates the attraction between

a. polar covalent bonds
b. ionic bonds
c. nonpolar covalent bonds
d. double covalent bonds

Figure 2-1 Chemical Bonding

In figures A & B below identify which one is the covalent bond and which one is the ionic bond. Show the direction of electron transfer with an arrow.

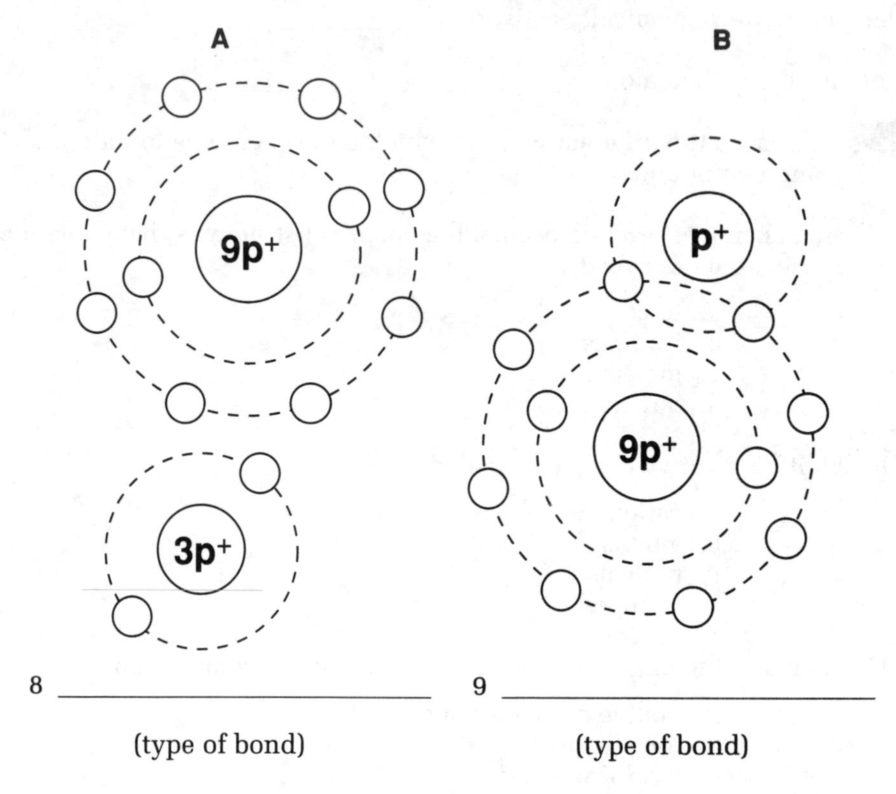

8 ______________________________ (type of bond)

9 ______________________________ (type of bond)

10. Atoms that complete their outer shells by sharing electrons with other atoms result in molecules held together by ______________________.

11. When one atom loses an electron and another accepts that electron the result is the for mation of a(n) ______________________.

Objective 3 Use chemical notations to symbolize chemical reactions.

_____ 1. Which one of the following statements about the reaction $H_2 + C_{12} \rightarrow 2HCl$ is not correct?

a. H_2 and C_{12} are the reactants
b. HCl is the product
c. one molecule of hydrogen contains two atoms
d. all of the above are correct

_____ 2. The symbol 2H means

a. one molecule of hydrogen
b. two molecules of hydrogen
c. two atoms of hydrogen
d. a, b, and c are correct

_____ 3. The symbol Na^+ refers to

a. one sodium ion (has gained an electron)
b. one sodium ion (has lost an electron)
c. one sodium ion (has gained a proton)
d. none of the above

_____ 4. The balanced equation $2H_2 + O_2 \rightarrow 2H_2O$ means that

a. one molecule of hydrogen and one molecule of oxygen have combined chemically to form one molecule of water.
b. two molecules of hydrogen and two molecules of oxygen have combined chemically to form two molecules of water.
c. two atoms of hydrogen and two atoms of oxygen have combined chemically to form four molecules of water.
d. two molecules of hydrogen and one molecule of oxygen have combined chemically to form two molecules of water.

_____ 5. The chemical notation that would indicate "one molecule of hydrogen composed of two hydrogen atoms" would be

a. $2H_2$
b. H_2
c. 2H
d. a, b, and c are correct

Objective 4 Distinguish among the major types of chemical reactions that are important for studying physiology.

_____ 1. When a molecule is broken down into smaller fragments the reaction is called

a. decomposition
b. reversible
c. exchange
d. synthesis

_____ 2. Which of the following is a synthesis reaction?

a. $AB \rightarrow A + B$
b. $A + B \leftrightarrow AB$
c. $AB + CD \rightarrow AD + CB$
d. $A+B \rightarrow AB$

_____ 3. Chemical reactions that require an input of energy, such as heat, are

a. exergonic
b. endergonic
c. decomposition reactions
d. in equilibrium

4. Chemical reaction that release energy are said to be _____________________.

5. When the rate of a synthesis reaction balances the rate of a decomposition reaction, the result is _____________________.

Objective 5 Describe the pH scale and the role of buffers in body fluids.

_____ 1. pH is a measure of the concentration of ____________________ in solution.

a. hydroxide ions
b. anions
c. hydrogen ions
d. sodium ions

_____ 2. A solution with a pH below 7 is called

a. alkaline
b. acidic
c. basic
d. neutral

_____ 3. Pure water has a pH of 7 because it contains

a. equal numbers of hydrogen and hydroxide ions
b. equal numbers of hydrogen and oxygen atoms
c. a variety of buffers
d. all of the above are correct

_____ 4. Which of the following substances would be the least acidic?

a. lemon juice with a pH of 2
b. white wine with a pH of 3
c. tomato juice with a pH of 4
d. urine with a pH of 6

_____ 5. If a substance has a pH greater than 7, it is

a. acidic
b. a salt
c. alkaline
d. buffered

_____ 6. To maintain homeostasis in the body, the normal pH range of the blood must remain at

a. 6.80 to 7.20
b. 7.35 to 7.45
c. 6.0 to 7.0
d. 6.80 to 7.80

7. Compounds in body fluids that maintain pH within normal limits are ____________________.

8.–10. Identify the numbered regions 8, 9, and 10 on the pH scale below.

Figure 2-2 pH Scale

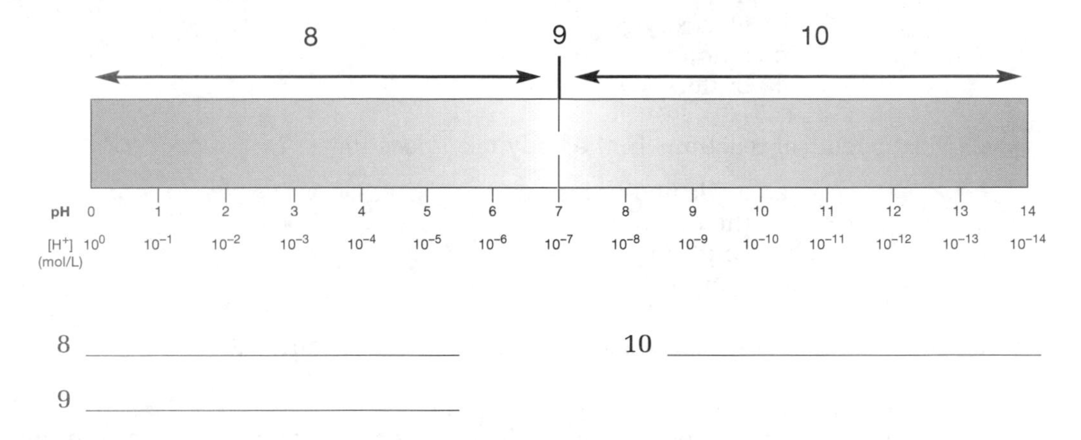

8 ______________________ 10 ______________________

9 ______________________

Objective 6 Distinguish between organic and inorganic compounds.

_____ 1. Which one of the following selections includes only *organic* compounds?

a. proteins, water
b. glucose, DNA
c. carbon dioxide, oxygen
d. fats, salts

_____ 2. The most important *inorganic* compound found in large quantity in the body is

a. DNA
b. ATP
c. carbohydrates
d. water

3. Compounds that contain the elements carbon, hydrogen, and usually oxygen are ______________ compounds.

4. Acids, bases, and salts are examples of ______________ compounds.

Objective 7 Explain how the chemical properties of water make life possible.

_____ 1. Which of the following statements about water is not correct?

a. 2/3 of the body's mass consists of water
b. water dissolves many compounds
c. water has a relatively low heat capacity
d. water remains liquid in cells over a wide range of environmental temperatures

_____ 2. The ideal medium for the absorption and transport of inorganic or organic compounds is

a. oil
b. water
c. blood
d. lymph fluid

_____ 3. During ionization, water molecules disrupt the ionic bonds of a solute and a mixture of ions is produced. These ions are called

a. cations
b. electrolytes
c. anions
d. buffers

_____ 4. Most chemical reactions in the body take place in

a. acids and bases
b. the blood
c. solution
d. lymph

5. A homogenous mixture containing a solvent and a solute is called a(n) ________________________.

6. Soluble inorganic compounds whose ions will conduct an electric current in solution are ________________________.

Objective 8 Describe the physiological roles of inorganic compounds.

_____ 1. A solute that dissociates to release hydrogen ions and causes a decrease in pH is

a. a base
b. a salt
c. water
d. an acid

_____ 2. A solute that removes hydrogen ions from a solution is

a. a salt
b. a base
c. an acid
d. a buffer

_____ 3. In the body, inorganic compounds

a. are the building blocks of proteins
b. are structural components of cells
c. serve as buffers
d. are the component parts of DNA

_____ 4. Inorganic ions are important in the function(s) of

a. blood clotting
b. muscle contractions
c. nerve impulse conduction
d. a, b, and c are correct

_____ 5. An example of a weak acid that serves as an effective buffer in the body is

a. hydrochloric acid
b. carbonic acid
c. sulfuric acid
d. acetic acid

_____ 6. A salt may best be described as

a. an organic molecule created by chemically altering an acid or base
b. an inorganic molecule that buffers solutions
c. an organic molecule used to flavor food
d. an inorganic molecule created by the reaction of an acid and a base

Objective 9 Discuss the structure and functions of carbohydrates, lipids, proteins, nucleic acids, and high-energy compounds.

_____ 1. Carbohydrates are most important to the body because they serve as primary sources of

a. tissue growth and repair
b. metabolites
c. energy
d. a, b, and c are correct

_____ 2. The most important metabolic fuel molecule in the body is

a. glucose
b. sucrose
c. starch
d. vitamins

_____ 3. The polysaccharide formed by stored glucose in the liver and muscle is

a. starch
b. cellulose
c. glycogen
d. sucrose

_____ 4. Most of the fat found in the human body is in the form of

a. steroids
b. triglycerides
c. phospholipids
d. monoglycerides

_____ 5. Butter, fatty meat, and ice cream are examples of sources of fatty acids that are said to be

a. dehydrogenated
b. monounsaturated
c. polyunsaturated
d. saturated

_____ 6. A steroid molecule is an example of a

a. high energy compound
b. carbohydrate
c. protein
d. lipid

_____ 7. The building blocks of proteins consist of chains of small molecules which are called

a. peptide bonds
b. amino acids
c. R groups
d. amino groups

_____ 8. Proteins differ from carbohydrates in that they

a. are not an energy nutrient
b. do not contain carbon, hydrogen, and oxygen
c. always contain nitrogen
d. are inorganic compounds

_____ 9. Each amino acid differs from others in the

a. location of the carbon atoms
b. number of carboxyl groups
c. size of the amino group
d. nature of the R group

_____ 10. Special proteins that are involved in metabolic regulation are called

a. enzymes
b. contractile proteins
c. transport proteins
d. structural proteins

_____ 11. The molecules that store and process information at the molecular level are the

a. metabolic steroids
b. nucleic acids
c. adenosines
d. a, b, and c are correct

_____ 12. The three basic components of a single nucleotide of a nucleic acid are

a. purines, pyrimidines, sugar
b. sugar, phosphate group, nitrogen base
c. guanine, cytosine, thymine
d. pentose, ribose, deoxyribose

_____ 13. In DNA and RNA, a nucleotide containing the nitrogen base cytosine would only base-pair with

a. uracil
b. adenine
c. guanine
d. thymine

_____ 14. The most important high energy compound in cells is

a. glucose
b. RNA
c. creatine phosphate
d. ATP

15. The attachment of a carboxylic acid group of one amino acid to the amino acid group of another forms a connection called a(n)_____________________.

16. The DNA contains the five carbon sugar _________________.

17. The nitrogen base found in RNA but not in DNA is ___________________.

18.–24. Locate and identify the numbered structures in the DNA molecule. Fill in the blanks below the drawing.

Figure 2-3 DNA Structure

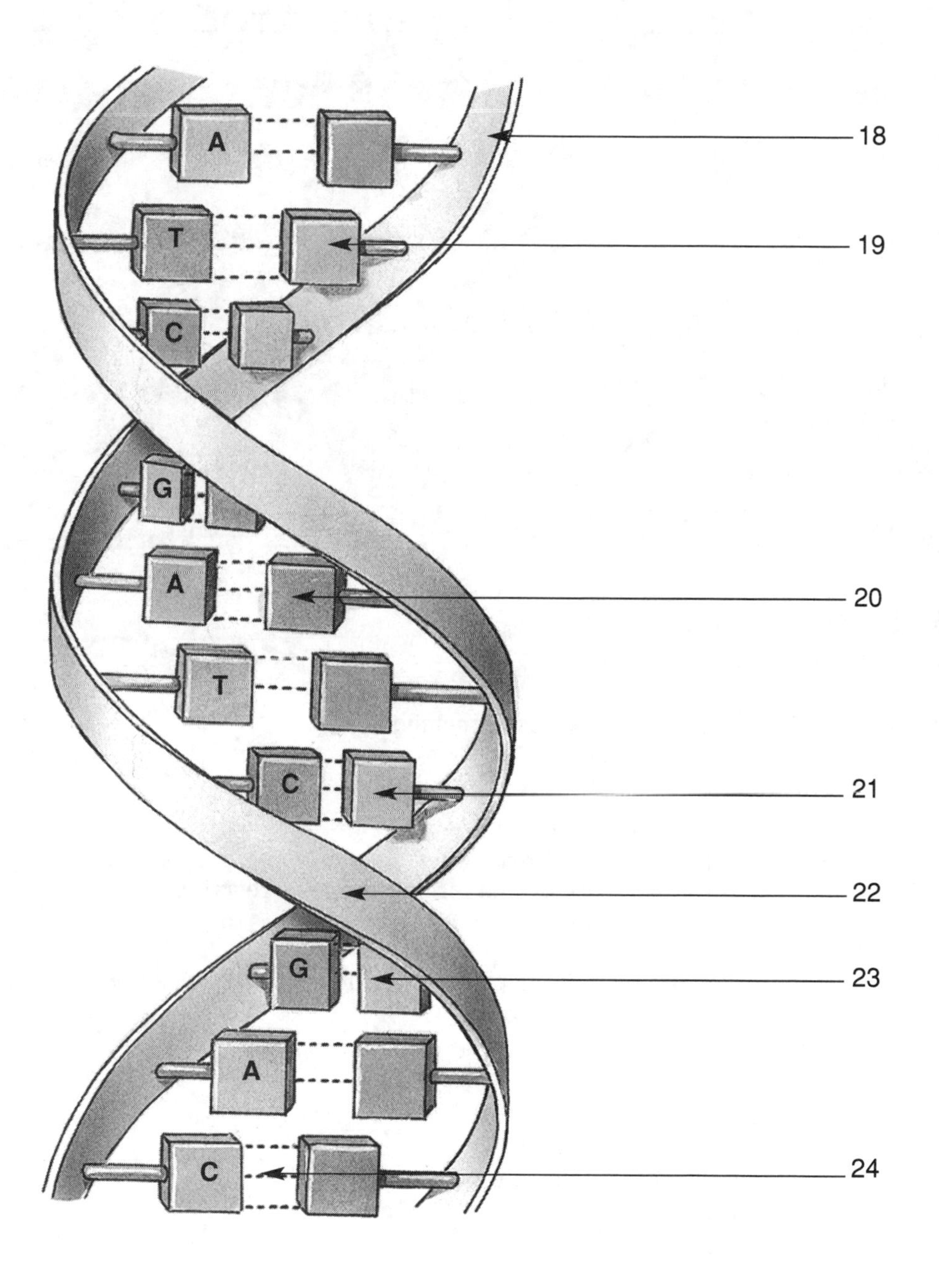

18 ____________________

19 ____________________

20 ____________________

21 ____________________

22 ____________________

23 ____________________

24 ____________________

25. Categorize the various classes of organic molecules. Place an X in the appropriate space to classify each type of organic molecule.

Table 2–1 Classification of Various Organic Molecules

Organic molecules	Carbohydrate	Lipid	Protein	Nucleic acid	High-energy compound
Amino Acid					
ATP					
Cholesterol					
Cytosine					
Disaccharide					
DNA					
Fatty acid					
Glucose					
Glycerol					
Glycogen					
Guanine					
Monosaccharide					
Nucleotide					
Phospholipid					
Polysaccharide					
RNA					
Starch					

Objective 10 Describe the role of enzymes in metabolism.

_____ 1. The presence of a specific enzyme affects only the

a. rate of a reaction
b. direction of the reaction
c. products that will be formed from the reaction
d. all of the above

_____ 2. Organic catalysts made by a living cell to promote a specific reaction are called

a. nucleic acids
b. buffers
c. enzymes
d. metabolites

_____ 3. Substrate molecules bind to enzymes at the

a. carboxyl group
b. active sites
c. amino groups
d. reactant sites

_____ 4. All enzymes are

a. high energy compounds
b. nucleic acids
c. inorganic molecules
d. proteins

_____ 5. In the reaction A + B → C, the letter(s) which represent the substrates in a reaction involving an enzyme is (are)

a. A and C
b. A and B
c. B and C
d. C only

Part II: Chapter Comprehensive Exercises

A. Word Elimination

Circle the term that does not belong in each of the following groupings.

1. protons neutrons electrons isotope subatomic
2. hydrogen compound oxygen nitrogen helium
3. decomposition exchange buffer reversible synthesis
4. water carbon dioxide salts glucose acids
5. carbonic acid carbohydrates lipids proteins nucleic acids
6. fatty acids glycogen glycerides steroids phospolipids
7. enzymes antibodies hormones keratin glycogen
8. uracil thymine adenine cytosine guanine
9. carbon hydrogen oxygen nitrogen carbon dioxide
10. sucrose monosaccharide lactose maltose disaccharide

B. Matching

Match the terms in Column "B" with the terms in Column "A." Use letters for answers in the spaces provided.

COLUMN A	COLUMN B
_____1. atom	A. an atom with a charge
_____2. atomic number	B. two atoms of same element with different number of neutrons.
_____3. buffer	C. liquid portion of solution
_____4. electron	D. acts as catalyst to speed up chemical reactions
_____5. enzyme	E. dissolved material in solution
_____6. ion	F. smallest unit of matter
_____7. isotope	G. reactants in chemical reaction
_____8. metabolism	H. resists changes in pH
_____9. solute	I. manufacture of compounds
_____10. solvent	J. has a negative charge
_____11. substrate	K. total of all chemical reactions in the body
_____12. synthesis	L. number of protons in an atom

C. Concept Map

This concept map summarizes and organizes some basic concepts covered in Chapter 2. Use the following terms to complete the map by filling in the spaces identified by the circled numbers.

Dissacharides	**DNA**	**Nitrogen Base**
Nucleotides	**Nucleic Acids**	**Lipids**
Fatty Acids	**Amino Acids**	

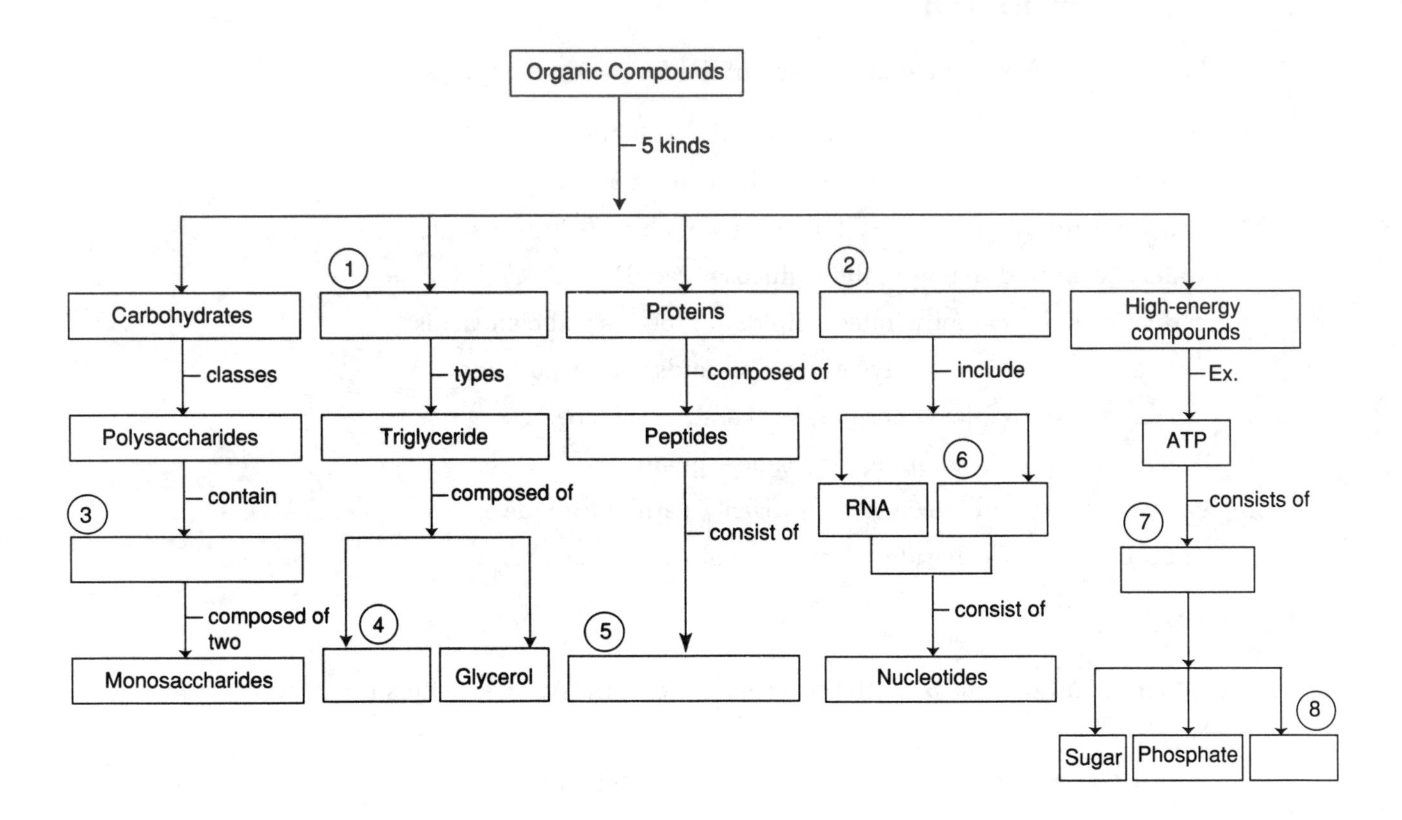

D. Crossword Puzzle

The following crossword puzzle reviews the material in Chapter 2. To complete the puzzle, you have to know the answers to the clues given, and you have to be able to spell the terms correctly.

ACROSS

1. The greater the concentration of ______ ions in solution, the lower the pH.
4. Neutrons and protons are located in the ______ of an atom.
5. Two atoms that have the same number of protons but a different number of neutrons are called _________.
8. A _____ consists of a chain of molecules called amino acids.
10. Fats and steroids are examples of _______.
11. The mass number of an atom changes if its number of ______ changes.

DOWN

2. The disaccharide sucrose consists of fructose bonded to _____.
3. A substance that helps stabilize pH is called a(n) ______.
5. An atom or molecule that has a positive or a negative charge is an ______.
6. When an atom loses an electron, it exhibits a ________ charge.
7. When phosphorus becomes an ion, it exhibits a _______ 3 charge.
9. The high-energy compound produced by the body is ________ triphosphate.

E. Short Answer Questions

Briefly answer the questions in the spaces provided below.

1. Suppose an atom has 8 protons, 8 neutrons, and 8 electrons. Construct a diagram of the atom and identify the subatomic particles by placing them in their proper locations.

2. Why are the elements helium, argon, and neon called *inert gases*?

3. In a water molecule (H_2O), the unequal sharing of electrons creates a *polar covalent bond*. Why?

4. List four (4) important characteristics of water that make life possible.

5. List the four (4) major classes of organic compounds found in the human body and give an example for each one.

6. What is the difference between a *saturated* and an *unsaturated* fatty acid?

7. Using the four (4) kinds of nucleotides that make up a DNA molecule, construct a model that will show the correct arrangement of the components which make up each nucleotide. *Name each nucleotide.*

8. What three (3) components make up one nucleotide of ATP?

CHAPTER

3

Cell Structure and Function

Overview

The basic unit of structure and function of all living things is the cell. In the living world a single cell can be a complete living thing or it can be one of billions of units which make up a complex living organism. In complex organisms such as humans, cells become specialized to form tissues and organs that perform specific functions. Cells differ in shape, size, and the roles they play in the human body. Cell structures called organelles perform the work of metabolism, are responsible for growth and reproduction, movement, responsiveness to internal and external stimuli, and maintenance of homeostasis.

The student activities and test questions in this chapter are designed to help you conceptualize, synthesize, and apply the principles of cell biology that provide the foundation for understanding how the body's organ systems work.

Review of Chapter Objectives

1. Discuss the basic concepts of the cell theory.
2. List the functions of the cell membrane and the structural features that enable it to perform those functions.
3. Describe the transport mechanisms that cells use to transport substances across the cell membrane.
4. Describe the organelles of a typical cell and indicate their specific functions.
5. Explain the functions of the cell nucleus.
6. Summarize the process of protein synthesis.
7. Describe the process of mitosis and explain its significance.
8. Define differentiation and explain its importance.

Part I: Objective Based Questions

Objective 1 Discuss the basic concepts of the cell theory.

_____ 1. The basic structural units of all plants and animals are

a. atoms
b. cells
c. molecules
d. protons, neutrons, electrons

_____ 2. The smallest functioning units of life are

a. tissues
b. organs
c. atoms
d. cells

_____ 3. Cells are produced only by the division of

a. pre-existing cells
b. tissues
c. molecules
d. organelles

_____ 4. The primary function of each cell in the body is to maintain ____________________.

Objective 2 List the functions of the cell membrane and the structural features that enable it to perform those functions.

_____ 1. The major components of the cell membrane are

a. carbohydrates, lipids, ions, vitamins
b. carbohydrates, fats, proteins, water
c. phospholipids, proteins, glycolipids, cholesterol
d. amino acids, fatty acids, carbohydrates, cholesterol

_____ 2. Most of the communication between the interior and exterior of the cell occurs by way of

a. the phospholipid bilayer
b. the peripheral proteins
c. integral protein channels
d. receptor sites

_____ 3. The cell membrane is

a. selectively permeable
b. differentially permeable
c. semi-permeable
d. all of the above

_____ 4. The structural feature that prevents the cell membrane from dissolving is the

a. tail (fatty-acid) portion is insoluble in water
b. membrane is a bilayered structure
c. protein molecules are too large to dissolve in water
d. all of the above

5. The molecular structure of the cell membrane consists of a(n) ________________.

6. The outer boundary of the intracellular material is called the ________________.

7. Locate and identify the following cell structures. Place your answers in the spaces below the drawing.

Figure 3-1 The Cell Membrane

Carbohydrate Chains	**Phospholipid Bilayer**	**Protein with Channels**
Proteins	**Cholesterol**	**Cell membrane**
Heads	**Tails**	**Cytoskeleton**

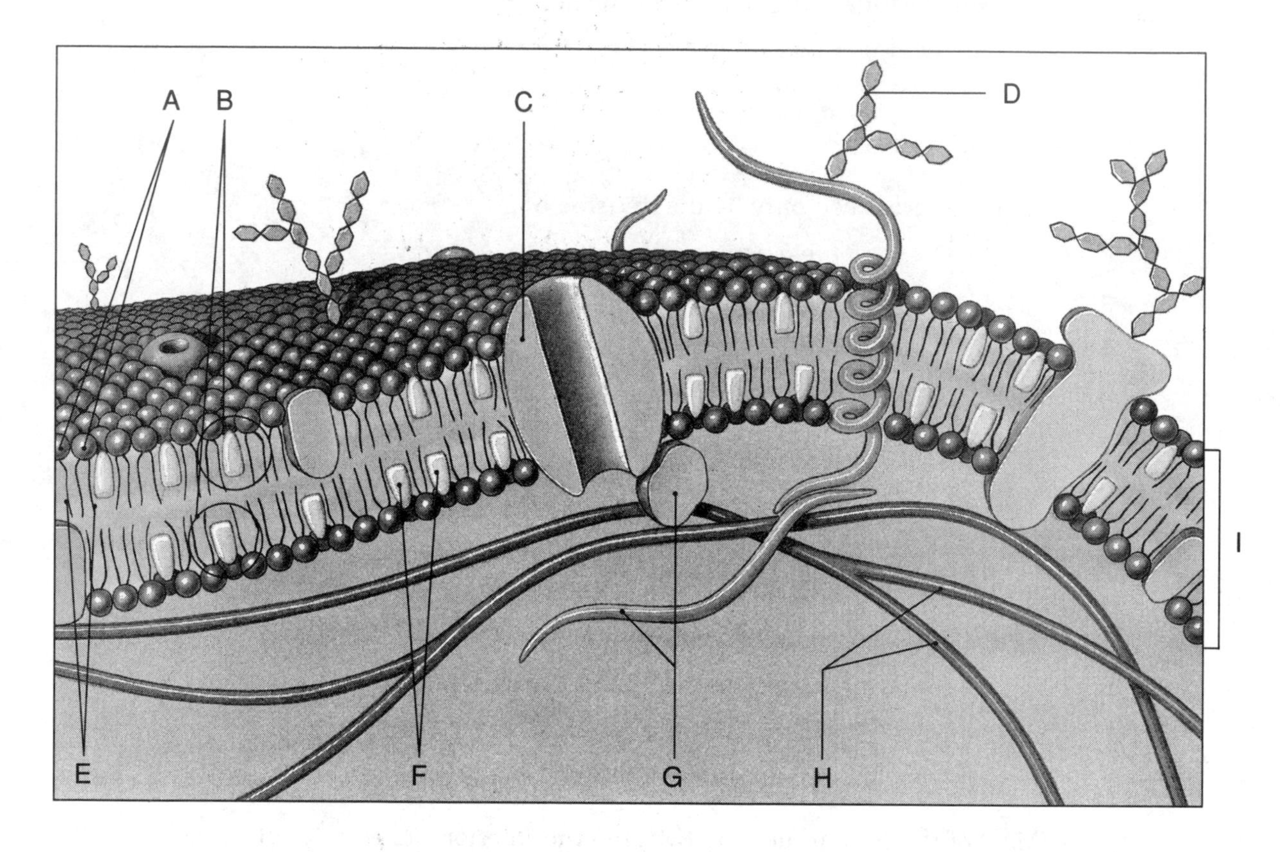

A ________________ F ________________

B ________________ G ________________

C ________________ H ________________

D ________________ I ________________

E ________________

Objective 3 Describe the transport mechanisms that cells use to transport substances across the cell membrane.

_____ 1. All transport through the cell membrane can be classified as

a. active or passive
b. diffusion or osmosis
c. pinocytosis or phagocytosis
d. permeable or impermeable

_____ 2. The mechanism by which glucose can enter the cytoplasm without expending ATP is via a

a. glycocalyx
b. recognition protein
c. carrier protein
d. catalyzed reaction

_____ 3. Ions and other small water-soluble materials cross the cell membrane only by passing through

a. ligands
b. channels
c. receptor proteins
d. membrane anchors

_____ 4. The rate that solute molecules are filtered depends on

a. their size
b. the force of the hydrostatic pressure
c. the rate at which water passes through the membrane
d. all of the above

_____ 5. The major difference between diffusion and bulk flow is that when molecules move by *bulk flow* they move

a. at random
b. as a unit in one direction
c. as individual molecules
d. slowly over long distances

_____ 6. The effect of *diffusion* in body fluids is that it tends to

a. increase the concentration gradient of the fluid
b. scatter the molecules and inactivate them
c. repel like charges and attract unlike charges
d. eliminate local concentration gradients

_____ 7. During *osmosis*, water will always flow across a membrane toward the solution that has the

a. highest concentration of solvents
b. highest concentration of solutes
c. equal concentration of solute
d. equal concentration of solvents

_____ 8. *Facilitated diffusion* differs from ordinary diffusion in that

a. ATP is expended during facilitated diffusion
b. molecules move against a concentration gradient
c. carrier proteins are involved
d. it is an active process utilizing carriers

9. The spreading of a drop of ink to color an entire glass of water illustrates the process of ___________.

10. A white blood cell engulfing a bacterium illustrates the process of ______________.

11. If a solution has the same solute concentration as the cytoplasm and will not cause a net movement of fluid in or out of the cells, the solution is said to be ________________.

12. Using the diagrams and criteria below and applying the principle of osmosis, identify the diagrams which represent a cell in a hypertonic solution, a hypotonic solution and an isotonic solution. Draw arrows to show the direction in which the net movement of water will occur.

Figure 3.2 Principles of Tonicity

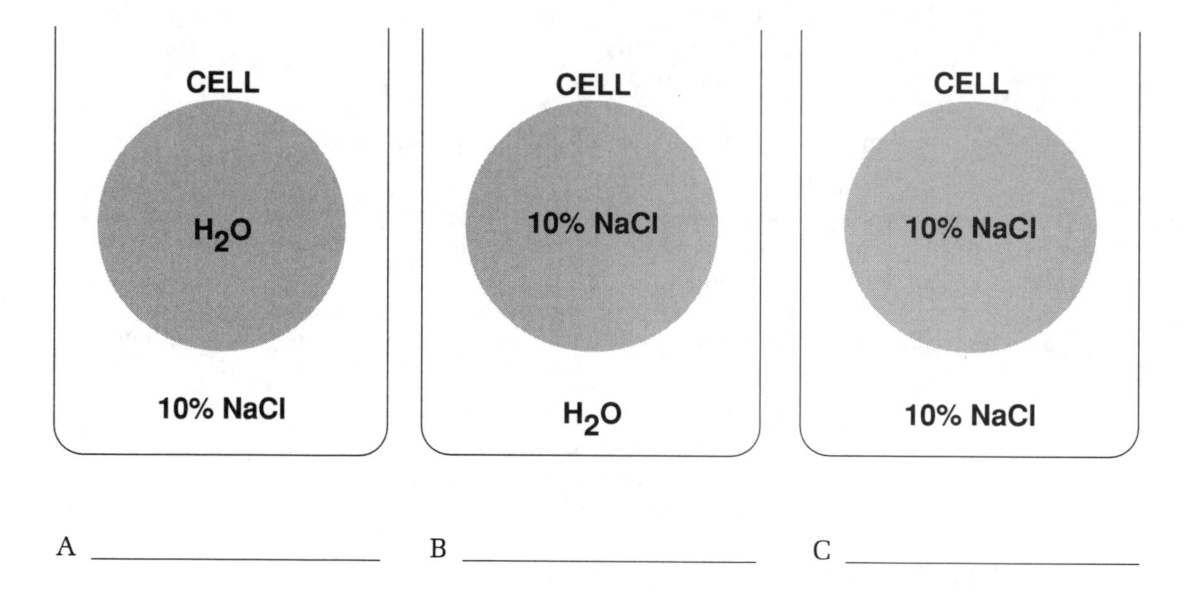

A ________________ B ________________ C ________________

Objective 4 Describe the organelles of a typical cell and indicate their specific functions.

_____ 1. Approximately 95 percent of the energy needed to keep a cell alive is generated by the activity of the

a. mitochondria
b. ribosomes
c. nucleus
d. microtubules

_____ 2. Nucleoli are nuclear organelles that

a. contain the chromosomes
b. are responsible for producing DNA
c. control nuclear operations
d. synthesize the components of the centrioles

_____ 3. The three major functions of the endoplasmic recticulum are

a. hydrolysis, diffusion, osmosis
b. synthesis, transport, storage
c. detoxification, packaging, modification
d. pinocytosis, phagocytosis, storage

_____ 4. The functions of the Golgi apparatus include

a. neutralization, absorption, assimilation, secretion
b. strength, movement, control, secretion
c. isolation, protection, sensitivity, organization
d. synthesis, storage, alteration, packaging

_____ 5. The major function of the ribosomes is to

a. phagocytize bacteria
b. produce protein
c. produce ATP
d. manufacture DNA

_____ 6. The major function of lysomes is to

a. alter amino acid sequences
b. produce carbohydrates
c. release digestive enzymes
d. produce protein

7. The areas of the endoplasmic reticulum that are called the rough ER contain _____________________.

8. The cell organelles responsible for the synthesis of protein using information provided by nuclear DNA are _____________________.

9. Using the terms below, identify the cell organelles in Figure 3-3. Place your answers in the spaces provided below the drawing.

Figure 3-3 Composite Cell

Centrioles	**Golgi Apparatus**	**Free Ribosomes**	**Cell Membrane**
Nuclear Envelope	**Nucleolus**	**Mitochondria**	**Lysosome**
Smooth E.R.	**Rough E.R.**	**Chromatin**	**Fixed Ribosomes**
Cytosol	**Secretory vesicles**	**Nuclear Pores**	**Cytoskeleton**
Microvilli	**Cilia**		

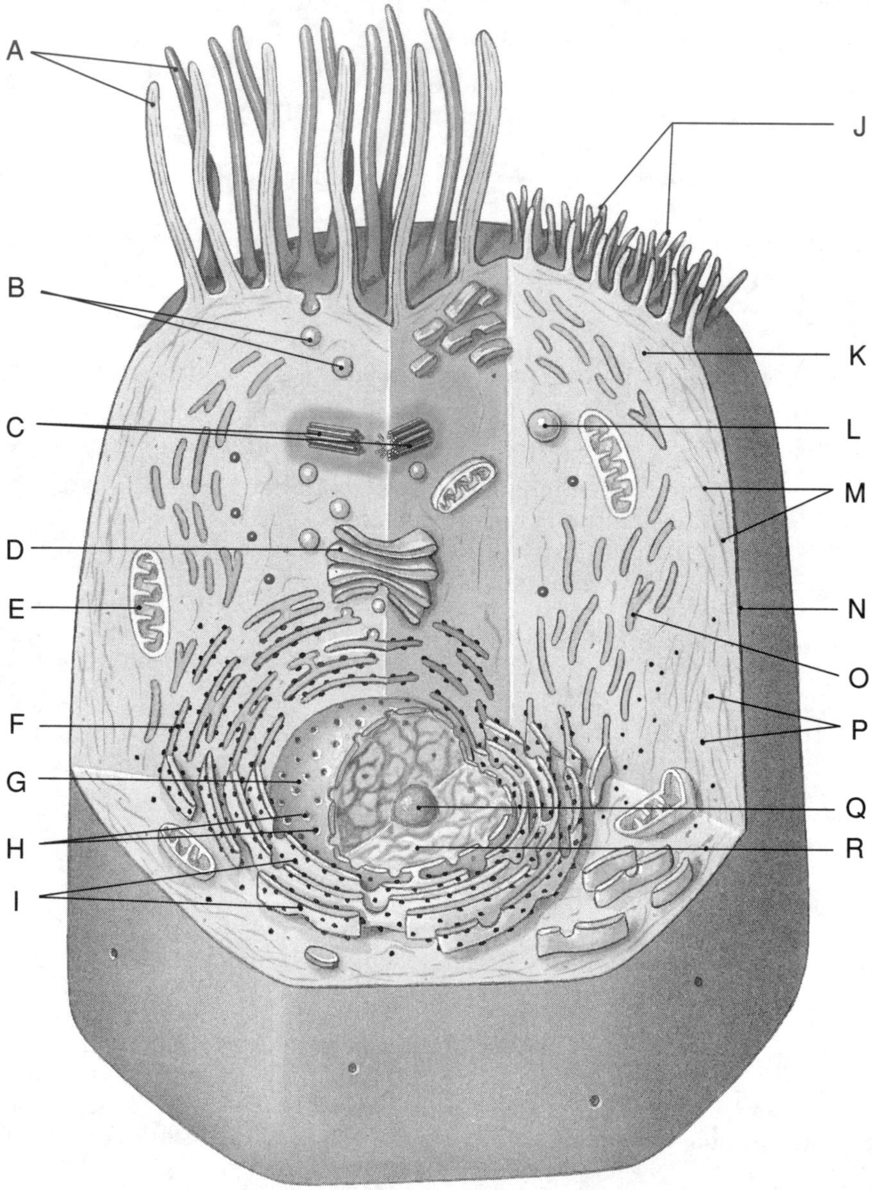
A
B
C
D
E
F
G
H
I
J
K
L
M
N
O
P
Q
R

A ____________ G ____________ M ____________

B ____________ H ____________ N ____________

C ____________ I ____________ O ____________

D ____________ J ____________ P ____________

E ____________ K ____________ Q ____________

F ____________ L ____________ R ____________

Objective 5 Explain the functions of the cell nucleus.

_____ 1. The major factor that allows the nucleus to control cellular operations is through its

a. location within the cell
b. regulation of protein synthesis
c. ability to communicate chemically through nucleus pores
d. "brain-like" sensory devices that monitor cell activity

_____ 2. The organelle that determines the structural and functional characteristics of the cell is the

a. nucleolus
b. nucleus
c. nuclear membrane
d. ribosome

_____ 3. There are _______________ chromosomes in the nucleus of a human cell, and these chromosomes consist of ____________ which are made of ____________.

a. 23, DNA, RNA
b. 46, DNA, genes
c. 23, genes, proteins
d. 46, genes, DNA

_____ 4. The genetic blueprint which determines our human characteristics is located in the

a. nucleolus
b. ribosomes
c. nucleus
d. centrioles

_____ 5. The nuclear pores provide an exit for _______________ into the cytosol so it can travel to the ribosomes.

a. RNA
b. DNA
c. ATP
d. chromosomes

Objective 6 Summarize the process of protein synthesis.

_____ 1. Which of the following sequences of events correctly describes protein synthesis?

a. DNA transcribes the coded message to RNA; RNA (messenger RNA) exits the nuclear pores; RNA translates the coded mes sage to ribosomes; ribosomes make protein.
b. DNA transcribes the coded message to RNA; RNA (transfer RNA) exits the nuclear pores; RNA carries the coded message to the ribosomes, where translation occurs; ribosomes make protein.
c. DNA translates the coded message to RNA; RNA exits the nuclear pores; RNA carries the coded message to the ribo somes, where transcription occurs; ribosomes make protein.
d. DNA transcribes the coded message to RNA; RNA (messenger RNA) exits the nuclear pores and translates the coded message to transfer RNA; tRNA carries the coded message to the ribo somes; ribosomes make protein.

_____ 2. The DNA transmits a chemical message to RNA. This step is called _____________.

a. codon
b. synthesis
c. transcription
d. translation

_____ 3. The RNA that exits the nuclear pores to the cytosol on its way to the ribosomes is called ________.

a. messenger RNA
b. ribosomal RNA
c. transcription RNA
d. transfer RNA

_____ 4. The process by which RNA gives the coded message to the ribosomes is called ___________.

a. codon
b. synthesis
c. transcription
d. translation

_____ 5. In order to make a protein according to the instructions coded in the DNA, ribosomes need amino acids. Where do the ribosomes get the amino acids?

a. from the cytosol
b. from the nucleus
c. from the SER
d. from within the ribosomes

_____ 6. If the DNA triplet is TAG, the corresponding codon on the messenger RNA would be _______.

a. ACT
b. AGC
c. ATC
d. AUC

_____ 7. If the messenger RNA has the codons CCC CGG UUA, the corresponding transfer RNA anticodons would be ___________.

a. GGG GCC AAU
b. CCC CGG TTA
c. GGG GCC AAT
d. CCC CGG TTU

8. A protein consists of a sequence of ______________ bonded together.

9. Identify the letters in Figure 3-4 that represent the following: DNA, mRNA, tRNA, ribosomes; nuclear pore, the process of transcription, and the process of translation. Place your answers in the spaces below the drawing.

Figure 3-4 Protein Synthesis

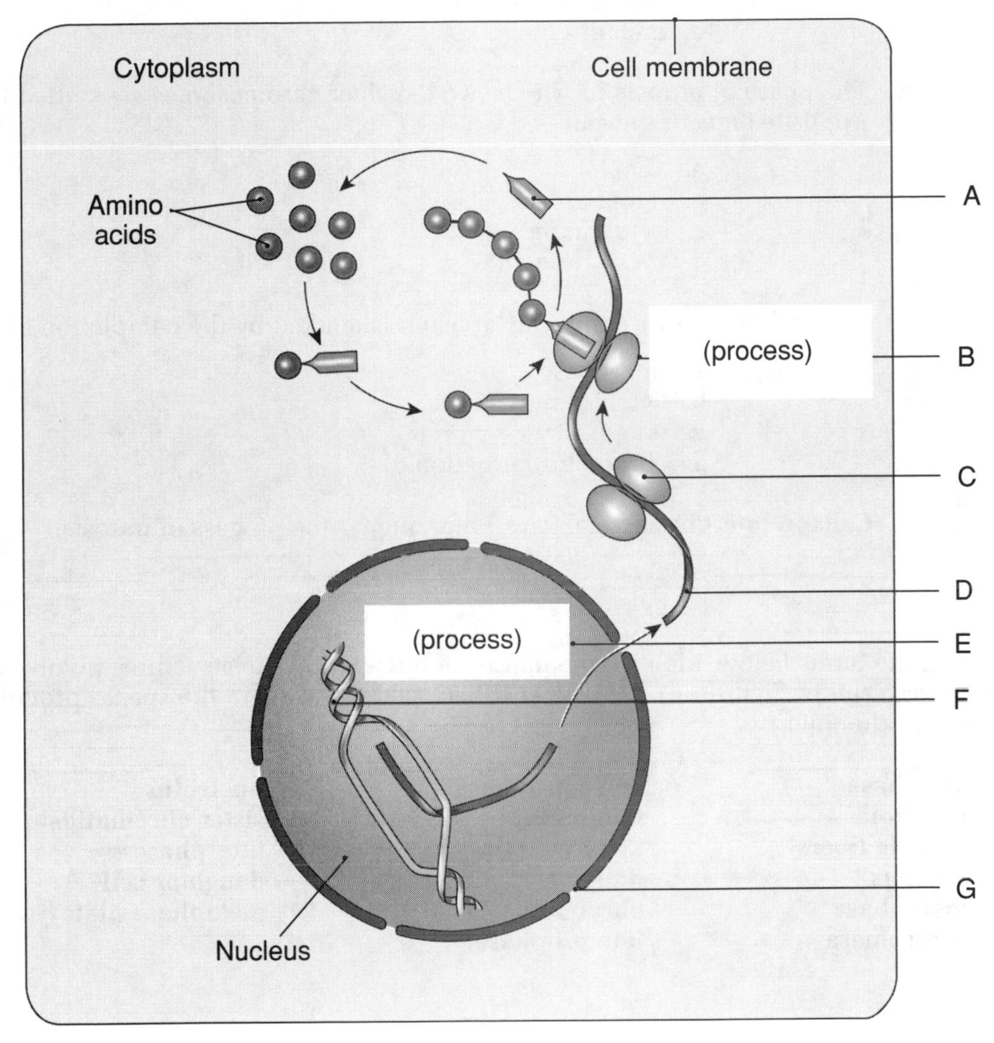

A ______________________

B ______________________

C ______________________

D ______________________

E ______________________

F ______________________

G ______________________

Objective 7 Describe the process of mitosis and explain its significance.

_____ 1. The four stages of mitosis in correct sequence are:

a. prophase, anaphase, metaphase, telophase
b. prophase, metaphase, telophase, anaphase
c. prophase, metaphase, anaphase, telophase
d. prophase, anaphase, telophase, metaphase

_____ 2. The phase of mitosis in which the chromatids move to a narrow central zone is called

a. telophase
b. prophase
c. anaphase
d. metaphase

_____ 3. The phase of mitosis in which two daughter chromosomes are pulled toward opposite ends of the cell is

a. anaphase
b. telophase
c. cytokinesis
d. metaphase

_____ 4. The end of the process of cell division is marked by the completion of

a. interphase
b. cytokinesis
c. telophase
d. chromatid formation

_____ 5. Cells in interphase are always preparing for the process of mitosis.

a. True
b. False

6. Using the terms below, identify the phases of mitosis and the structures involved in the process, seen in Figure 3–5 on page 47. Place your answers in the spaces provided beneath the figure.

anaphase
centrioles
spindle fibers
nucleus
metaphase
centromere
early prophase
telophase
cytokinesis
daughter chromosomes
chromatin
late prophase
nucleolus
sister chromatids
interphase
daughter cell
metaphase plate

Figure 3-5 Mitosis

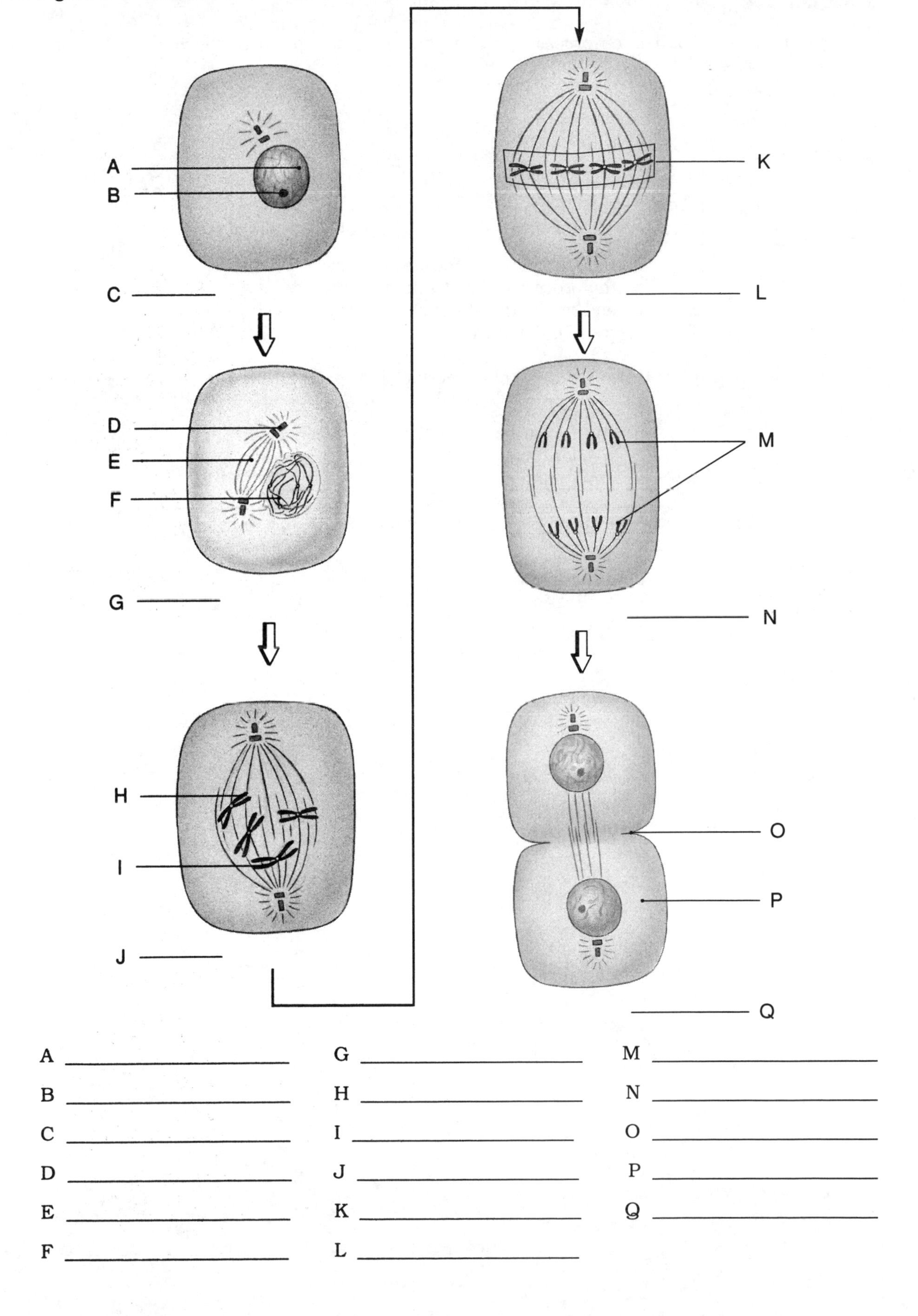

A ______________________ G ______________________ M ______________________

B ______________________ H ______________________ N ______________________

C ______________________ I ______________________ O ______________________

D ______________________ J ______________________ P ______________________

E ______________________ K ______________________ Q ______________________

F ______________________ L ______________________

Objective 8 Define differentiation and explain its importance.

_____ 1. The specialization process which produces cells to perform specific functions is called

a. selection
b. metastasis
c. differentiation
d. replication

_____ 2. The process of differentiation, which causes cells to have different characteristics, involves

a. gene activation and deactivation
b. the process of fertilization
c. embryonic development
d. protein denaturation

_____ 3. The process that restricts a cell to performing specific functions is known as

a. maturation
b. specialization
c. speciation
d. replication

_____ 4. A single cell with all of its genetic potential intact is produced by

a. mitosis
b. differentiation
c. replication
d. fertilization

Part II: Chapter Comprehensive Exercises

A. Word Elimination

Circle the term that does not belong in each of the following groupings.

1. extracellular interstitial organelle cytosol intracellular
2. microvilli lysosomes cilia cytoskeleton centrosome
3. anchoring recognition receptor carrier glycocalyx
4. distance diffusion filtration vesicular carrier-mediated
5. isotonic hypertonic hyperosmotic saturation hypotonic
6. diffusion endocytosis phagocytosis pinocytosis exocytosis
7. enzymes ribosomes suicide packets lysosome autolysis
8. nucleoli DNA chromatin nucleoplasm mitochondria
9. codon mRNA DNA anticodon tRNA
10. prophase interphase metaphase telophase anaphase

B. Matching

Match the terms in Column B with the terms in Column A. Use letters for answers in the spaces provided.

COLUMN A	COLUMN B
_____ 1. active transport	A. protein "factories"
_____ 2. sodium ions	B. carrier mediated
_____ 3. potassium ions	C. high concentration in cytosol
_____ 4. ribosomes	D. DNA nitrogen base
_____ 5. Golgi apparatus	E. chromosomes
_____ 6. DNA strands	F. nuclear division
_____ 7. thymine	G. RNA nitrogen base
_____ 8. uracil	H. cell specialization
_____ 9. mitosis	I. high concentration in extracellular fluid
_____ 10. differentiation	J. membrane turnover

C. Concept Map

The concept map summarizes and organizes some basic concepts in Chapter 3. Use the following terms to complete the map by filling in the spaces identified by the circled numbers.

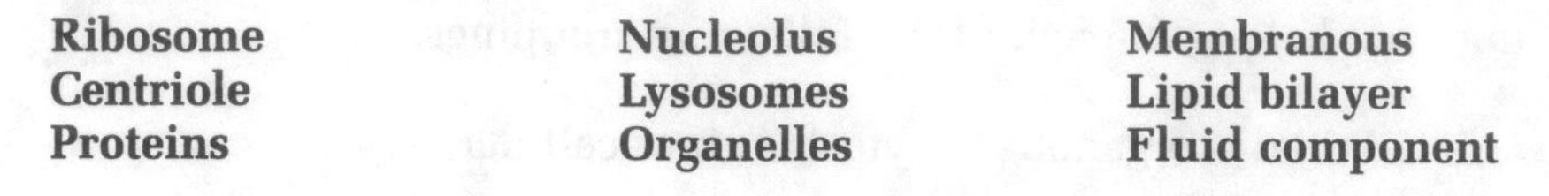

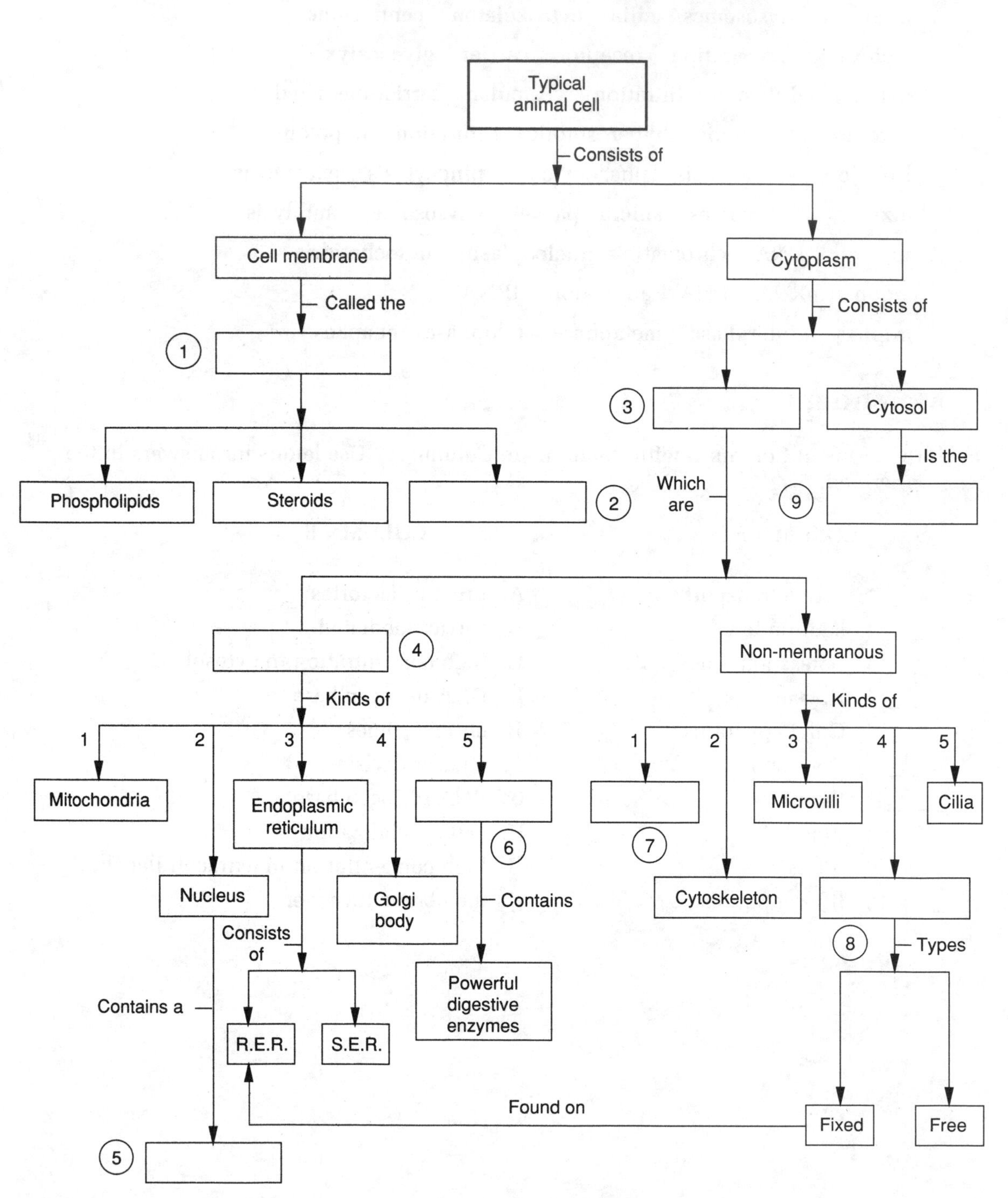

D. Crossword Puzzle

The following crossword puzzle reviews the material in Chapter 3. To complete the puzzle, you have to know the answers to the clues given, and you have to be able to spell the terms correctly.

ACROSS

5. A solution that has a lower concentration of solutes than that within the cell.
7. The second phase of mitosis.
9. Molecules that bond together to produce a protein.
10. A type of RNA that delivers information to the ribosomes.
11. A cell structure that contains DNA.

DOWN

1. A cell organelle that digests incoming material.
2. The organelle responsible for making ATP.
3. Cell organelles that produce protein.
4. A genetic change in the DNA.
5. A solution that has a greater concentration of solutes than that within the cell.
6. The process by which a cell engulfs particles.
7. The movement of water from an area of high concentration to an area of low concentration.

E. Short Answer Questions

Briefly answer the following questions in the spaces provided.

1. Confirm your understanding of cell specialization by citing five (5) systems in the human body and naming a specialized cell found in each system.

2. List four (4) general functions of the cell membrane.

3. List three (3) ways in which the cytosol differs chemically from the extracellular fluid.

4. What *organelles* would be necessary to construct a functional "typical" cell? (Assume the presence of cytosol and a cell membrane.)

5. What are the *functional* differences among centrioles, cilia, and flagella?

6. What three (3) major factors determine whether a substance can diffuse across a cell membrane?

CHAPTER

4

The Tissue Level of Organization

Overview

No single cell contains the metabolic machinery and organelles needed to perform all the functions of the human body. Individual cells of similar structure and function join together to form groups called tissues, which are identified on the basis of their origin, location, shape and function.

This chapter is an introduction to the study of tissues called Histology. Emphasis is placed on the four major types: epithelial, connective, muscle, and nervous tissues. These types are identified according to the organ system in which they are found or the function that they perform. Knowledge of tissue structure and function is important in understanding how individual cells are organized to form tissues and how tissues are arranged to form the organ systems of the complete organism. There are interrelationships between tissue types and their functions and between the tissues in an organ and the organ's function. The study of tissues requires a great deal of memorization and visualization. Your textbook is an excellent source of reference to help you complete and master the exercises in this chapter which are designed to locate, identify, and learn the function of each type of tissue. You will find, after studying these exercises, that all the tissues must function together to maintain homeostasis.

Review of Chapter Objectives

1. Identify the four major tissues of the body and their roles.
2. Describe the types and functions of epithelial cells.
3. Describe the relationship between form and function for each epithelial type.
4. Compare the structures and functions of the various types of connective tissues.
5. Explain how epithelial and connective tissues combine to form four different types of membranes and specify the functions of each.
6. Describe the three types of muscle tissue and the special structural features of each type.
7. Discuss the basic structure and role of neural tissue.
8. Explain how tissues respond in a coordinated manner to maintain homeostasis.
9. Describe how aging affects tissues of the body.

Part I: Objective Based Questions

Objective 1 Identify the four major tissues of the body and their roles.

_____ 1. The four primary tissue types found in the human body are

a. squamous, cuboidal, columnar, glandular
b. adipose, elastic, reticular, cartilage
c. skeletal, cardiac, smooth, muscle
d. epithelial, connective, muscle, neural

_____ 2. Each cell specializes to perform a relatively restricted range of functions through the process of

a. fertilization
b. differentiation
c. speciation
d. translation

_____ 3. Collections of specialized cells and cell products that perform a limited range of functions are called

a. systems
b. organs
c. tissues
d. organelles

Objective 2 Describe the types and functions of epithelial cells.

_____ 1. The type of tissue that covers exposed surfaces and lines internal passageways and body cavities is

a. muscle
b. neural
c. epithelial
d. connective

_____ 2. The two types of layering recognized in epithelial tissues are

a. cuboidal and columnar
b. squamous and cuboidal
c. columnar and stratified
d. simple and stratified

_____ 3. The types of cells that form glandular epithelium that secrete enzymes and buffers in the pancreas and salivary glands are

a. simple squamous epithelium
b. simple cuboidal epithelium
c. stratified cuboidal epithelium
d. transitional epithelium

_____ 4. The type of epithelial tissue found only along the ducts that drain sweat glands is

a. transitional epithelium
b. simple squamous epithelium
c. stratified cuboidal epithelium
d. transitional epithelium

5. The type of tissue that makes up the surface of the skin is _____________________.

6. In general, the primary function of epithelial tissue is that it provides ________________ from the external and internal environment.

Objective 3 Describe the relationship between form and function for each epithelial type.

_____ 1. If epithelial cells are classified according to their cell shape, the classes include

a. simple, stratified, pseudostratified
b. squamous, cuboidal, columnar
c. simple, squamous, stratified
d. pseudostratified, stratified, columnar

_____ 2. If epithelial cells are classified according to their function, the classes would include those involved with

a. support, storage, transport
b. defense, support, storage
c. lining, covering, secreting
d. protection, defense, transport

_____ 3. Simple epithelial cells are characteristic of regions where

a. mechanical or chemical stresses occur
b. support and flexibility are necessary
c. padding and elasticity are necessary
d. secretion and absorption occur

_____ 4. A single layer of epithelium covering a basement membrane is termed

a. simple epithelium
b. stratified epithelium
c. squamous epithelium
d. cuboidal epithelium

5. Several types of epithelial cells are shown in Figure 4-1. Match each structure in the Cell Name column with the name and function of that cell type in the Cell Function column.

Figure 4-1 Epithelial Cells

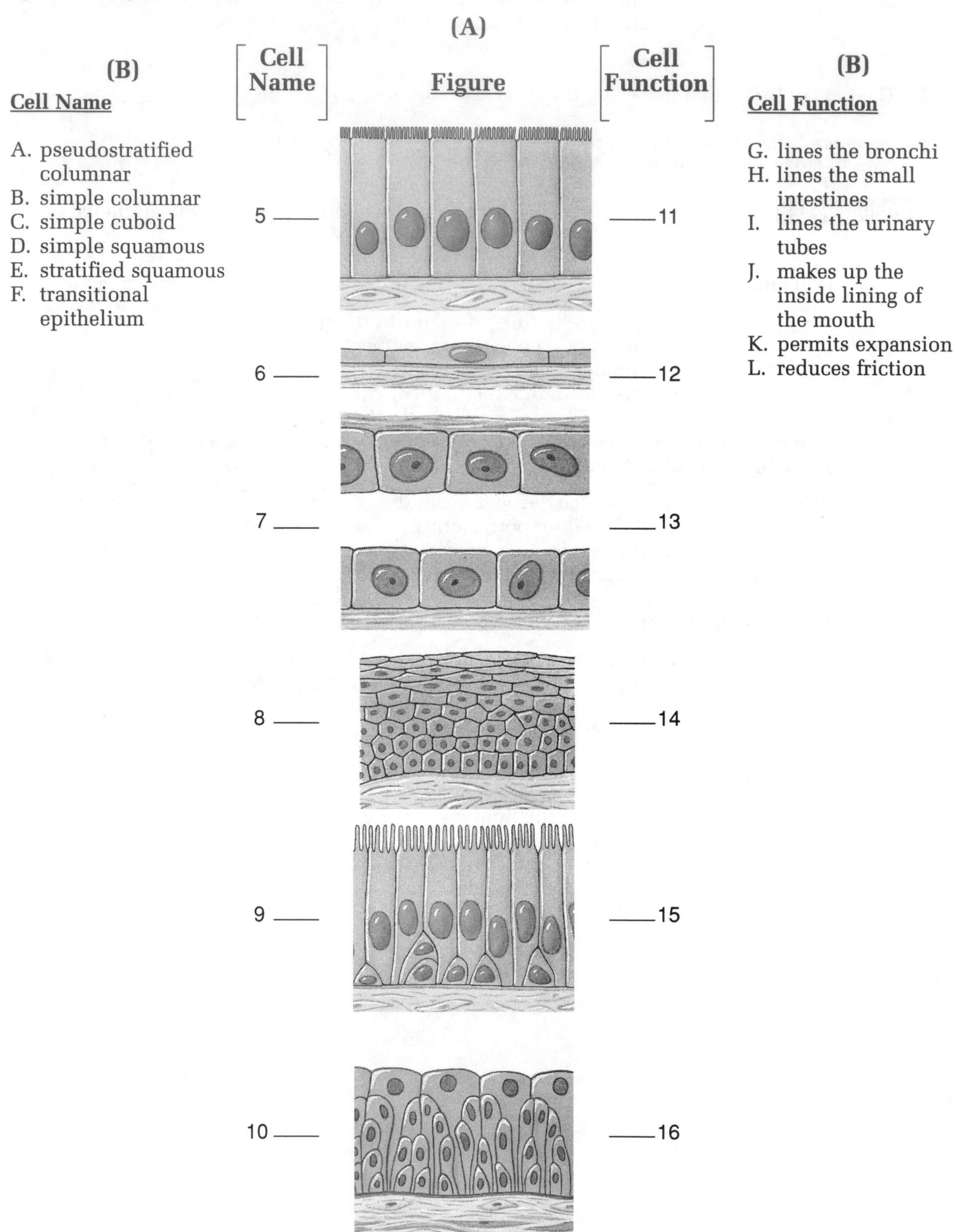

Objective 4 Compare the structures and functions of the various types of connective tissues.

_____ 1. The three basic components of all connective tissues are:

a. free exposed surface, exocrine secretions, endocrine secretions
b. fluid matrix, cartilage, osteocytes
c. specialized cells, extracellular protein fibers, ground substance
d. satellite cells, cardiocytes, osteocytes

_____ 2. The three classes of connective tissue based on structure and function are:

a. fluid, supporting, and connective tissue proper
b. cartilage, bone, and blood
c. collagenic, reticular, and elastic
d. adipose, reticular, and ground

_____ 3. The two major cell populations found in connective tissue proper are:

a. fibroblasts and adipocytes
b. mast cells and lymphocytes
c. melanocytes and mesenchymal cells
d. fixed cells and wandering cells

_____ 4. Most of the volume in loose connective tissue is made up of:

a. elastic fibers
b. ground substance
c. reticular fibers
d. collagen fibers

_____ 5. The major purposes of adipose tissue in the body are:

a. strength, flexibility, elasticity
b. support, connections, conduction
c. padding, cushioning, insulating
d. absorption, compression, lubrication

_____ 6. The three major subdivisions of the extracellular fluid in the body are:

a. blood, water, and saliva
b. plasma, interstitial fluid, and lymph
c. blood, urine, and saliva
d. spinal fluid, cytosol, and blood

_____ 7. The two types of supporting connective tissue found in the body are:

a. regular and irregular connective tissue
b. collagen and reticular fibers
c. proteoglycans and chondrocytes
d. cartilage and bone

_____ 8. The three major types of cartilage found in the body are:

a. collagen, reticular, and elastic
b. regular, irregular, and dense
c. hyaline, elastic, and fibrocartilage
d. interstitial, appositional and calcified

_____ 9. The pads that lie between the vertebrae in the vertebral column contain:

a. elastic fibers
b. fibrocartilage
c. hyaline cartilage
d. bone

_____ 10. One difference between bone and cartilage is that:

a. bone is highly vascular while cartilage is not
b. bone repairs easily while cartilage does not
c. oxygen demand is high in bone while it is low in cartilage
d. a, b, and c are correct

11. Of the four primary types, the tissue that stores energy in bulk quantities is _______________.

12. The most common fibers in connective tissue proper are _______________.

13. The least specialized connective tissue in the adult body is _______________.

14. Connective tissue fibers forming a branching, interwoven framework that is tough but flexible, describes __________________ fibers.

15. Using the terms below, identify the following cells and/or structures in the examples in Figure 4-2. Place your answers in the spaces below the drawings.

Figure 4-2 Examples of Connective Tissues

adipose tissue	**bone tissue**	**dense connective tissue**
central canal	**collagen fibers**	**canaliculi**
hyaline cartilage	**loose connective tissue**	**matrix**

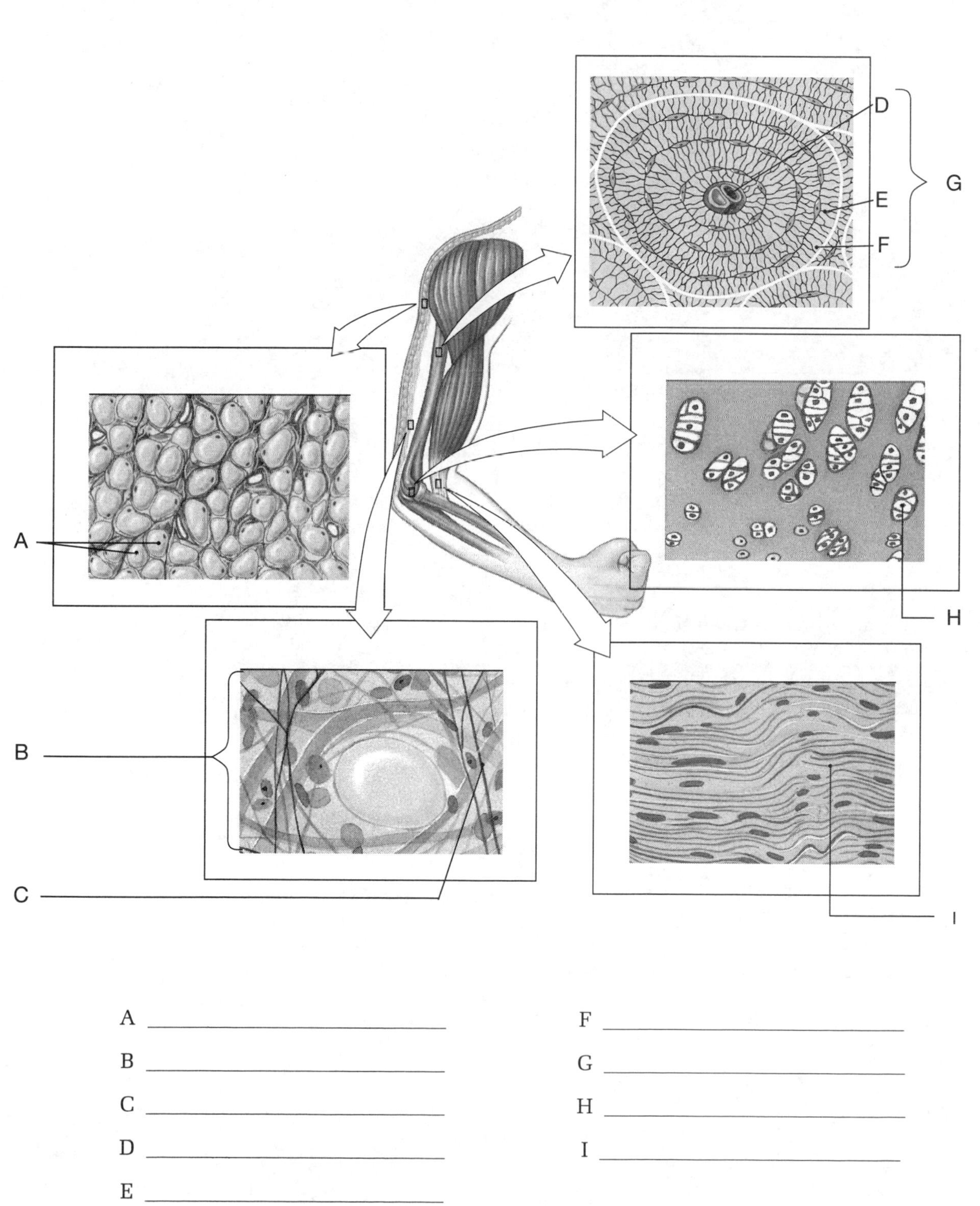

A ______________________

B ______________________

C ______________________

D ______________________

E ______________________

F ______________________

G ______________________

H ______________________

I ______________________

Objective 5 Explain how epithelial and connective tissues combine to form four different types of membranes and specify the functions of each.

_____ 1. The type of membranes that line cavities which communicate with the exterior of the body are

a. mucous membranes
b. serous membranes
c. cutaneous membranes
d. synovial membranes

_____ 2. The reduction of friction between the parietal and visceral surfaces of an internal cavity is the function of

a. cutaneous membranes
b. mucous membranes
c. serous membranes
d. synovial membranes

_____ 3. The pleura, peritoneum, and pericardium are examples of

a. mucous membranes
b. serous membranes
c. synovial membranes
d. cutaneous membranes

_____ 4. The mucous membranes that are lined by simple epithelia perform the functions of

a. digestion and circulation
b. respiration and excretion
c. absorption and secretion
d. a, b, and c are correct

_____ 5. Which of the following membranes consist of epithelial tissue and loose connective tissue?

a. mucous, serous, cutaneous, synovial
b. cutaneous and synovial
c. mucous and cutaneous
d. serous, mucous, synovial

_____ 6. The visceral and parietal pleura are made of

a. epithelial and loose connective tissue
b. adipose and areolar cells
c. squamous cells and loose connective tissues
d. serous membrane

7. The loose connective tissue of a mucous membrane is called the _______________.

8. The membranes associated with freely-moveable joints are _______________.

9. Identify the type of membrane that is located within each of the body regions identified by the arrows in Figure 4-3. Write your answers in the spaces provided below.

Figure 4-3 Body Membranes

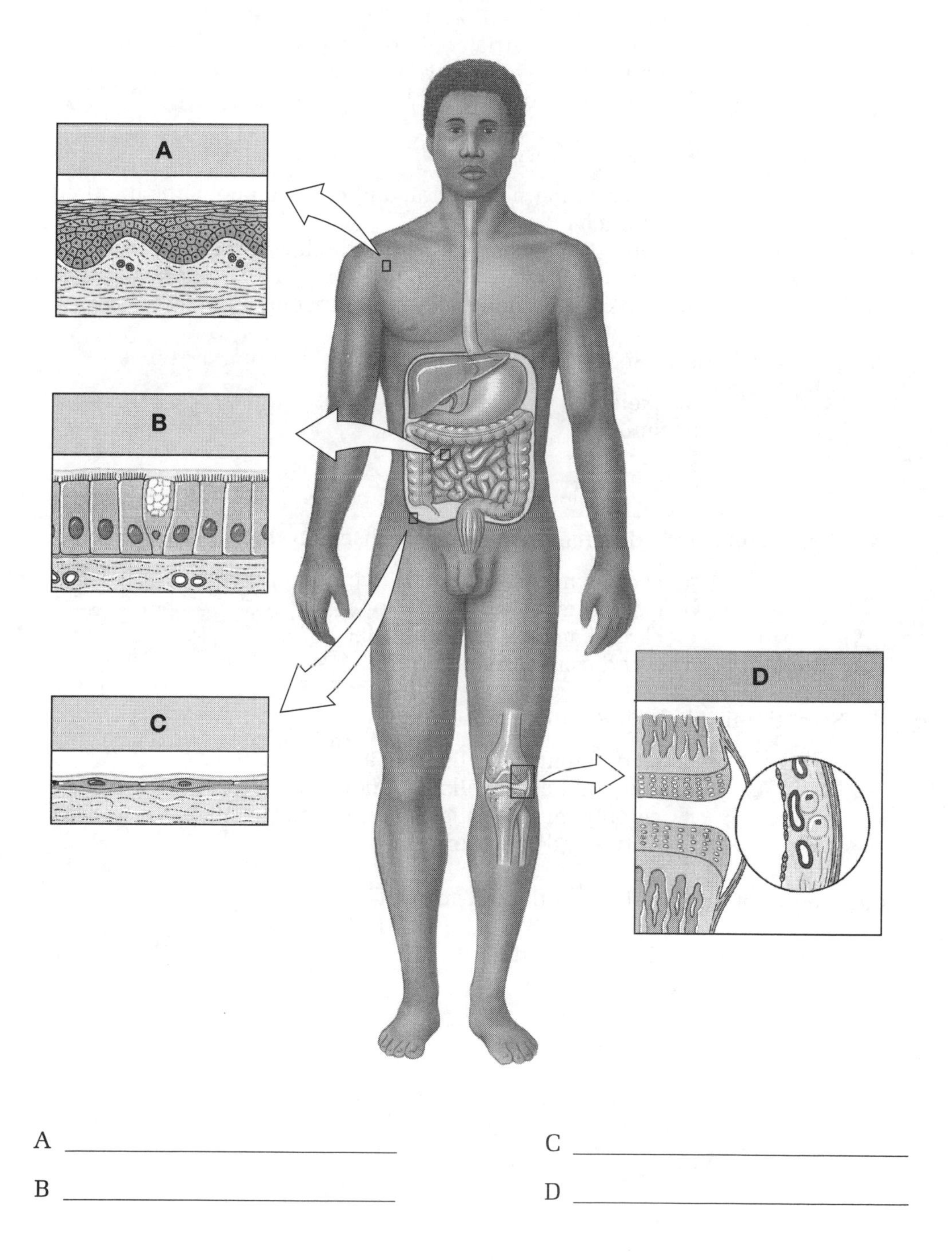

A ______________________

B ______________________

C ______________________

D ______________________

Objective 6 Describe the three types of muscle tissue and the special structural features of each type.

_____ 1. The three types of muscle tissue found in the body are

a. elastic, hyaline, fibrous
b. striated, non-striated, fibrous
c. voluntary, involuntary, nonstriated
d. skeletal, cardiac, smooth

_____ 2. Skeletal muscle fibers are very unusual because they may be

a. a foot or more in length, and each cell contains hundreds of nuclei
b. subject to pacemaker cells, which establish contraction rates
c. devoid of striations, spindle-shaped with a single nucleus
d. unlike smooth muscle cells, capable of division

_____ 3. The muscle tissue located in the walls of hollow internal organs is

a. skeletal
b. smooth
c. cardiac
d. voluntary

_____ 4. Cardiac muscle is different than skeletal muscle in that

a. cardiac muscle is under involuntary control
b. cardiac muscle has striations
c. cardiac muscle has intercalated discs
d. a and c are correct

_____ 5. Smooth muscle is like cardiac muscle in that

a. smooth muscle cells have a single nucleus
b. smooth muscle cells are under involuntary control
c. smooth muscle cells have intercalated discs
d. only a and b are correct

6. A unique feature of muscle tissue is that it is capable of _______________.

7. Striated, involuntary, multinucleated, describes the structural and functional characteristics of ____________ muscle.

8. Identify the following muscle tissue types and structures in the drawings below. Place your answers in the spaces below the drawing.

Figure 4-4 Muscle Tissue

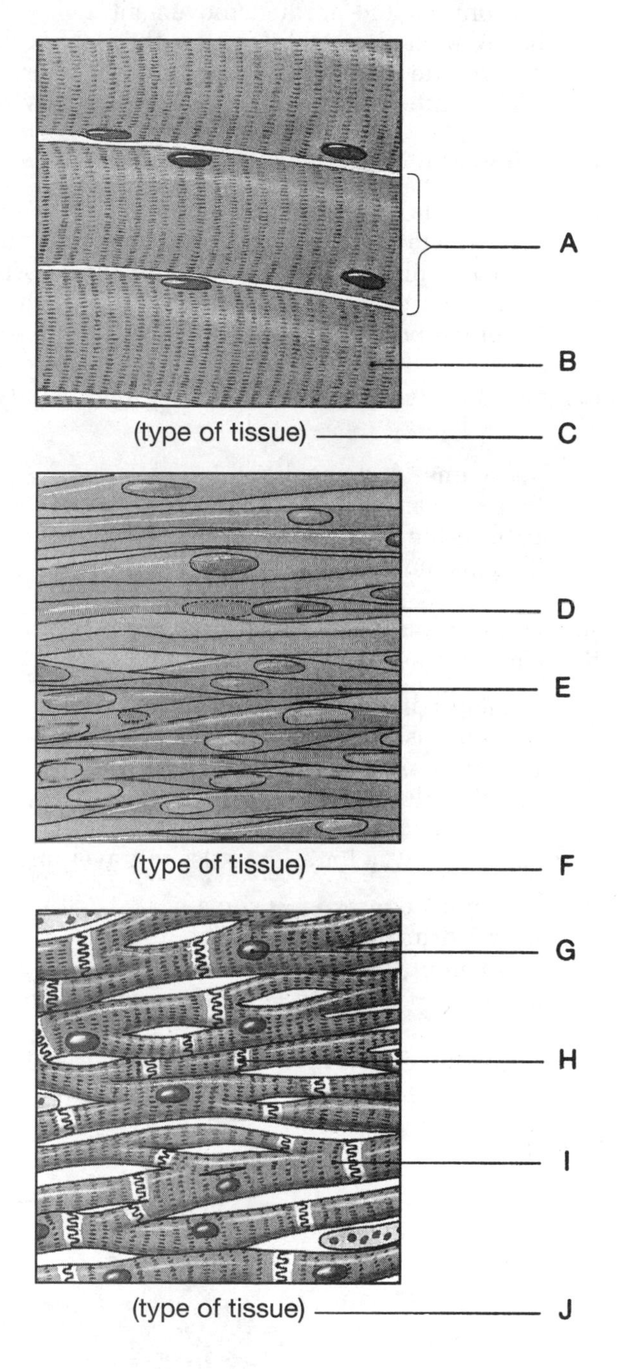

A ______________________ F ______________________

B ______________________ G ______________________

C ______________________ H ______________________

D ______________________ I ______________________

E ______________________ J ______________________

Objective 7 Discuss the basic structure and role of neural tissue.

_____ 1. Neural tissue is specialized to

a. contract and produce movement
b. conduct electrical impulses throughout the body
c. provide structural support and fill internal spaces
d. line internal passageways and body cavities

_____ 2. The major function of neurons in neural tissue is to

a. provide a supporting framework for neural tissue
b. regulate the composition of the interstitial fluid
c. act as phagocytes that defend neural tissue
d. transmit signals that take the form of changes in the transmembrane potential

_____ 3. Structurally, neurons are unique because they are the only cells in the body that have

a. lacunae and canaliculi
b. axons and dendrites
c. satellite cells and neuroglia
d. soma and stroma

_____ 4. Cells of the nervous system that function to protect, provide nourishment and support the neural tissue are

a. neuroglia
b. neurons
c. nephrons
d. all of the above

_____ 5. The unidirectional pathway for an impulse to travel through a neuron is

a. dendrite → axon → soma
b. dendrite → soma → axon
c. axon → soma → dendrite
d. soma → dendrite → axon

6. Identify the structures in the figure below. Place the answers in the spaces next to the letters.

Figure 4-5 A Typical Neuron

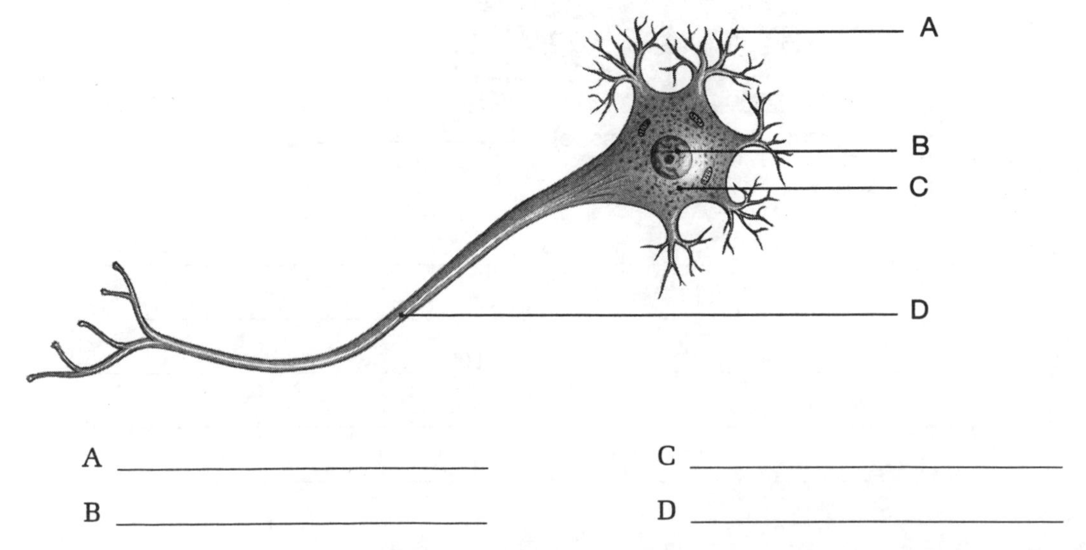

A _______________________ C _______________________

B _______________________ D _______________________

Objective 8 Explain how tissues respond in a coordinated manner to maintain homeostasis.

_____ 1. The restoration of homeostasis after an injury involves two related processes, which are

a. necrosis and fibrosis
b. infection and immunization
c. inflammation and regeneration
d. isolation and reconstruction

_____ 2. The release of histamine by mast cells at an injury site produces the following responses

a. redness, warmth, and swelling
b. bleeding, clotting, healing
c. necrosis, fibrosis, scarring
d. hematoma, shivering, retraction

_____ 3. The first evidence of the tissue repair process is

a. restoration
b. regeneration
c. isolation
d. inflammation

_____ 4. The process of tissue repair is called

a. cloning
b. regeneration
c. reconstruction
d. fibrosis

Objective 9 Describe how aging affects the tissues of the body.

_____ 1. The two primary requirements for maintaining tissue homeostasis over time are

a. exercise and supplements
b. hormonal therapy and adequate nutrition
c. metabolic turnover and adequate nutrition
d. supplements and hormonal therapy

_____ 2. Tissue changes with age include

a. thinner epithelial tissue
b. more brittle connective tissue
c. inability to repair cardiac and nerve cells
d. all of the above

_____ 3. Tissue changes with age can be the result of

a. hormonal changes
b. improper nutrition
c. inadequate amount of activity
d. all of the above

Part II: Chapter Comprehensive Exercises

A. Word Elimination

Circle the term that does not belong in each of the following groupings.

1. epithelial connective adipose muscle neural
2. protection permeability sensation storage secretion
3. microvilli gap junction tight junction desmosomes intercellular cement
4. microvilli stereocilia cilia flagella villus
5. squamous cuboidal columnar connective transitional
6. support covering transporting storing defending
7. fibroblasts macrophages matrix adipocytes mast cells
8. lymph canaliculi lacunae osteocytes matrix
9. mucous serous visceral cutaneous synovial
10. soma neuroglia dendrites nucleus axon

B. Matching

Match the terms in column "B" with the terms in column "A." Use letters for answers in the spaces provided.

COLUMN A	COLUMN B
___ 1. microvilli	A. connects bone to bone
___ 2. tendons	B. wandering cells
___ 3. ligaments	C. movement
___ 4. fibroblasts	D. found in certain joints
___ 5. mast cells	E. voluntary
___ 6. synovial membrane	F. connects muscle to bone
___ 7. muscle tissue	G. intercellular junction
___ 8. skeletal muscle	H. repair process
___ 9. synapse	I. absorption
___ 10. regeneration	J. fixed cells

C. Concept Map

This concept map organizes and summarizes the concepts presented in Chapter 4. Use the following terms to complete the map filling in the boxes identified by the circled numbers, 1-10.

adipose	**blood**	**cartilage**
columnar	**connective**	**ligaments**
loose	**neurons**	**skeletal**
immunity		

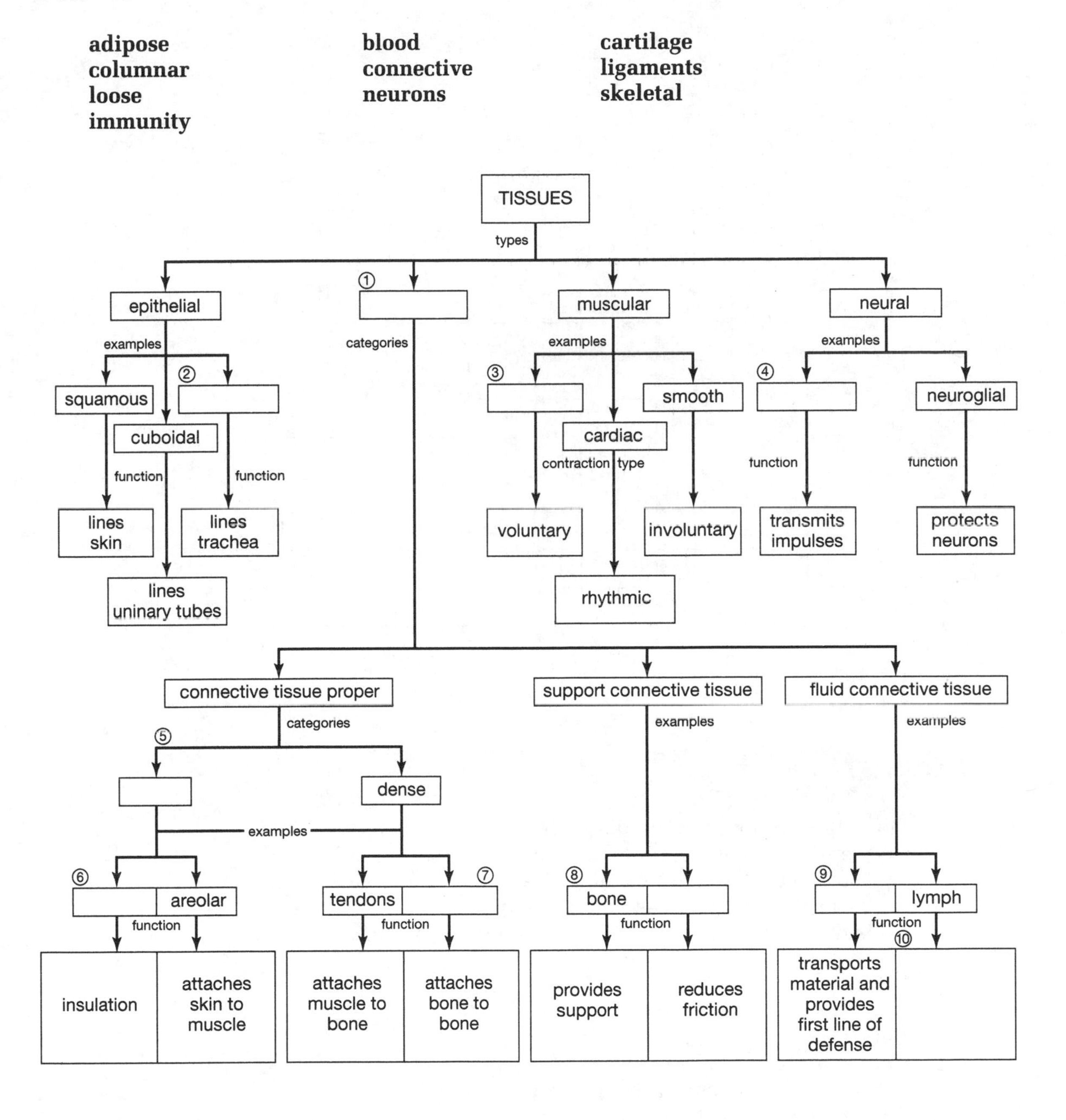

D. Crossword Puzzle

The following crossword puzzle reviews the material in Chapter 4. To complete the puzzle, you must know the answers to the clues given, and must be able to spell the terms correctly.

ACROSS

5. Discs that are found in cardiac cells.
10. Tissue that has the ability to contract.
11. Type of tissue that conducts impulses.
12. An elongated cell that can conduct impulses.
13. Cells that appear to be flat and that line some organs.

DOWN

1. Cells that connect skin to muscle.
2. Cells that have a liquid matrix.
3. The portion of a neuron that contains the organelles.
4. Tissues that connect muscle to bone.
6. The tissue that makes up the lining of many organs.
7. The tissue that has a matrix associated with the cells.
8. Tissues that connect bone to bone.
9. Cells that form concentric rings around blood vessels.

E. Short Answer Questions

Briefly answer the following questions in the spaces provided below.

1. What are the four primary tissue types in the body?

2. Summarize the four essential functions of epithelial tissue.

3. What is the functional difference between microvilli and cilia on the exposed surfaces of epithelial cells?

4. How do the processes of merocrine, apocrine, and holocrine secretions differ?

5. List the types of exocrine glands in the body and identify their secretions.

6. What three basic components are found in all connective tissues?

7. What three classifications are recognized to classify connective tissues?

8. What three basic types of fibers are found in connective tissue?

9. What four kinds of membranes consisting of epithelial and connective tissues that cover and protect other structures and tissues are found in the body?

10. What are the three types of muscle tissue?

11. What two types of cell populations make up neural tissue, and what is the primary function of each type?

CHAPTER

5

The Integumentary System

Overview

Do you know what the largest organ in the human body is? If you answered "skin" you are correct. The skin, which makes up the greatest part of the Integumentary System, is considered to be the largest structurally integrated organ in the body. In addition to the skin, the integumentary system consists of associated structures, including hair, nails and a variety of glands. Of all the body systems the integument is the only one that is seen every day. Millions of dollars are spent each year for skin, hair, and nail care to enhance their appearance and prevent disorders which may alter desirable structural features on or below the surface of the skin. All of the four tissue types studied in Chapter 4 are found in the structural makeup of the skin, each contributing to the many functions performed by the skin and its associated structures. These functions include: protection, excretion, secretion, absorption, synthesis, storage, sensitivity, and temperature regulation.

Completion and mastery of the exercises in Chapter 5 will increase your awareness of the structural and functional features of the integument and the important roles the skin, hair, and nails play in our lives.

Review of Chapter Objectives

1. Describe the general functions of the integumentary system.
2. Describe the main structural features of the epidermis and explain their functional significance.
3. Explain what accounts for individual and racial differences in skin, such as skin color.
4. Describe how the integumentary system helps to regulate body temperature.
5. Discuss the effects of ultraviolet radiation on the skin and the role played by melanocytes.
6. Discuss the functions of the skin's accessory structures.
7. Explain the mechanisms that produce hair and determine hair texture and color.
8. Explain how the skin responds to injury and repairs itself.
9. Summarize the effects of the aging process on the skin.

Part I: Objective Based Questions

Objective 1 Describe the general functions of the integumentary system.

_____ 1. The two functional components of the integument include

a. dermis and epidermis
b. hair and skin
c. cutaneous membrane and accessory structures
d. elastin and keratin

_____ 2. The skin maintains normal body temperature by regulating

a. heat gain or loss to the environment
b. the loss of body fluids
c. large reserves of lipids
d. information to the nervous system

_____ 3. The structure(s) of the integumentary system involved in protection is/are

a. hair
b. skin
c. nails
d. all of the above

_____ 4. All of the following are functions of the integumentary system except

a. protection of underlying tissue
b. synthesis of vitamin A
c. maintenance of body temperature
d. excretion

_____ 5. The link between the integument and the nervous system is in the form of

a. stimuli from the environment
b. chemicals from metabolism
c. body temperature
d. receptors in the skin

Objective 2 Describe the main structural features of the epidermis and explain their functional significance.

_____ 1. The layers of the epidermis, beginning with the deepest layer and proceeding outwardly, include the stratum

a. corneum, granulosum, spinosum, germinativum
b. granulosum, spinosum, germinativum, corneum
c. spinosum, germinativum, corneum, granulosum
d. germinativum, spinosum, granulosum, corneum

_____ 2. The layers of the epidermis where mitotic divisions occur are

a. germinativum and spinosum
b. corneum and germinativum
c. spinosum and coreum
d. mitosis occurs in all layers

_____ 3. Epidermal cells in the stratum spinosum and germinativum function as chemical factories in that they can covert

a. steroid precursors to vitamin D when exposed to sunlight
b. eleidin to keratin
c. keratohyalin to eleidin
d. a and c only

_____ 4. The lower epidermal layers of the skin are responsible for

a. capillary and nerve supply
b. support and attachment
c. nutrient and oxygen supply
d. vitamin D production

5. Keratin, a fibrous protein, would be found primarily in the _______________.

6. The area where the skin is thick, such as the palms of the hands and the soles of the feet, is called the _______________.

7. Identify the various components of the Integumentary System in Figure 5-1. Place your answers in the spaces provided below the drawing.

Figure 5-1 Components of the Integumentary System

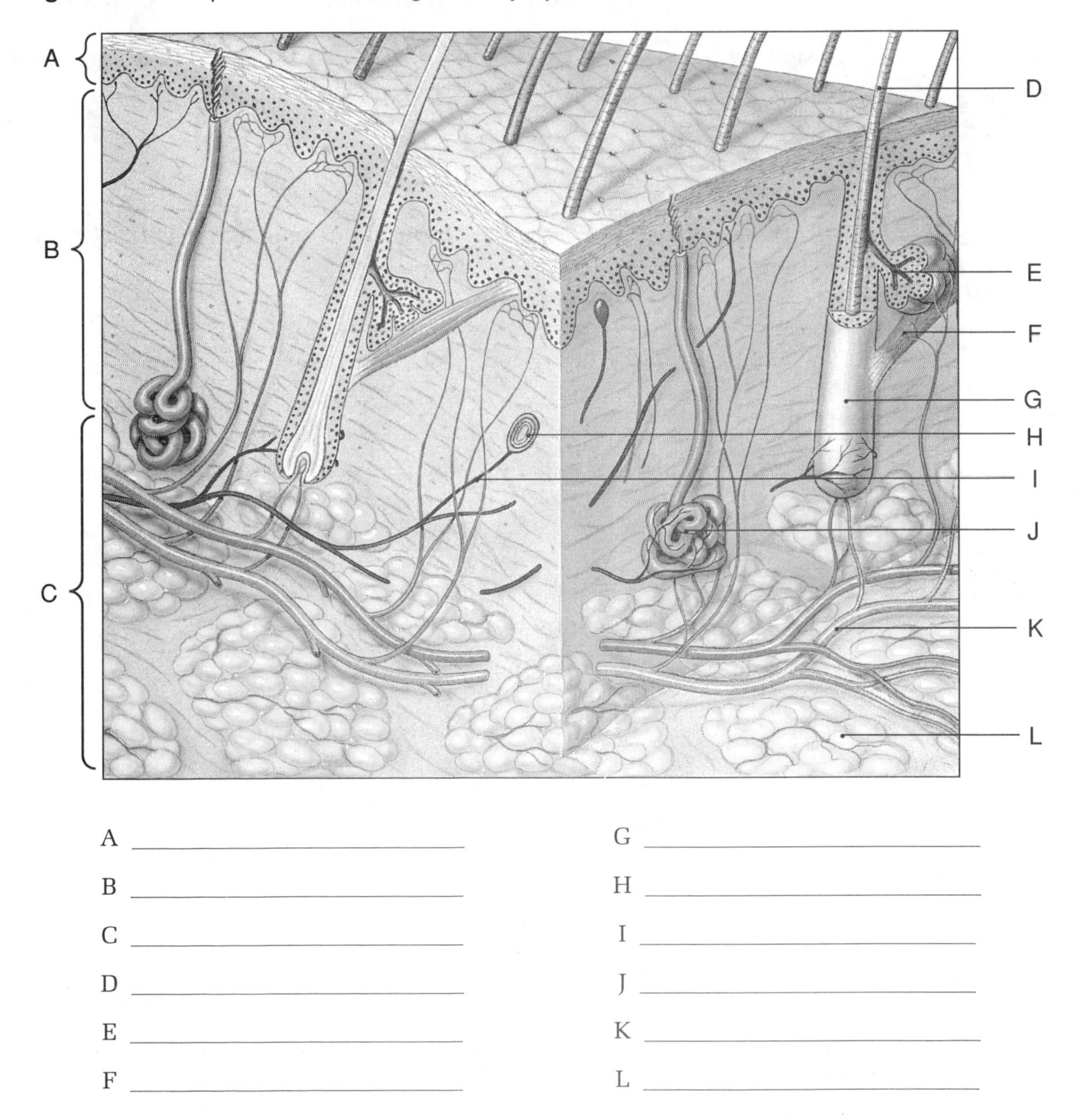

A _______________

B _______________

C _______________

D _______________

E _______________

F _______________

G _______________

H _______________

I _______________

J _______________

K _______________

L _______________

Objective 3 Explain what accounts for individual and racial differences in skin, such as skin color.

_____ 1. Differences in skin color between individuals and races reflect

a. numbers of melanocytes
b. melanocyte distribution patterns
c. levels of melanin synthesis
d. U.V. responses and nuclear activity

_____ 2. The two basic factors interacting to produce skin color are

a. sunlight and ultraviolet radiation
b. the presence of carotene and melanin
c. melanocyte production and oxygen supply
d. circulatory supply and pigment concentration and composition

_____ 3. Darker skin color is produced as a result of

a. active melanocytes
b. excessive blood supply to the skin
c. exposure to ultraviolet radiation
d. excessive numbers of melanocytes

_____ 4. Some people only "burn" when exposed to the sun. The reason they don't tan is

a. they have the gene for albinism
b. their melanocytes are inactive
c. they don't have a sufficient number of melanocytes
d. all of the above

5. The pigment produced by melanocytes is ____________________.

6. The pigment which absorbs ultraviolet radiation before it can damage mitochondrial DNA is ______________.

_____ 7. Albinos have

a. no melanocytes
b. fewer melanocytes than non-albinos
c. the same number of melanocytes as non-albinos
d. overactive melanocytes

Objective 4 Describe how the integumentary system helps regulate body temperature.

_____ 1. The glands responsible for assisting in cooling the body are

a. apocrine
b. eccrine
c. sebaceous
d. sweat

_____ 2. If body temperature drops below normal, heat is conserved by a(n) ________________ in the diameter of dermal blood vessels.

a. increase
b. the blood vessels are not affected
c. decrease
d. none of the above

_____ 3. When the body temperature becomes abnormally high, thermoregulatory homeostasis is maintained by

a. an increase in sweat gland activity and blood flow to the skin
b. a decrease in blood flow to the skin and sweat gland activity
c. an increase in blood flow to the skin and a decrease in sweat gland activity
d. an increase in sweat gland activity and a decrease in blood flow to the skin

_____ 4. The sweat glands that communicate with hair follicles are called

a. arrector pilli
b. apocrine
c. sebaceous
d. merocrine

_____ 5. Perspiration is a secretion discharged directly onto the surface of the skin by

a. apocrine sweat glands
b. sebaceous glands
c. arrector pilli
d. merocrine sweat glands

Objective 5 Discuss the effects of ultraviolet radiation on the skin and the role played by melanocytes.

_____ 1. Melanin prevents skin damage due to U. V. light by

a. covering and protecting the epidermal layers
b. absorbing U. V. light
c. protecting the nuclei of epidermal cells
d. b and c are correct

_____ 2. Excessive exposure to U.V. light may damage

a. cellular DNA resulting in mutations
b. connective tissue and cause wrinkling
c. chromosomes and causing cancer
d. all of the above

_____ 3. Excessive exposure to U.V. light may cause

a. a decrease in the number of melanocytes
b. an increase in the number of melanocytes
c. a decrease in vitamin D production
d. damage to the DNA in cells in the stratum germinativum area

4. Melanocytes prevent skin damage due to U.V. light by protecting the _____________ within the nuclei of epidermal cells.

5. Melanocytes begin producing melanin when they are exposed to _______________.

Objective 6 Discuss the functions of the skin's accessory structures.

_____ 1. Accessory structures of the skin include

a. dermis, epidermis, hypodermis
b. cutaneous and subcutaneous layers
c. hair follicles, sebaceous and sweat glands
d. blood vessels, macrophages, neurons

_____ 2. Sensible perspiration released by the eccrine sweat glands serves to

a. cool the surface of the skin
b. reduce body temperature
c. dilute harmful chemicals
d. all of the above

_____ 3. Nail production occurs at an epithelial fold not visible from the surface called the

a. eponychium
b. cuticle
c. nail root
d. lunula

_____ 4. Natural body odor is produced by the _____________ glands.

a. apocrine
b. eccrine
c. sebaceous
d. sweat

_____ 5. The smooth muscles found in the dermis of the skin are

a. arrector gracili
b. arrector pili
c. arrector brachii
d. none of the above

_____ 6. Glands associated with acne production are the

a. eccrine
b. sebaceous
c. sweat
d. apocrine

7. Protection for the tips of the fingers and toes is provided by the _____________.

8. Hair develops from a group of epidermal cells at the base of a tube-like depression called a(n) _____________.

9. The accessory structures that prevent the entry of foreign particles into the eye are the _____________.

10. The arrector pili are muscles of the integument involved in creating _____________.

11. Identify the structure in drawings (a) and (b). Place your answers in the spaces provided below the drawings.

Figure 5-2 Nail Structure **(a)** Nail Surface **(b)** Sectional View

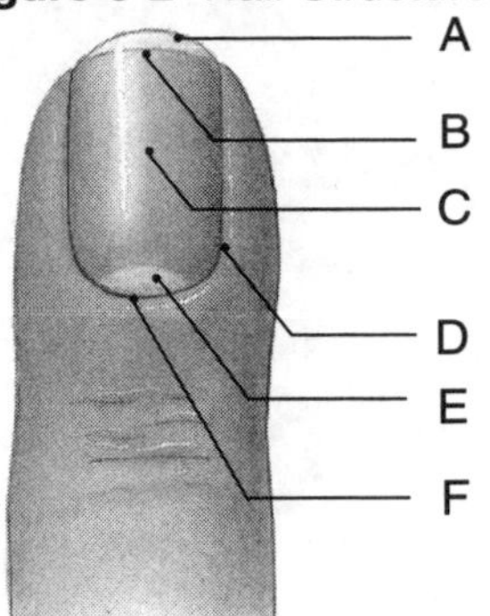

(a)

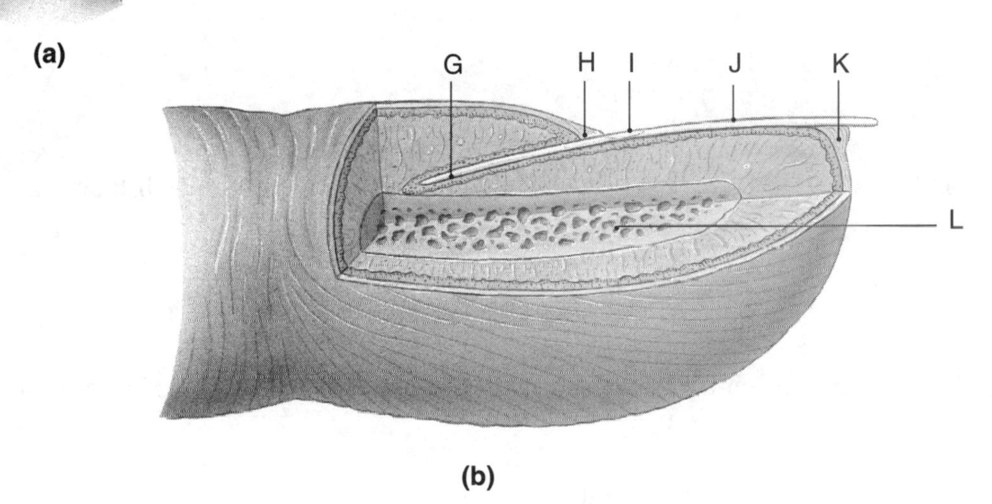

(b)

A ______________________ G ______________________

B ______________________ H ______________________

C ______________________ I ______________________

D ______________________ J ______________________

E ______________________ K ______________________

F ______________________ L ______________________

Objective 7 Explain the mechanisms that produce hair and determine hair texture and color.

_____ 1. Hair production occurs in the

a. reticular layers of the dermis
b. papillary layer of the dermis
c. hypodermis
d. stratum germinativum of the epidermis

_____ 2. Except for red hair, the natural factor responsible for varying shades of hair color is

a. number of melanocytes
b. amount of carotene production
c. the type of pigment present
d. all of the above

_____ 3. The development of gray hair is due to the

a. death of hair follicles
b. production of air bubbles in the hair
c. production of gray pigments
d. reduction of melanocyte activity

_____ 4. The various types of hair are due to the

a. arrector pili
b. follicles
c. hair papilla
d. melanocytes

_____ 5. The shaft of the hair is stiff due to the presence of a protein substance called

a. elastin
b. collagen
c. keratin
d. vellus

_____ 6. The fine "peach fuzz" hairs formed over much of the body surface are called

a. vellus
b. lunula
c. arrector pili
d. eccrines

Objective 8 Explain how the skin responds to injury and repairs itself.

_____ 1. The immediate response by the skin to an injury is

a. bleeding occurs and mast cells trigger an inflammation response
b. the epidermal cells are immediately replaced
c. fibroblasts in the dermis create scar tissue
d. the formation of a scab

_____ 2. The practical limit to the healing process of the skin is the formation of inflexible, fibrous, noncellular

a. scabs
b. skin grafts
c. ground substance
d. scar tissue

_____ 3. A second-degree burn is readily identified by the appearance of

a. charring
b. blisters
c. inflammation
d. tenderness

_____ 4. Skin regeneration to replace lost epidermal and dermal cells can occur because of the activity of

a. melanocytes
b. merocrine glands
c. stem cells
d. the cuticle

_____ 5. An essential part of the healing process during which the edges of a wound are pulled closer together is called

a. cyanosing
b. regressing
c. regeneration
d. contraction

Objective 9 Summarize the effects of the aging process in the skin.

_____ 1. Dangerously high body temperatures occur sometimes in the elderly due to

a. reduction in the number of Langerhan cells
b. decreased blood supply to the dermis
c. decreased sweat gland activity
d. b and c only

_____ 2. A factor which causes increased skin damage and infection in the elderly is

a. decreased sensitivity of the immune system
b. decreased vitamin D production
c. a decline in melanocyte activity
d. a decline in glandular activity

_____ 3. Hair turns gray or white due to

a. a decline in glandular activity
b. a decrease in the number of Langerhan cells
c. decreased melanocyte activity
d. decreased blood supply to the dermis

_____ 4. Sagging and wrinkling of the integument occurs from

a. the decline of germinativum cell activity in the epidermis
b. a decrease in the elastic fiber network of the dermis
c. a decrease in vitamin D production
d. deactivation of sweat glands

5. In older Caucasians, the skin becomes very pale because of a decline in _____________ activity.

6. In older adults, dry and scaly skin is usually a result of a decrease in ___________ activity.

Part II: Chapter Comprehensive Exercises

A. Word Elimination

Circle the term that does not belong in each of the following groupings.

1. protection fat storage excretion secretion cutaneous
2. dermis germinativum spinosum granulosum lucidum
3. touch pain secretion pressure temperature
4. hair follicles melanocytes sebaceous glands sweat glands nails
5. protect cushion stabilize insulate guard
6. sebaceous holocrine acne sebum apocrine
7. apocrine merocrine sebaceous eccrine sweat
8. cuticle arrector pili lunula nail root eponychium
9. increased immunity dry skin gray hair wrinkling weak muscles
10. papillary reticulear collagen dermis epidermis

B. Matching

Match the terms in Column "B" with the terms in Column "A." Use letters for answers in the spaces provided.

COLUMN A	COLUMN B
___ 1. cutaneous membrane	A. fibrous protein
___ 2. melanin	B. perspiration
___ 3. ultraviolet radiation	C. subcutaneous layer
___ 4. hypodermis	D. acne
___ 5. keratin	E. skin
___ 6. merocrine glands	F. body odor
___ 7. cuticle	G. synthesis of vitamin D
___ 8. apocrine glands	H. skin pigment
___ 9. regeneration	I. stem cells
___ 10. sebaceous glands	J. eponychium

C. Concept Map

This concept map summarizes and organizes some of the ideas in Chapter 5. Use the following terms to complete the map by filling in the boxes identified by the circled numbers, 1-6.

dermis	**glands**	**lubrication**
production of secretions	**sensory reception**	**vitamin D synthesis**

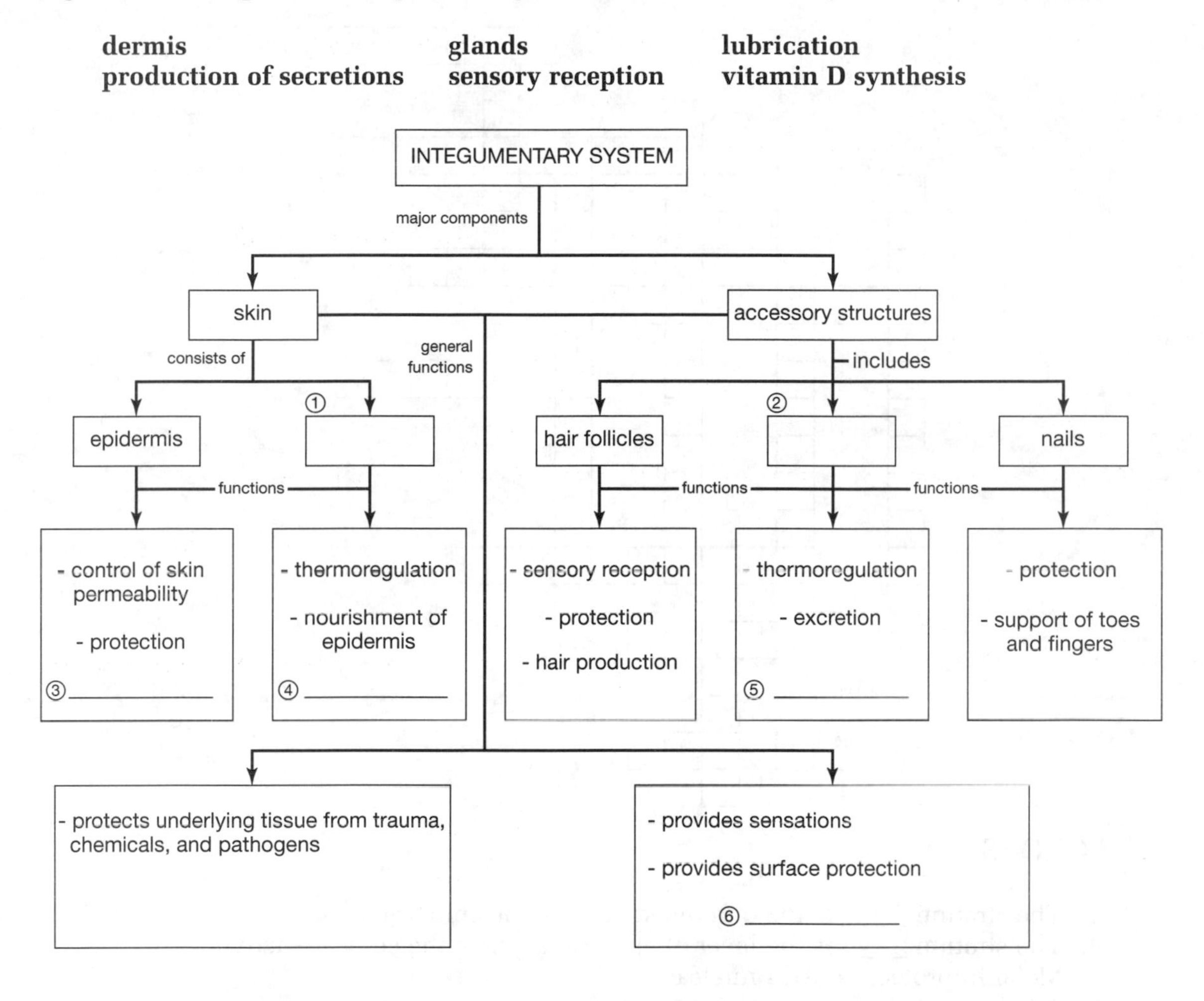

D. Crossword Puzzle

This crossword puzzle reviews the material in Chapter 5. To complete the puzzle, you must know the answers to the clues given, and you must be able to spell the terms correctly.

ACROSS

1. The stratum _____ is the outermost layer of the epidermis.
4. The stratum _____ is the layer of epidermis where the cells are actively growing.
7. Melanin protects a cell's nuclear ____.
8. The layer of skin that contains most of the accessory structures.
9. The ____ pili muscles are responsible for goosebumps.
11. Freckles are spot concentrations of ____.
13. The stratum corneum layer is a part of this layer of skin.
14. A blocked sebaceous gland can result in this skin condition.
15. The integumentary system includes the skin, hair, nails, and ____.

DOWN

2. Skin is the largest ____ of the body.
3. Albinos have the same number of ____ as non-albinos.
5. A term that refers to skin and its accessory structures.
6. The skin becomes pale if ____ are underactive.
10. The gland that helps maintain proper body temperature.
12. The gland that produces "natural body odor."

E. Short Answer Questions

Briefly answer the following questions in the spaces provided below.

1. A friend says to you, "Don't worry about what you say to her; she is thick-skinned." Anatomically speaking, what areas of the body would your friend be referring to? Why are these areas thicker?

2. Two females are discussing their dates. One of the girls says, "I liked everything about him except he had body odor." What is the cause of body odor?

3. A hypodermic needle is used to introduce drugs into the loose connective tissue of the hypodermis. Beginning on the surface of the skin in the region of the thigh, list, in order, the layers of tissue the needle would penetrate to reach the hypodermis.

4. The general public associates a tan with good health. What is wrong with this assessment?

5. Many shampoo advertisements list the ingredients, such as honey, kelp extracts, beer, vitamins and other nutrients as being beneficial to the hair. Why could this be considered false advertisement?

6. Two teenagers are discussing their problems with acne. One says to the other, "Sure wish I could get rid of these whiteheads." The other replies, "At least you don't have blackheads like I do." What is the difference between a "whitehead" and a "blackhead?"

CHAPTER

6

The Skeletal System

Overview

Can you imagine what the human body would be like if it were devoid of bones or some other form of supporting framework? Picture a "blob" utilizing amoeboid movement. Ugh! The skeletal system consists of bones and related connective tissues which include cartilage, tendons, and ligaments. Bone is a living tissue and is functionally dynamic. It provides a supportive framework for vital body organs, serves as areas for muscle attachment, articulates at joints for stability and movement and assists in respiratory movements. In addition, it provides areas of storage for substances such as calcium and lipids, and blood cell formation occurs within the cavities containing bone marrow.

The skeletal system consists of 206 bones, 80 of which are found in the axial division, and 126 which make up the appendicular division. Many of the bones of the body, especially those of the appendicular skeleton, provide a system of levers used in movement, and are utilized in numerous ways to control the environment that surrounds you every second of your life. Few people relate the importance of movement as one of the factors necessary to maintain life, however the body doesn't survive very long without the ability to produce movements.

The student study and review for this chapter includes microscopic and macroscopic features of bone, bone development and growth, location and identification of bones, joint classification, and the structure of representative articulations.

Review of Chapter Objectives

1. Describe the functions of the skeletal system.
2. Compare the structures and functions of compact and spongy bones.
3. Discuss the processes by which bones develop and grow and account for variations in their internal structure.
4. Describe the remodeling and repair of the skeleton and discuss homeostatic mechanisms responsible for regulating mineral deposition and turnover.
5. Name the components of the axial and appendicular skeletons and their functions.
6. Identify the bones of the skull.
7. Discuss the differences in the structure and function of the various vertebrae.
8. Relate the structural differences between the pectoral and pelvic girdles to their various functional roles.

9. Distinguish among different types of joints and link structural features to joint functions.
10. Describe the dynamic movements of the skeleton and the structure of representative articulations.
11. Explain the relationship between joint structure and mobility, using specific examples.
12. Discuss the functional relationships between the skeletal system and other body systems.

Part I: Objective Based Questions

Objective 1 Describe the functions of the skeletal system.

_____ 1. The function(s) of the skeletal system is/are

a. a storage area for calcium and lipids
b. it is involved in blood cell formation
c. it provides structural support for the entire body
d. all of the above

_____ 2. Storage of lipids that represent an important energy reserve in bone occur in areas of

a. red marrow
b. yellow marrow
c. bone matrix
d. ground substance

_____ 3. Of the five major functions of the skeleton, the two that depend on the dynamic nature of bone are

a. storage and support
b. blood cell formation and lipid storage
c. storage of lipids and calcium
d. support and blood cell formation

_____ 4. The support tissues in the body consist of

a. bone and muscle
b. collagen and elastin
c. bone and cartilage
d. all of the above

Objective 2 Compare the structures and functions of compact and spongy bones.

_____ 1. One of the basic histological differences between compact and spongy bone is that in compact bone

a. the basic functional unit is the Haversian System
b. there is a lamella arrangement
c. there are plates or struts called trabeculae
d. osteons are not present

_____ 2. Compact bone is usually found where

a. bones are not heavily stressed
b. stresses arrive from many directions
c. trabeculae are aligned with extensive cross-bracing
d. stresses arrive from a limited range of directions

_____ 3. Spongy and cancellous bone, unlike compact bone, resembles a network of bony struts separated by spaces that are normally filled with

a. osteocytes
b. lacunae
c. bone marrow
d. lamella

_____ 4. Spongy bone is found primarily at the ____________ of long bones.

a. bone surfaces, except inside joint capsules
b. expanded ends of long bones, where they articulate with other skeletal elements
c. axis of the diaphysis
d. exterior region of the bone shaft to withstand forces applied at either end

5. The basic functional unit of compact bones is the ______________________.

6. The expanded region of a long bone consisting of spongy bone is called the ___________________.

7. Identify the following structures in a long bone. Place your answers in the spaces below the drawing.

Figure 6-1 Structure of a Long Bone

compact bone	**distal epiphysis**	**marrow cavity**
blood vessels	**spongy bone**	**proximal epiphysis**
articular cartilage	**endosteum**	**diaphysis**
periosteum		

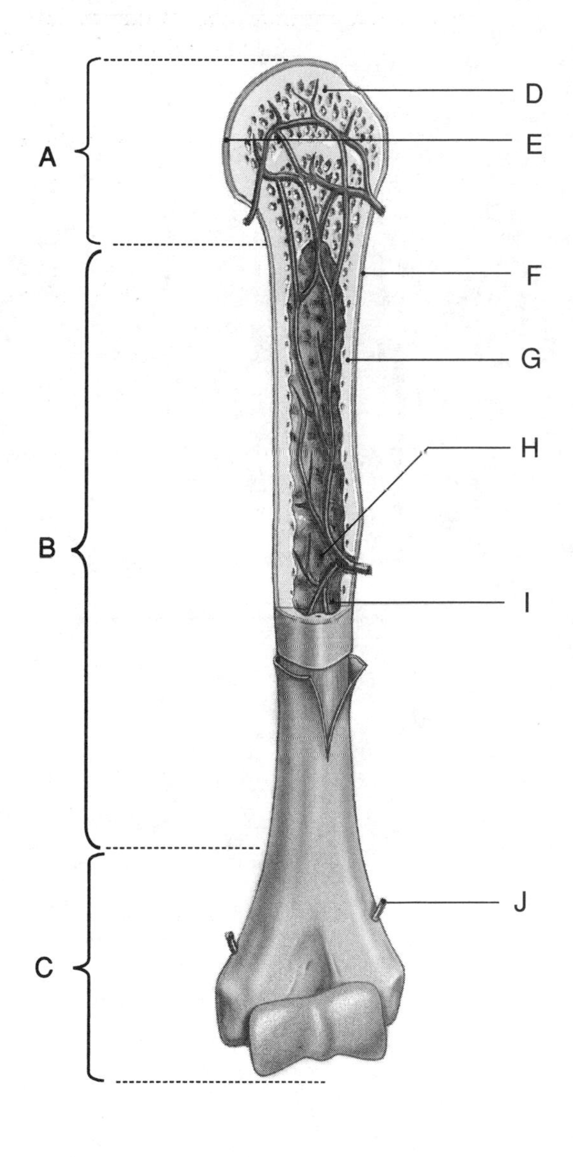

A ______________________ F ______________________

B ______________________ G ______________________

C ______________________ H ______________________

D ______________________ I ______________________

E ______________________ J ______________________

8. Identify the following structures in a typical bone. Place your answers in the spaces below the drawing.

Figure 6-2 Structure of a Typical Bone

osteons	**vein**	**spongy bone**
artery	**compact bone**	**central canal**

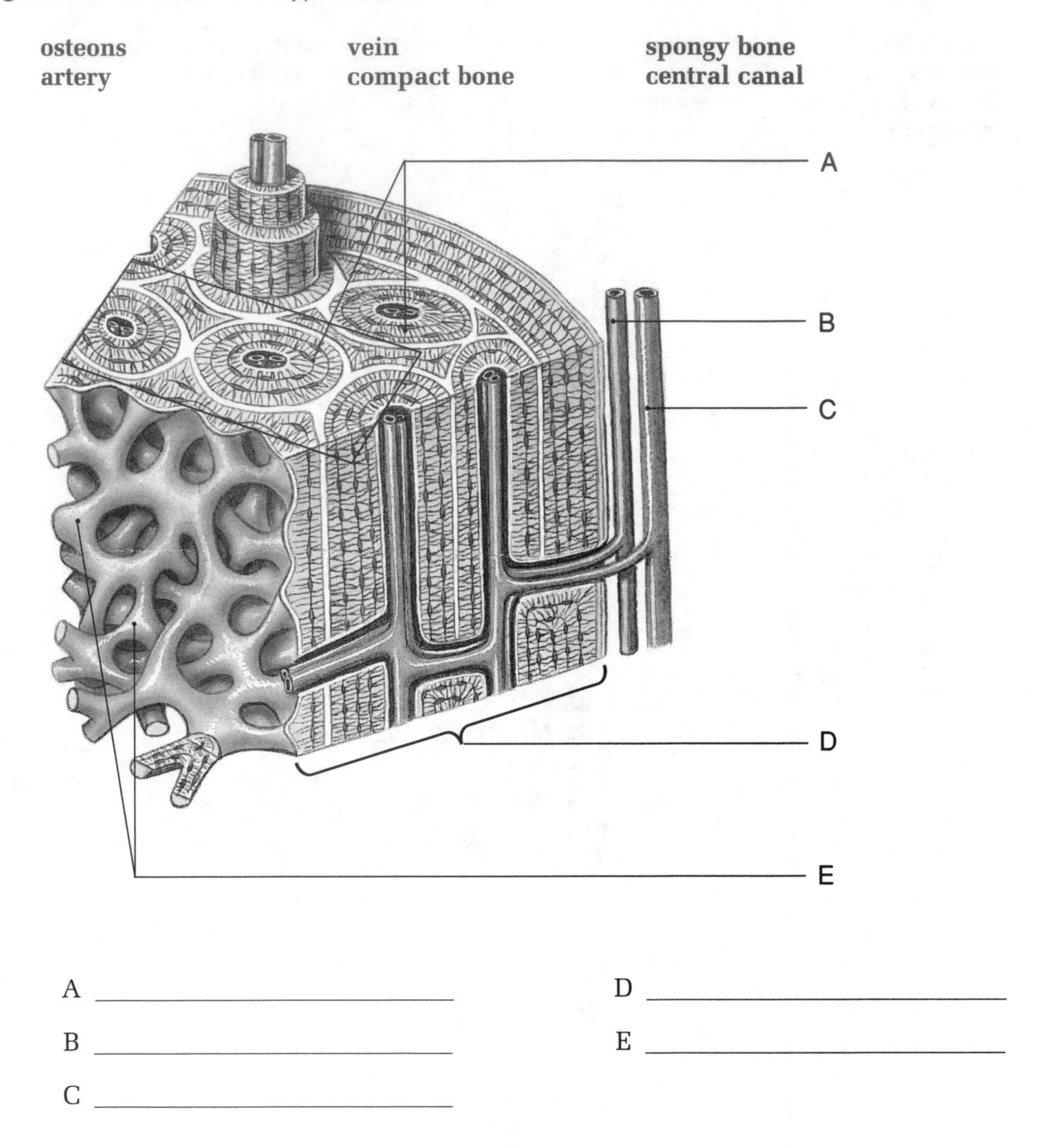

A ______________________

B ______________________

C ______________________

D ______________________

E ______________________

Objective 3 Discuss the processes by which bones develop and grow and account for variations in their internal structures.

_____ 1. The ossification process first occurs in the

a. diaphysis
b. distal ends of the bone
c. epiphysis
d. proximal end of the bone

_____ 2. From the following steps identify the correct sequence in the process of endochondral ossification.

(1) blood vessels invade the perichondrium
(2) osteoclasts create a marrow-cavity
(3) chondrocytes enlarge and calcify
(4) osteoblasts replace calcified cartilage with spongy bone
(5) the perichondrium is converted into a periosteum and the inner layer produces bone.

a. 1, 5, 3, 4, 2
b. 3, 1, 5, 4, 2
c. 1, 3, 5, 4, 2
d. 2, 3, 1, 5, 4

_____ 3. Secondary ossification centers occur

a. in the medullary cavity of the diaphysis
b. at the outer surface of the diaphysis
c. in the center of the epiphysis
d. at the surface of the epiphysis

_____ 4. When sexual hormone production increases, bone growth

a. slows down
b. increases, but only in thickness
c. accelerates rapidly
d. is not affected

_____ 5. Endochondral ossification begins with the formation of a

a. cartilage model
b. calcified model
c. membranous mode
d. a fibrous connective tissue model

_____ 6. In intramembranous ossification

a. osteoblasts differentiate within a connective tissue
b. osteoblasts cluster together in ossification centers
c. spicules radiate out from ossification centers to join with neighboring spicules
d. both a and c are correct

_____ 7. The presence of an epiphyseal line indicates

a. epiphyseal growth has ended
b. epiphyseal growth is beginning
c. growth in bone diameter is beginning
d. the bone is fractured at that location

8. Identify the following structures in Figure 6-3. Place your answers in the spaces provided below the drawings.

Figure 6-3 Endochondral Ossification

disintegrating chondrocytes
blood vessels
marrow cavity
diaphysis
enlarging chondrocytes
epiphyseal plate
epiphysis

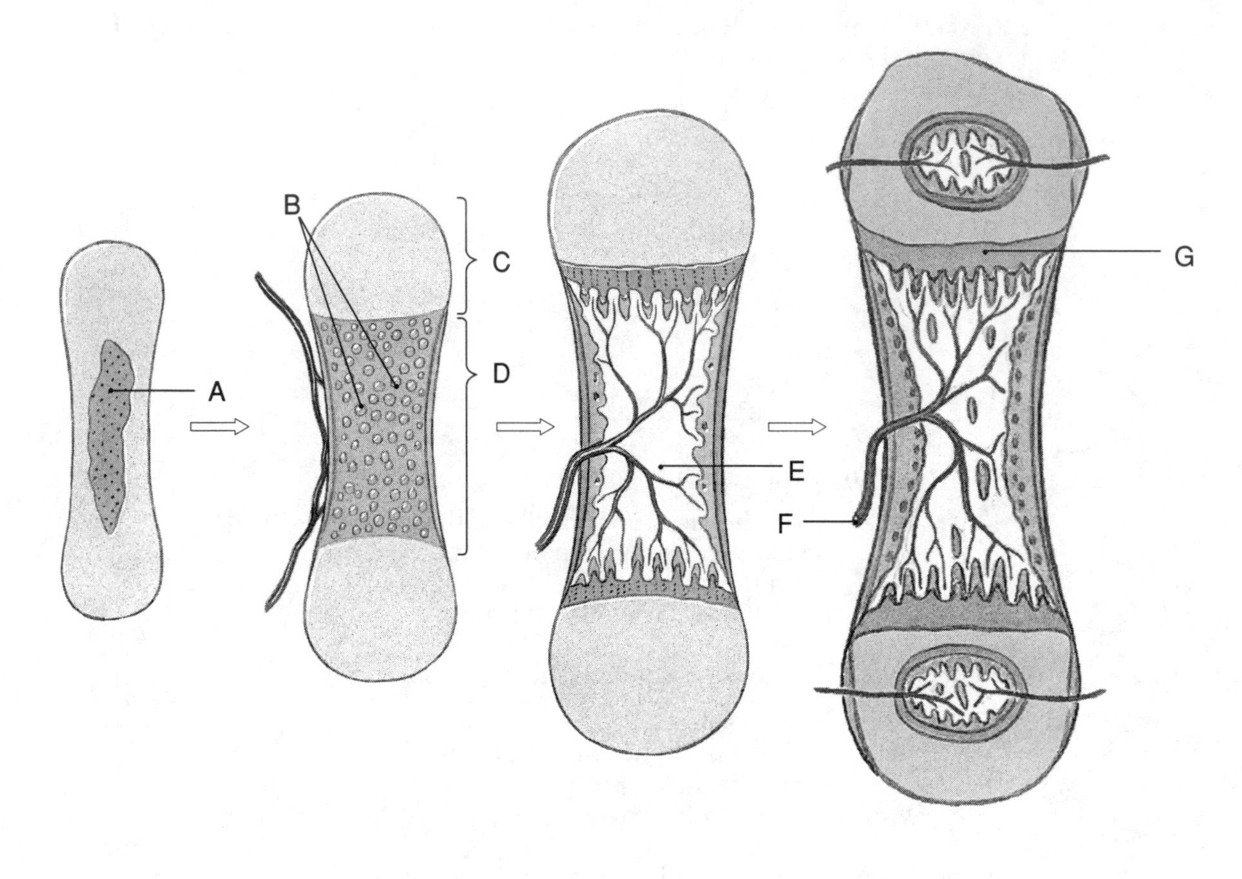

A ______________________________

B ______________________________

C ______________________________

D ______________________________

E ______________________________

F ______________________________

G ______________________________

Objective 4 Describe the remodeling and repair of the skeleton and discuss homeostatic mechanisms responsible for regulating mineral deposition and turnover.

_____ 1. The organic and mineral components of the bone matrix are continually being recycled and renewed through the process of

a. calcification
b. ossification
c. remodeling
d. regeneration

_____ 2. During bone renewal, as one osteon forms through the activity of osteoblasts, another is destroyed by

a. osteocytes
b. osteoclasts
c. chondrocytes
d. calcification

_____ 3. The condition that produces a reduction in bone mass sufficient to compromise normal function is

a. osteopenia
b. osteitis deformans
c. osteomyelitis
d. osteoporosis

_____ 4. In order for bone to repair itself

a. osteoblast activity must be greater than osteoclast activity
b. osteocyte activity must increase
c. osteoclast activity must be greater than osteoblast activity
d. osteoblast activity must be greater than osteocyte activity

_____ 5. Elevated levels of calcium ions in the blood stimulates the secretion of the hormone

a. calciton
b. parathyroid hormone
c. growth hormone
d. thyroxine

_____ 6. Of the following selections, the one that describes a homeostatic mechanism of the skeleton is

a. as one osteon forms through the activity of osteoblasts, another is destroyed by osteoclasts
b. mineral absorption from the mother's bloodstream during pre natal development
c. Vitamin D stimulating the absorption and transport of calcium and phosphate ions
d. a, b, and c are correct

_____ 7. Vitamin D is necessary for

a. collagen formation
b. reducing osteoblast activity
c. increasing osteoclast activity
d. absorption and transport of calcium and phosphate ions

_____ 8. The parathyroid hormone

a. stimulates osteoclast activity
b. increases the rate of calcium absorption
c. decreases the rate of calcium excretion
d. all of the above

_____ 9. A lack of exercise causes bones to become

a. thicker
b. thin and brittle
c. longer
d. all of the above

_____ 10. The hormone calcitonin functions to

a. decrease the rate of calcium excretion
b. decrease the rate of calcium absorption
c. stimulate osteoclast activity
d. decrease the level of calcium in the blood.

Objective 5 Name the components of the axial and appendicular skeletons and their functions.

_____ 1. The axial skeleton can be recognized because it

a. includes the bones of the arms and legs
b. forms the longitudinal axis of the body
c. includes the bones of the pectoral and pelvic girdles
d. a, b, and c are correct

_____ 2. Of the following selections, the one that includes bones found exclusively in the axial skeleton is

a. ear ossicles, scapula, clavicle, sternum, hyoid
b. vertebral, ischium, ilium, skull, ribs
c. skull, vertebrae, ribs, sternum, hyoid
d. sacrum, ear ossicles, skull, scapula, ilium

_____ 3. The axial skeleton creates a framework that supports and protects organ systems in the

a. dorsal and ventral body cavities
b. pleural cavity
c. abdominal cavity
d. pericardial cavity

_____ 4. The bones that make up the appendicular division of the skeleton consist of the

a. bones which form the longitudinal axis of the body
b. rib cage and vertebral column
c. skull and the arms and legs
d. pectoral and pelvic girdles, and the upper and lower limbs

_____ 5. One of the major functional differences between the appendicular and axial divisions is that the appendicular division

a. serves to adjust the position of the head, neck, and trunk
b. protects organ systems in the dorsal and ventral body cavities
c. makes you an active, mobile individual
d. assists directly in respiratory movements

_____ 6. A composite structure that includes portions of both the appendicular and axial skeleton is the

a. pelvis
b. pectoral girdle
c. pelvic girdle
d. a, b, and c are correct

_____ 7. The unique compromise of the articulations in the appendicular skeleton is

a. the stronger the joint, the less restricted the range of motion
b. the weaker the joint, the more restricted the range of motion
c. the stronger the joint, the more restricted the range of motion
d. the strength of the joint and range of motion are unrelated

8. The bones of the skeleton provide an extensive surface area for the attachment of _______________.

9. The appendicular skeleton includes the bones of the pectoral and pelvic girdles and the upper and lower _______________.

10. The only direct connection between the pectoral girdle and the axial skeleton is the _______________.

11. Identify all the bones of the skeleton designated by leader lines. Place your answers in the spaces provided below the drawing.

Figure 6-4 The Skeleton - Axial and Appendicular Divisions

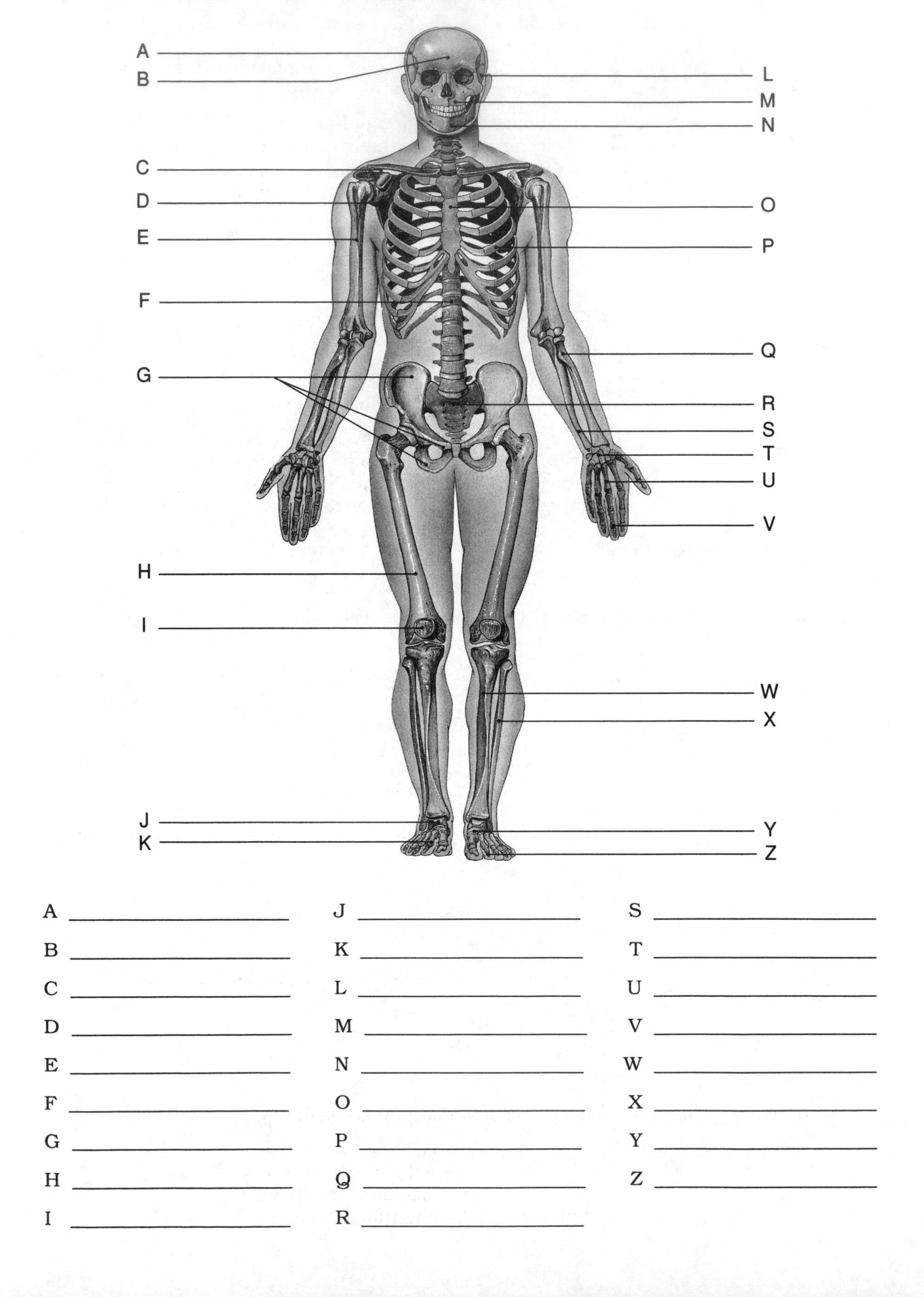

A ______________________ J ______________________ S ______________________

B ______________________ K ______________________ T ______________________

C ______________________ L ______________________ U ______________________

D ______________________ M ______________________ V ______________________

E ______________________ N ______________________ W ______________________

F ______________________ O ______________________ X ______________________

G ______________________ P ______________________ Y ______________________

H ______________________ Q ______________________ Z ______________________

I ______________________ R ______________________

Objective 6 Identify the bones of the skull.

_____ 1. At birth, the bones of the skull can be distorted without damage because of the

a. cranial foramina
b. fontanels
c. alveolar process
d. cranial ligaments

_____ 2. The most significant growth in the skull occurs before age five when the

a. brain stops growing and cranial sutures develop
b. brain development is incomplete until maturity
c. cranium of a child is larger than that of an adult
d. ossification and articulation process is completed

_____ 3. The bones of the cranium that exclusively represent single, unpaired bones are

a. occipital, parietal, frontal, temporal
b. occipital, frontal, sphenoid, ethmoid
c. frontal, temporal, parietal, sphenoid
d. ethmoid, frontal, parietal, temporal

_____ 4. The paired bones of the cranium are the

a. ethmoid and sphenoid
b. frontal and occipital
c. occipital and parietal
d. parietal and temporal

_____ 5. The associated bones of the skull include the

a. mandible and maxilla
b. nasal and lacrimal
c. hyoid and auditory ossicles
d. vomer and palatine

_____ 6. The sutures that articulate the bones of the skull are the

a. parietal, occipital, frontal, temporal
b. calvaria, foramen, condyloid, lacerum
c. posterior, anterior, laternal, dorsal
d. lambdoidal, sagittal, coronal, squamosal

_____ 7. Foramina, located on the bones of the skull, serve primarily as passageways for

a. airways and ducts for secretions
b. sight and sound
c. nerves and blood vessels
d. muscle fibers and nerve tissue

_____ 8. Areas of the head that are involved in the formation of the skull are called

a. fontanels
b. craniocephalic
c. craniulums
d. ossification centers

_____ 9. The sinuses or internal chambers in the skull are found in

a. sphenoid, ethmoid, vomer, lacrimal bones
b. sphenoid, frontal, ethmoid, maxillary bones
c. ethmoid, frontal, lacrimal, maxillary bones
d. lacrimal, vomer, ethmoid, frontal bones

10. Identify the following structures in Figure 6-5. Place your answers in the spaces provided below the drawing.

Figure 6-5 Lateral View of the Skull

Frontal	**Parietal**	**Mandible**
Mastoid process	**Sphenoid**	**Ethmoid**
Temporal	**Zygomatic bone**	**Maxilla**
Occipital	**Styloid process**	**Nasal**
Lacrimal	**Zygomatic arch**	

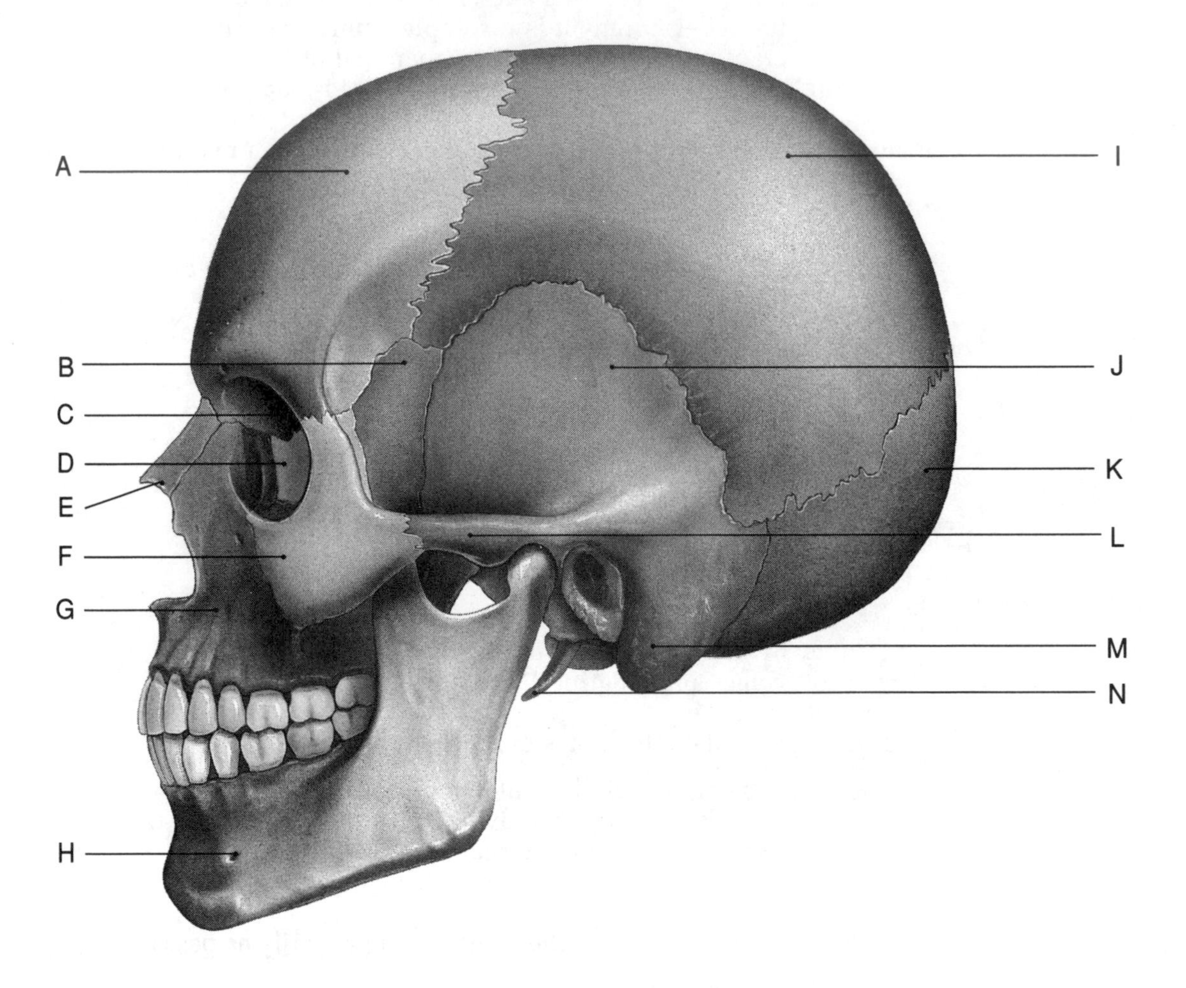

A ______________________ H ______________________

B ______________________ I ______________________

C ______________________ J ______________________

D ______________________ K ______________________

E ______________________ L ______________________

F ______________________ M ______________________

G ______________________ N ______________________

11. Identify the following structures in Figure 6-6. Place your answers in the spaces provided below the drawing.

Figure 6-6 Anterior View of the Skull

Frontal	**Temporal**	**Zygomatic**
Vomer	**Parietal**	**Ethmoid**
Maxilla	**Nasal**	**Sphenoid**
Lacrimal	**Mandible**	**Nasal Septum**

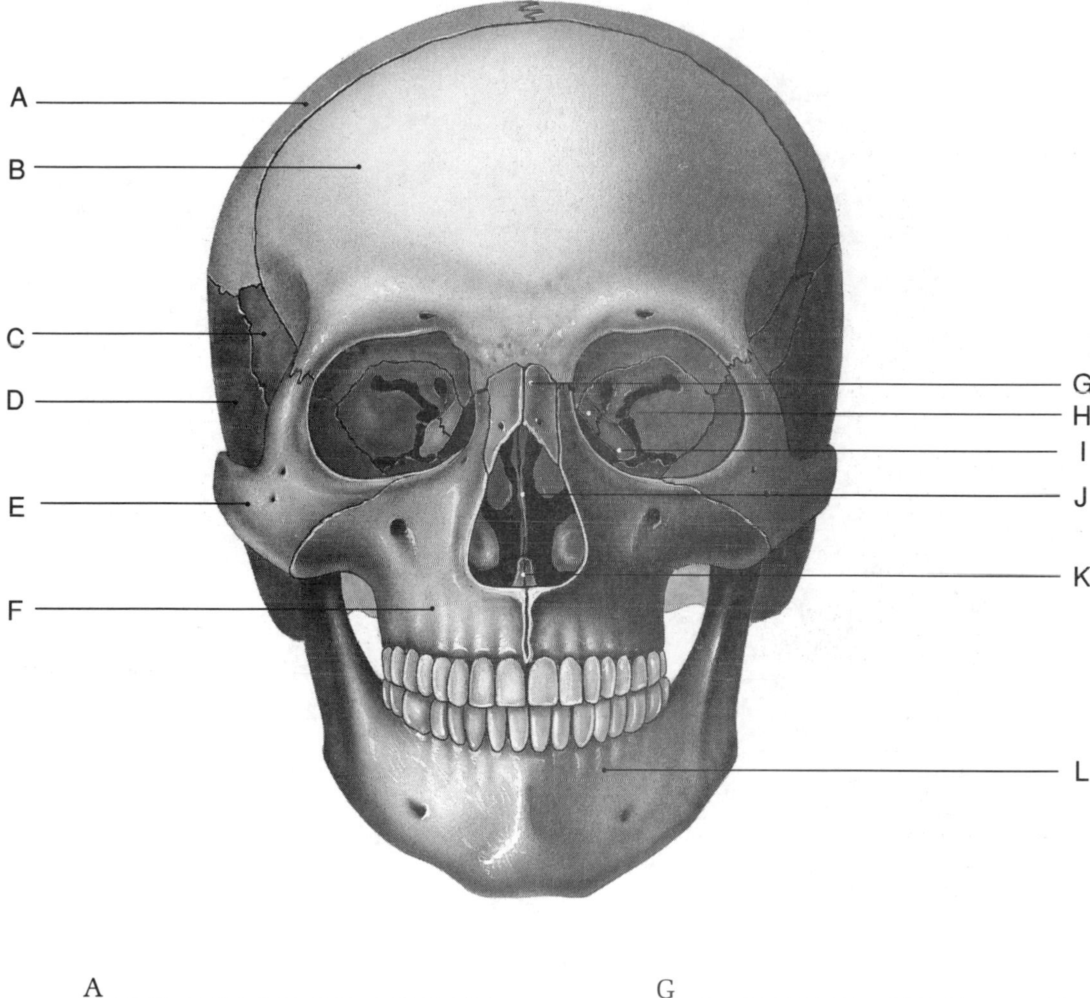

A ______________________ G ______________________

B ______________________ H ______________________

C ______________________ I ______________________

D ______________________ J ______________________

E ______________________ K ______________________

F ______________________ L ______________________

12. Identify the following structures in Figure 6-7. Place your answers in the spaces provided below the drawing.

Figure 6-7 Inferior View of the Skull

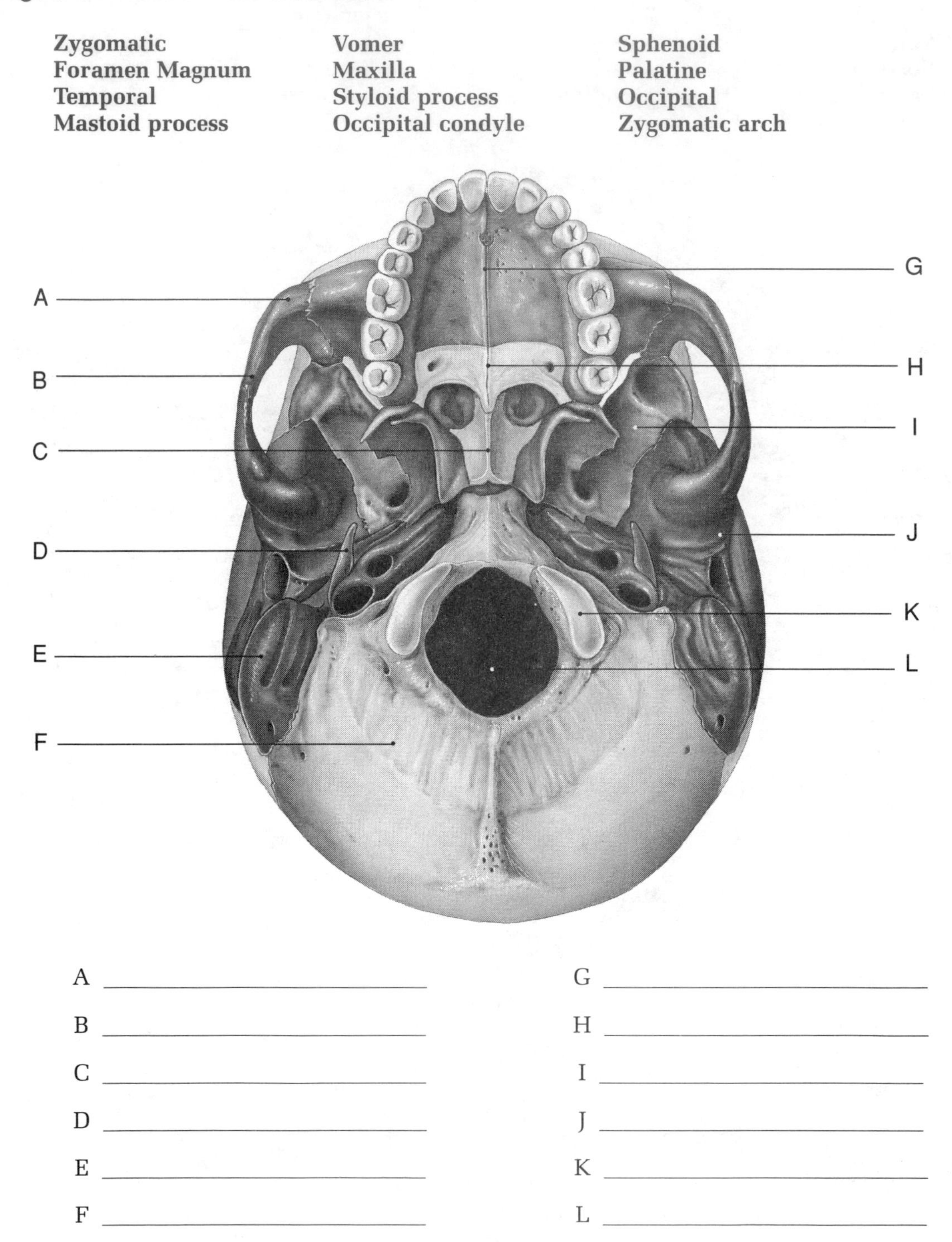

13. Identify the following structures in Figure 6-8. Place your answers in the spaces provided below the drawings.

Figure 6-8 Infant Skull

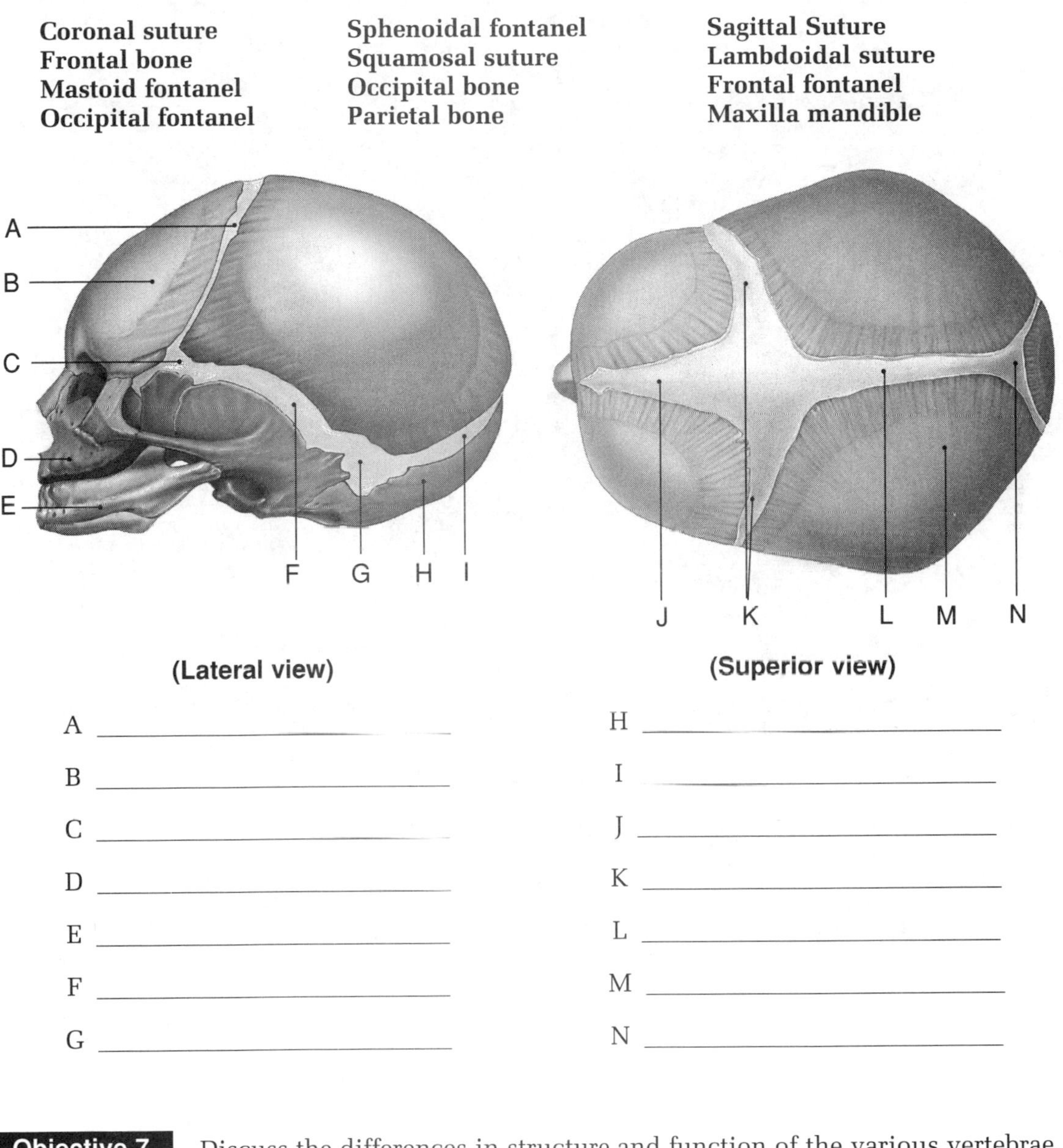

Objective 7 Discuss the differences in structure and function of the various vertebrae.

_____ 1. The vertebrae that indirectly effect changes in the volume of the rib cage are the

a. cervical vertebrae
b. thoracic vertebrae
c. lumbar vertebrae
d. sacral vertebrae

_____ 2. The most massive and least mobile of the vertebrae are the

a. thoracic
b. cervical
c. lumbar
d. sacral

_____ 3. Of the following selections, the one which correctly identifies the sequence of the vertebrae from superior to inferior is

a. thoracic, cervical, lumbar, coccyx, sacrum
b. cervical, lumbar, thoracic, sacrum, coccyx
c. cervical, thoracic, lumbar, sacrum, coccyx
d. cervical, thoracic, sacrum, lumbar, coccyx

_____ 4. C1 and C2 have specific names which are the

a. sacrum and coccyx
b. atlas and axis
c. cervical and costal
d. sacrum and coccyx

_____ 5. The secondary curves that appear several months after birth are the

a. sacral and lumbar
b. thoracic and lumbar
c. cervical and lumbar
d. thoracic and sacral

_____ 6. The vertebral column contains __________ cervical vertebrae, __________ thoracic vertebrae and __________ lumbar vertebrae

a. 7, 12, 5
b. 5, 12, 7
c. 7, 5, 12
d. 12, 7, 5

_____ 7. Cervical vertebrae can usually be distinquished from other vertebrae by the presence of

a. transverse processes
b. transverse foramina
c. large spinous processes
d. facets for articulations of the ribs

_____ 8. The odontoid process is found in the

a. sacrum
b. coccyx
c. axis
d. atlas

_____ 9. Costal processes are located on the __________ vertebrae

a. thoracic
b. cervical
c. lumbar
d. sacral

_____ 10. Thoracic vertebrae can be distinquished from other vertebrae by the presence of

a. transverse processes
b. transverse foramina
c. facets for the articulation of ribs
d. costal cartilages

11. An attachment site for a muscle that closes the anal opening is the primary purpose of the ________.

12. The medium, heart-shaped, flat face which serves as a facet for rib articulation in the thoracic vertebrae is called the ____________.

13. The vertebrae that stabilize relative positions of the brain and spinal cord are the ____________ vertebrae.

14. Identify the following structures in Figure 6-9. Place your answers in the spaces provided below the drawing.

Figure 6-9 The Vertebral Column

Cervical
Thoracic
Sacrum
Coccygeal
Lumbar

A
B
C
D
E

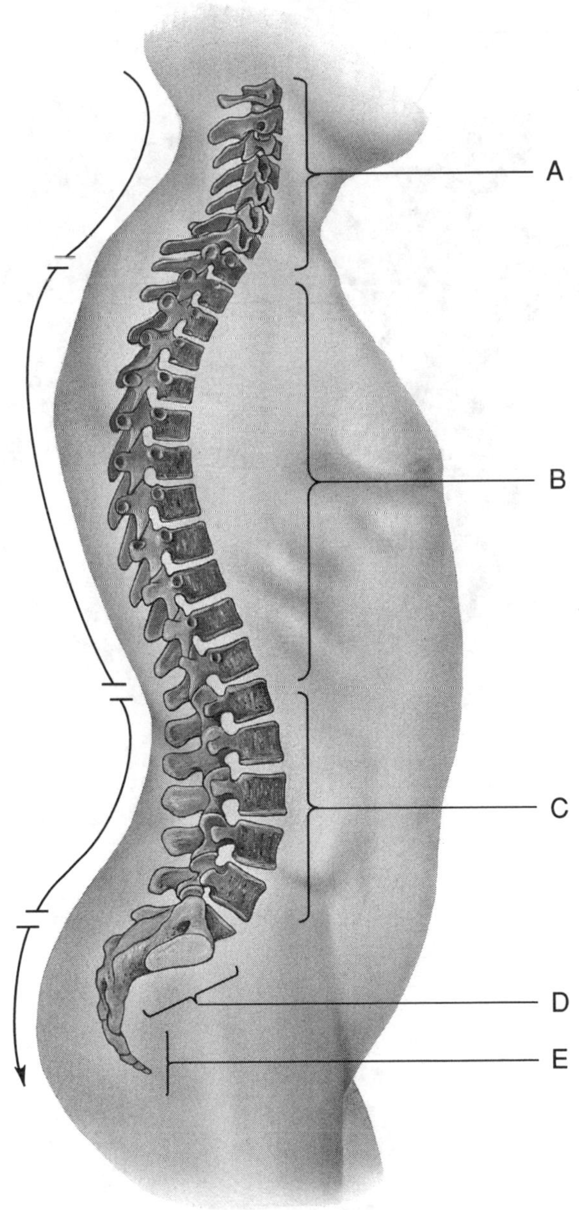

A ______________________

B ______________________

C ______________________

D ______________________

E ______________________

15. Identify the following structures in Figure 6-10. Place your answers in the spaces provided below the drawing.

Figure 6-10 A Typical Vertebrae

Vertebral body	**Spinous process**	**Vertebral foramen**
Transverse process	**Pedicle**	**Lamina**

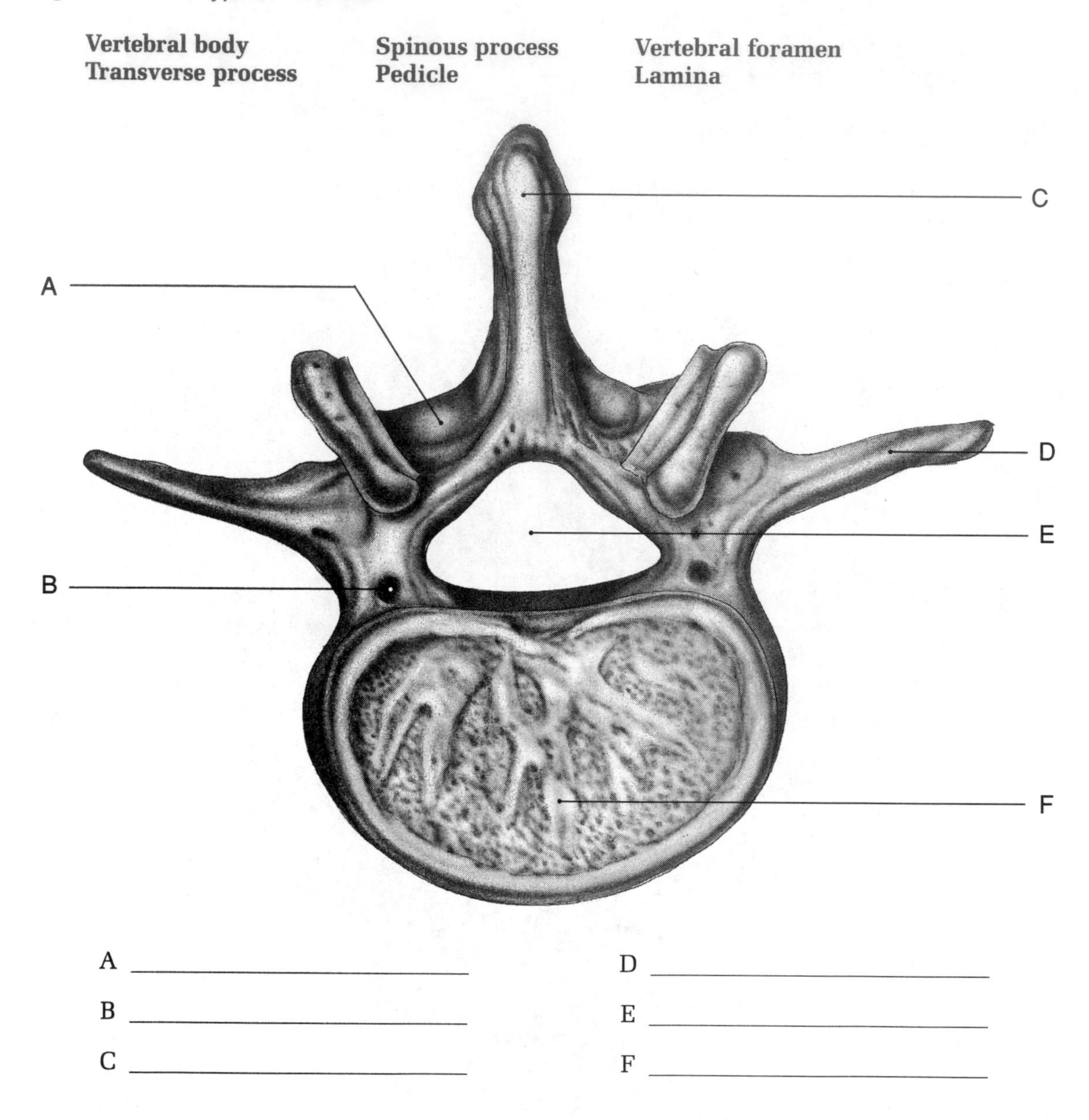

A ______________________ D ______________________

B ______________________ E ______________________

C ______________________ F ______________________

16. Identify the following structures in Figure 6-11. Place your answers in the spaces provided below the drawing.

Figure 6-11 The Ribs and Sternum

True ribs	**Manubrium**	**False ribs**
Xiphoid process	**Sternum**	**Body of sternum**
Costal cartilages	**Floating ribs**	

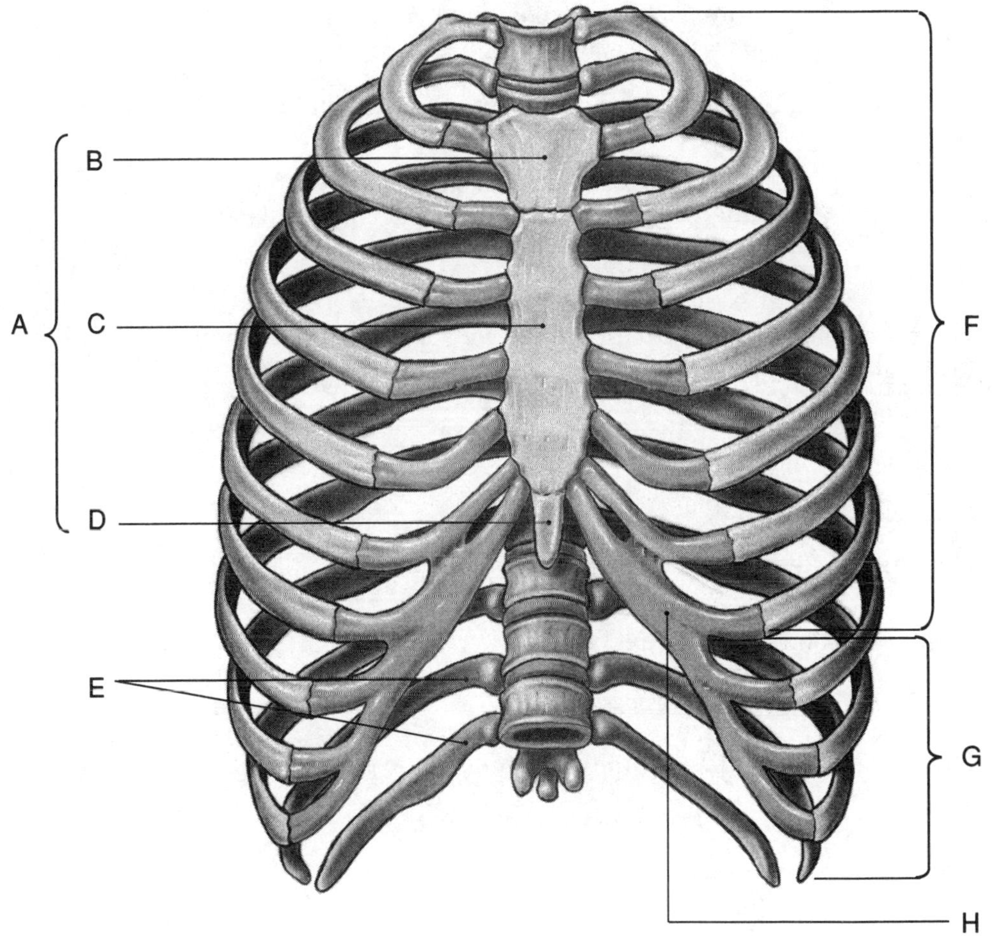

A ______________________ E ______________________

B ______________________ F ______________________

C ______________________ G ______________________

D ______________________ H ______________________

Objective 8 Relate the structural differences between the pectoral and pelvic girdles to their various functional roles.

_____ 1. The bones of the pectoral girdle include the

a. clavicle and scapula
b. ilium and ishcium
c. humerus and femur
d. ulna and radius

_____ 2. The primary function of the pectoral girdle is to

a. protect the organs of the thorax
b. provide areas for articulation with the vertebral column
c. position the shoulder joint and provide a base for arm movement
d. support and maintain the position of the skull

_____ 3. The bones of the pelvic girdle include the

a. tibia and fibula
b. ilium, pubis, ischium
c. ilium, ischium, acetabulum
d. coxa, patella, acetabulum

_____ 4. The pelvic girdle's role of bearing the weight of the body is possible because it has

a. heavy bones and strong, stable joints
b. a high degree of flexibility at the hip joint
c. a wide range of motion
d. all of the above

_____ 5. The primary type of tissue responsible for stabilizing, positioning and bracing the pectoral girdle is

a. tendons
b. ligaments
c. cartilage
d. muscles

_____ 6. The large posterior process on the scapula is the

a. coracoid process
b. acromion process
c. olecranon fossa
d. styloid process

_____ 7. The bone that articulates with the scapula at the glenoid fossa is the

a. radius
b. ulna
c. femur
d. humerus

_____ 8. The parallel bones that support the forearm are the

a. ulna and radius
b. humerus and femur
c. tibia and fibula
d. scapula and clavicle

_____ 9. The articulation which limits movement between the two pubic bones is the

a. obturator foramen
b. greater sciatic notch
c. pubic symphysis
d. pubic tubercle

_____ 10. The bone that articulates with the coxa at the acetabulum is the

a. humerus
b. femur
c. sacrum
d. tibia

_____ 11. The large medial bone of the lower leg is the

a. femur
b. fibula
c. tibia
d. humerus

_____ 12. The general appearance of the pelvis of the female compared to the male is that the female pelvis is

a. heart-shaped
b. robust, heavy, rough
c. relatively deep
d. broad, light, smooth

_____ 13. The bones of the wrist are called the

a. carpals
b. metacarpals
c. tarsals
d. metatarsals

_____ 14. The bones of the foot are called the

a. metacarpals
b. carpals
c. tarsals
d. metatarsals

_____ 15. The anatomical name for the heel bone is the

a. navicular
b. calcaneus
c. talus
d. cuboid

_____ 16. The bones of the fingers and toes are collectively referred to as

a. tarsals and metatarsals
b. carpals and tarsals
c. phalanges
d. carpals and metacarpals

17. The only direct connection between the pectoral girdle and the axial skeleton is the ____________.

18. The pelvic girdle consists of six bones collectively referred to as the ______________.

19. The ulna and radius both have long shafts that contain like processes called ______________ processes.

20. The process that the tibia and fibula have in common that acts as a shield for the ankle is the ________________.

21. At the hip joint to either side, the head of the femur articulates with the ________________.

22. The popliteal ligaments are responsible for reinforcing the back of the ________________.

23. An enlarged pelvic outlet in the female is an adaptation for ________________.

24. The bone which cannot resist strong forces but provides the only fixed support for the pectoral girdle is the ________________.

25. The longest and heaviest bone in the body is the _______________.

26. Identify the following structures in Figure 6-12. Place your answers in the spaces provided below the drawing.

Figure 6-12 The Scapula

Medial border	**Lateral border**	**Superior border**
Coracoid process	**Glenoid cavity**	**Acromion process**
Spine	**Body**	

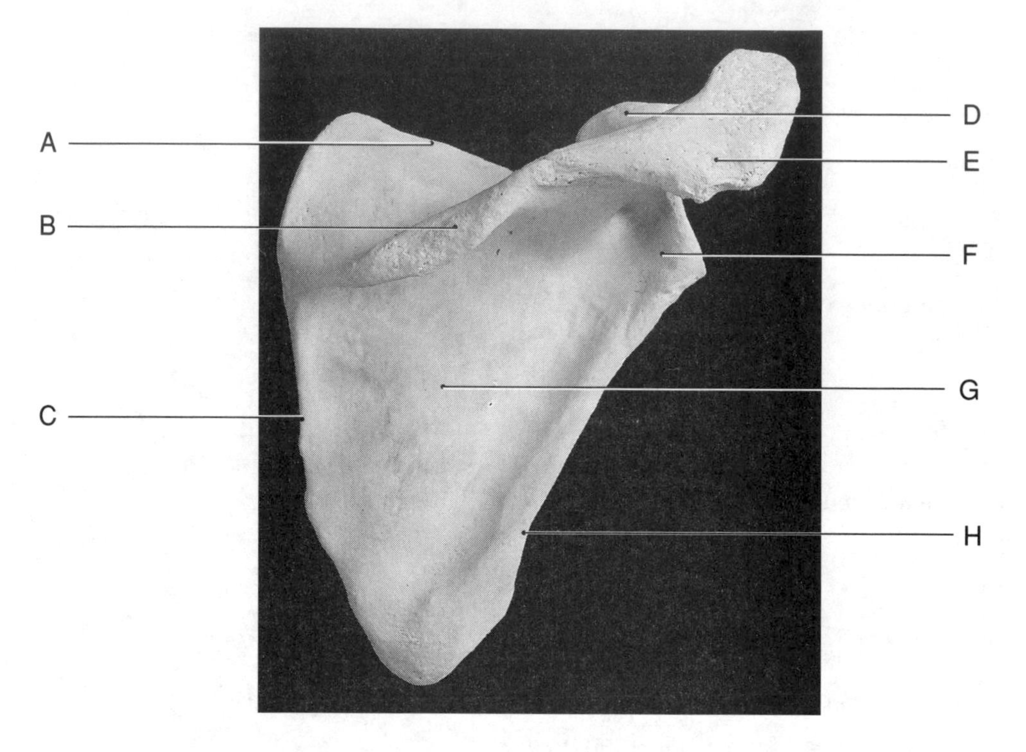

A _______________ E _______________

B _______________ F _______________

C _______________ G _______________

D _______________ H _______________

27. Identify the following components of the pelvis. Place your answers in the spaces provided below the drawing.

Figure 6-13 The Pelvis - Anterior View

Ilium	**Sacrum**	**Iliac crest**
Ischium	**Coccyx**	**Obturator foramen**
Pubis	**Symphysis pubis**	**Acetabulum**

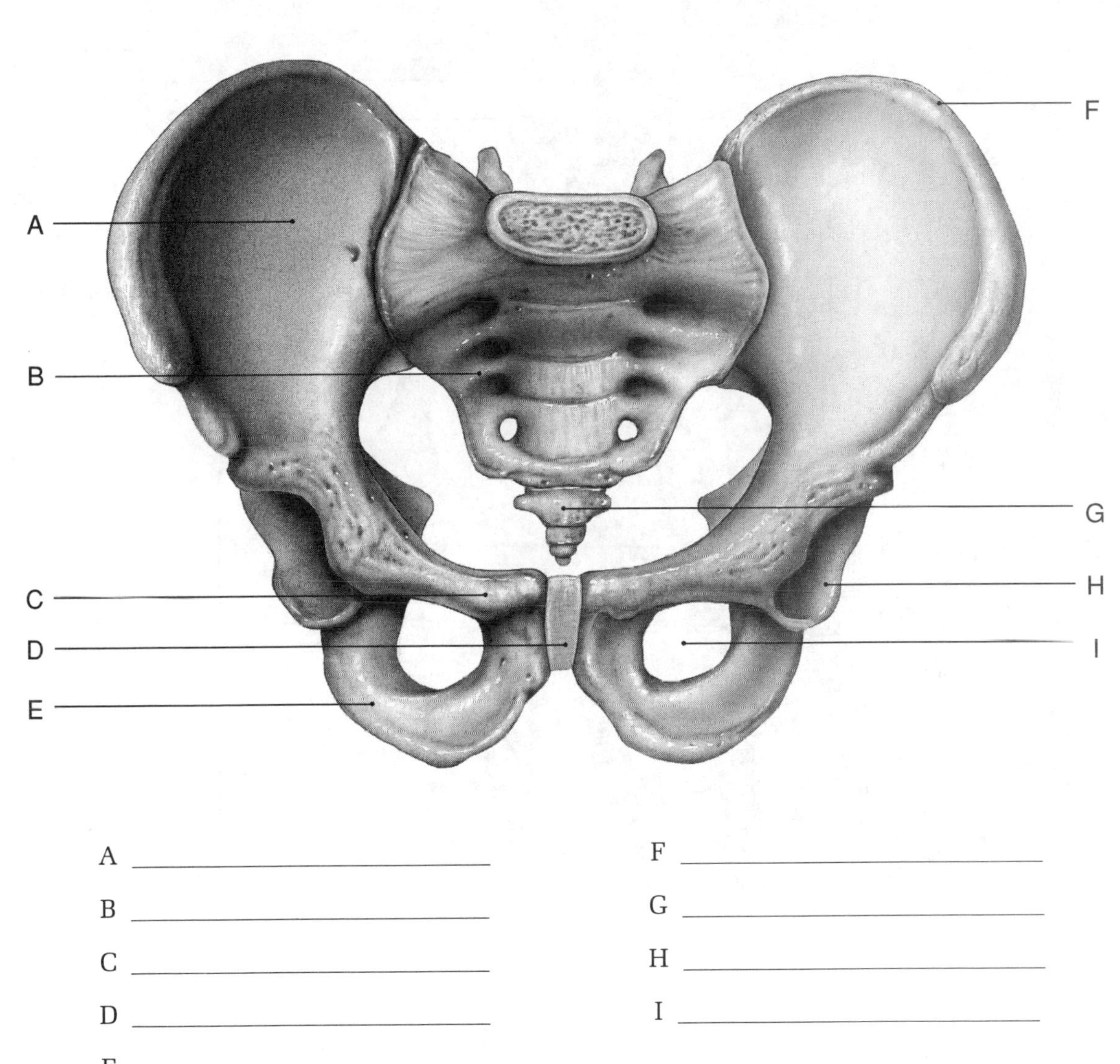

A ____________________ F ____________________

B ____________________ G ____________________

C ____________________ H ____________________

D ____________________ I ____________________

E ____________________

28. Identify the following bones in the wrist and hand. Place your answers in the spaces provided below the drawing.

Figure 6-14 Bones of the Wrist and Hand

Ulna **Radius** **Carpals**
Metacarpals **Phalanges**

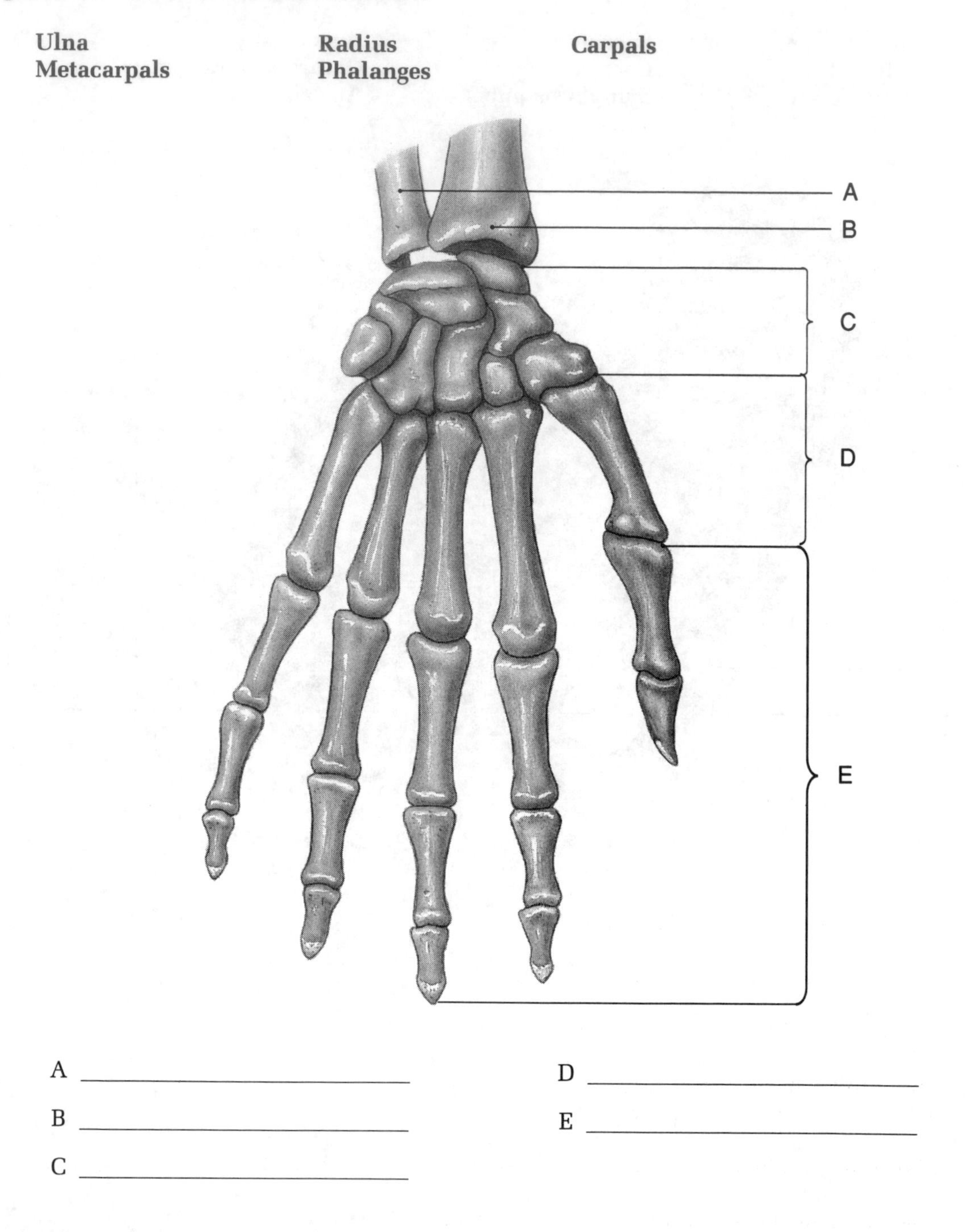

A ______________________

B ______________________

C ______________________

D ______________________

E ______________________

Objective 9 Distinguish among different types of joints and link structural features to joint functions.

_____ 1. An immovable joint is a(n)

a. synarthrosis
b. diarthrosis
c. amphiarthrosis
d. syndesmosis

_____ 2. A slightly movable joint is a(n)

a. gomphosis
b. synarthrosis
c. amphiarthrosis
d. synostosis

_____ 3. A freely movable joint is a(n)

a. syndesmosis
b. amphiarthrosis
c. synarthrosis
d. diarthrosis

_____ 4. The joint that permits the greatest range of motion of any joint in the body is the

a. hip joint
b. shoulder joint
c. elbow joint
d. knee joint

_____ 5. The joint that is correctly matched with the type of joint indicated is

a. symphysis pubis – fibrous
b. knee – synovial
c. sagittal suture – cartilaginous
d. intervertebral disc – synovial

_____ 6. The function(s) of synovial fluid that fills the joint cavity is/are

a. it nourishes the chondrocytes
b. it provides lubrication
c. it acts as a shock absorber
d. all of the above

_____ 7. The primary function(s) of menisci in synovial joints is to

a. subdivide a synovial cavity
b. channel flow of synovial fluid
c. allow for variations in the shapes of articular surfaces
d. all of the above

8. A synarthrotic joint found only between the bones of the skull is a _______________.

9. A totally rigid immovable joint resulting from fusion of bones is a _______________.

10. The amphiarthrotic joint where bones are separated by a wedge or pad of fibrocartilage is a _____________.

11. Diarthrotic joints that permit a wide range of motion are called _____________ joints.

12. The extremely stable joint that is almost completely enclosed in a bony rocket is the _____________ joint.

13. The joint that resembles three separate joints with no single unified capsule or common synovial cavity is the ____________ joint.

14. The radial collateral, annular, and ulnar collateral ligaments provide stability for the _____________ joint.

15. Identify the following structures in the knee joint. Place your answers in the spaces provided below the drawing.

Figure 6-15 A Sectional View of the Knee (Synovial) Joint

Femur	**Tibia**	**Knee muscle**
Tendon	**Fat pad**	**Intracapsular ligament**
Bursa	**Meniscus**	**Patellar ligament**
Patella	**Joint cavity**	**Joint capsule**

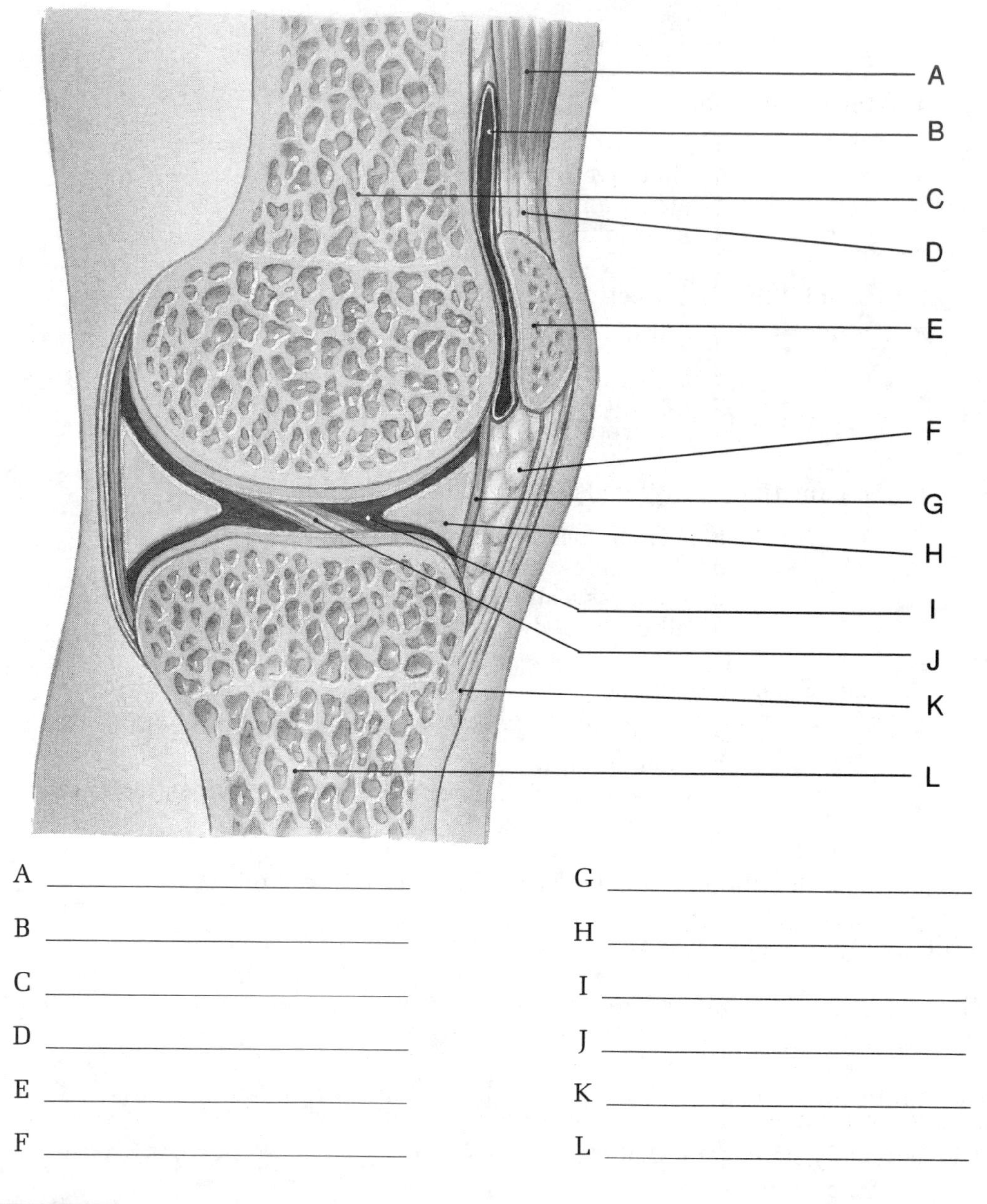

A __________________ G __________________

B __________________ H __________________

C __________________ I __________________

D __________________ J __________________

E __________________ K __________________

F __________________ L __________________

Objective 10 Describe the dynamic movements of the skeleton and the structure of representative articulations.

_____ 1. Decreasing the angle between bones is called

a. flexion
b. extension
c. abduction
d. adduction

_____ 2. The movement that allows you to gaze at the ceiling is

a. rotation
b. circumduction
c. hyperextension
d. elevation

_____ 3. Movements of the vertebral column are limited to

a. flexion and extension
b. lateral flexion
c. rotation
d. all of the above

_____ 4. Movement in the wrist and hand in which the palm of the hand is turned forward is

a. supination
b. pronation
c. dorsal flexion
d. plantar flexion

_____ 5. A movement towards the midline of the body is called

a. abduction
b. inversion
c. eversion
d. adduction

_____ 6. The movement of the thumb that allows for grasping is

a. inversion
b. opposition
c. supination
d. retraction

_____ 7. Twiddling your thumbs during a lecture demonstrates the action that occurs at a

a. hinge joint
b. ball and socket joint
c. saddle joint
d. gliding joint

_____ 8. Contraction of the biceps brachii muscle produces

a. pronation of the forearm and extension of the elbow
b. supination of the forearm and extension of the elbow
c. supination of the forearm and flexion of the elbow
d. pronation of the forearm and flexion of the elbow

_____ 9. Movements such as dorsiflexion and plantar flexion involve moving the

a. leg
b. hip
c. arm
d. foot

_____ 10. To do a split, the initial movement of the legs is

a. abduction
b. adduction
c. extension
d. flexion

11. Identify the body movements in the following illustrations. Place your answers in the spaces provided below the drawings.

Figure 6-16 Body Movements

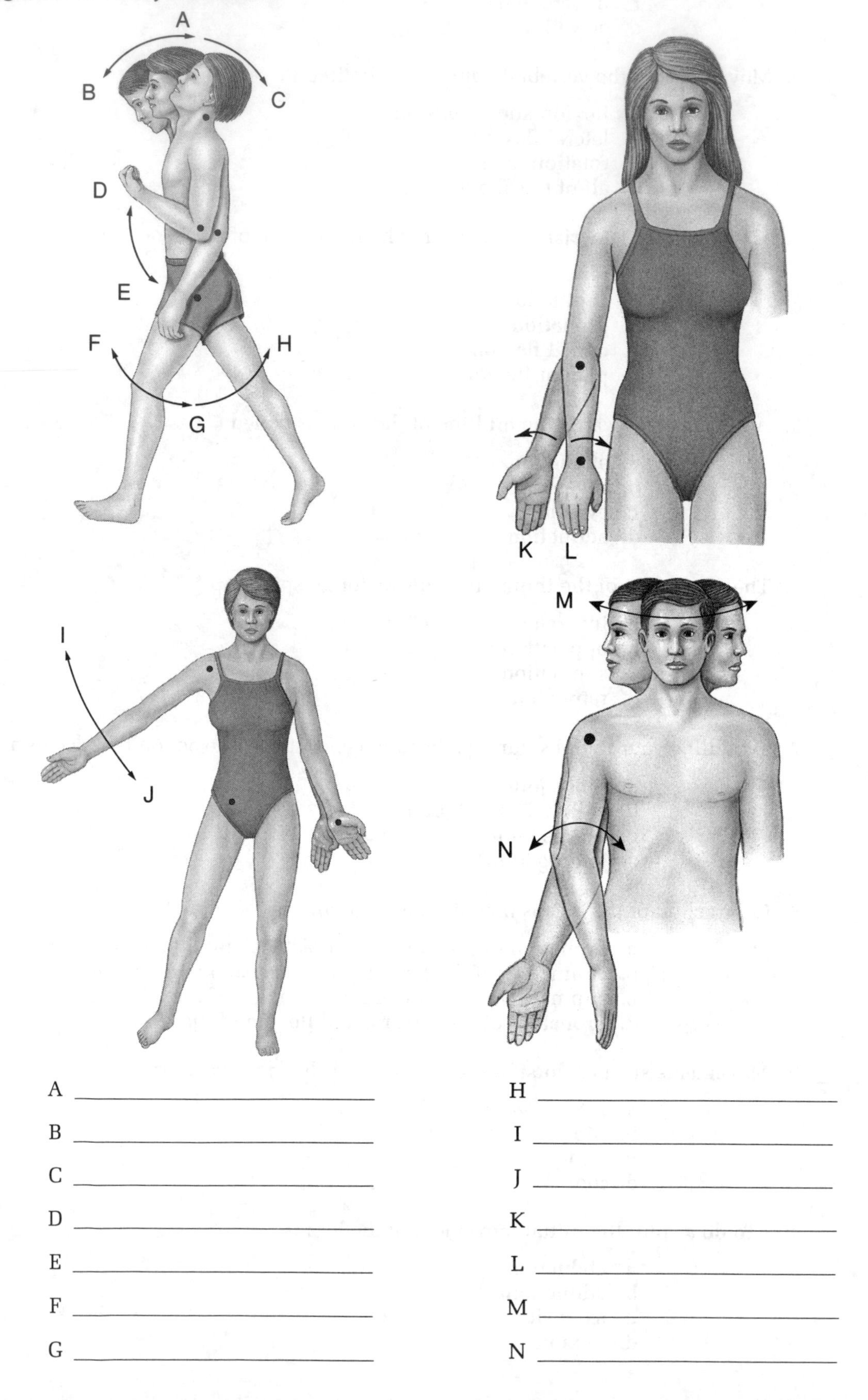

A ______________________________ H ______________________________

B ______________________________ I ______________________________

C ______________________________ J ______________________________

D ______________________________ K ______________________________

E ______________________________ L ______________________________

F ______________________________ M ______________________________

G ______________________________ N ______________________________

12. Identify the special movements in the following illustrations. Place your answers in the spaces provided below the drawings.

Figure 6-17 Special Movements

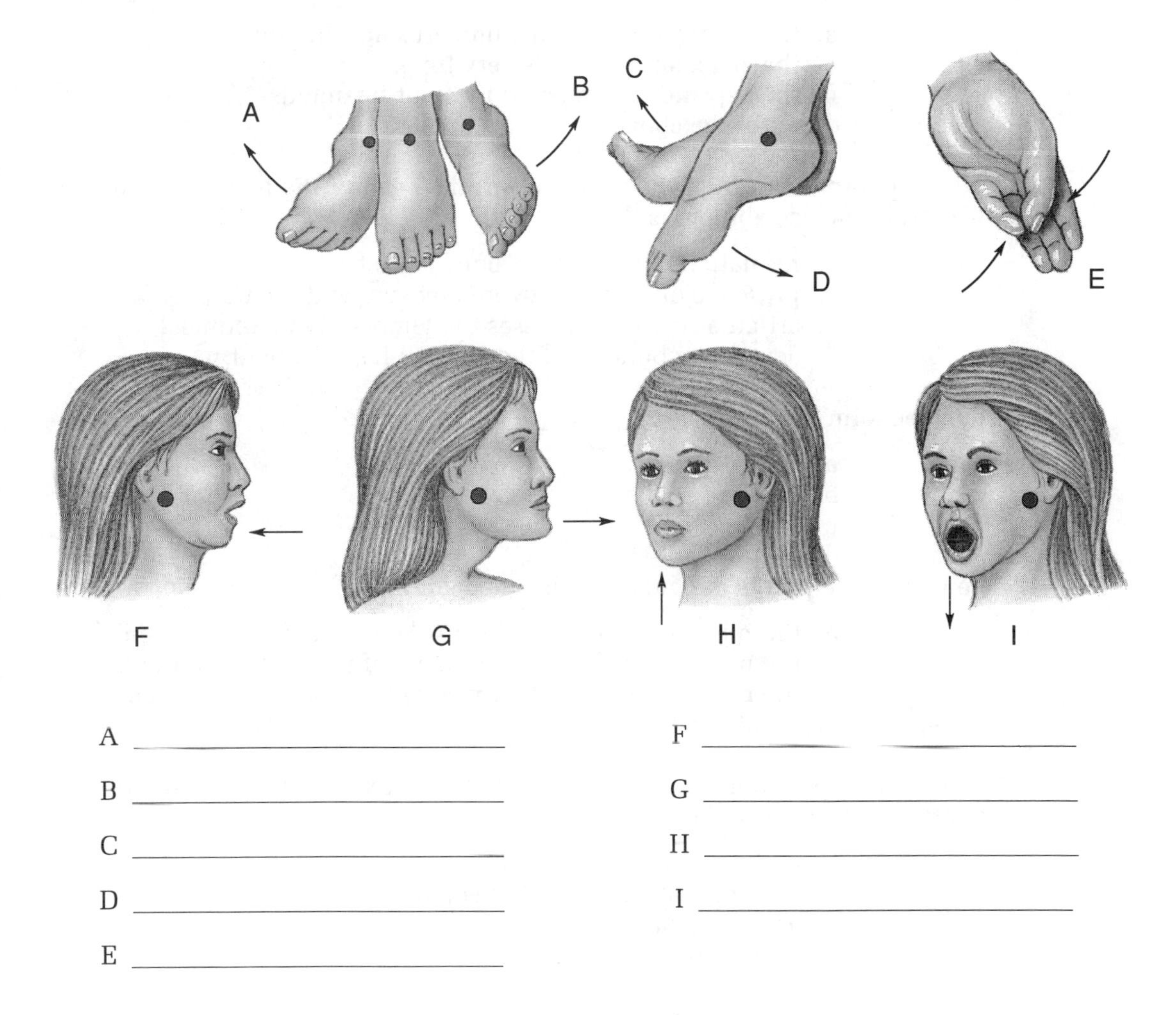

A ______________________

B ______________________

C ______________________

D ______________________

E ______________________

F ______________________

G ______________________

H ______________________

I ______________________

Objective 11 Explain the relationship between join structure and mobility, using specific examples.

_____ 1. The elbow joint is quite stable because

a. the bony surfaces of the humerus and the ulna lock
b. the articular capsule is very thick
c. the capsule is reinforced by stout ligaments
d. all of the above

_____ 2. In the hip joint, the arrangement that keeps the head of the femur from moving away from the acetabulem is the

a. formation of a complete bony socket
b. presence of fat pads covered by synovial membranes
c. articular capsule encloses the femoral head and neck
d. acetabular bones and the femoral head fit tightly

_____ 3. The knee joint functions as a _______________ joint

a. hinge
b. ball and socket
c. saddle
d. gliding

_____ 4. The reason the points of contact in the knee joint are constantly changing is

a. there is no single unified capsule or a common synovial cavity
b. the menisci conform to the shape of the surface of the femur
c. the rounded femoral condyles roll across the top of the tibia
d. all of the above

_____ 5. The reason opposition can occur between the hallux (first toe) and the first metatarsal in the articulation is

a. a hinge rather than a saddle joint
b. a saddle rather than a hinge joint
c. a gliding diarthrosis
d. like a ball-and-socket joint

_____ 6. The function(s) of the intervertebral discs is/are to

a. act as shock absorbers
b. prevent bone-to-bone contact
c. allow for flexion and rotation of the vertebral column
d. a, b, and c are correct

7. Using the following selections, identify the types of synovial joints, seen in Figure 6-18 on page 115. Place your answers in the spaces below each type of joint.

Hinge joint	**Ball-and-socket joint**	**Pivot joint**
Saddle joint	**Ellipsoidal joint**	**Gliding joint**

Figure 6-18 Types of Synovial Joints

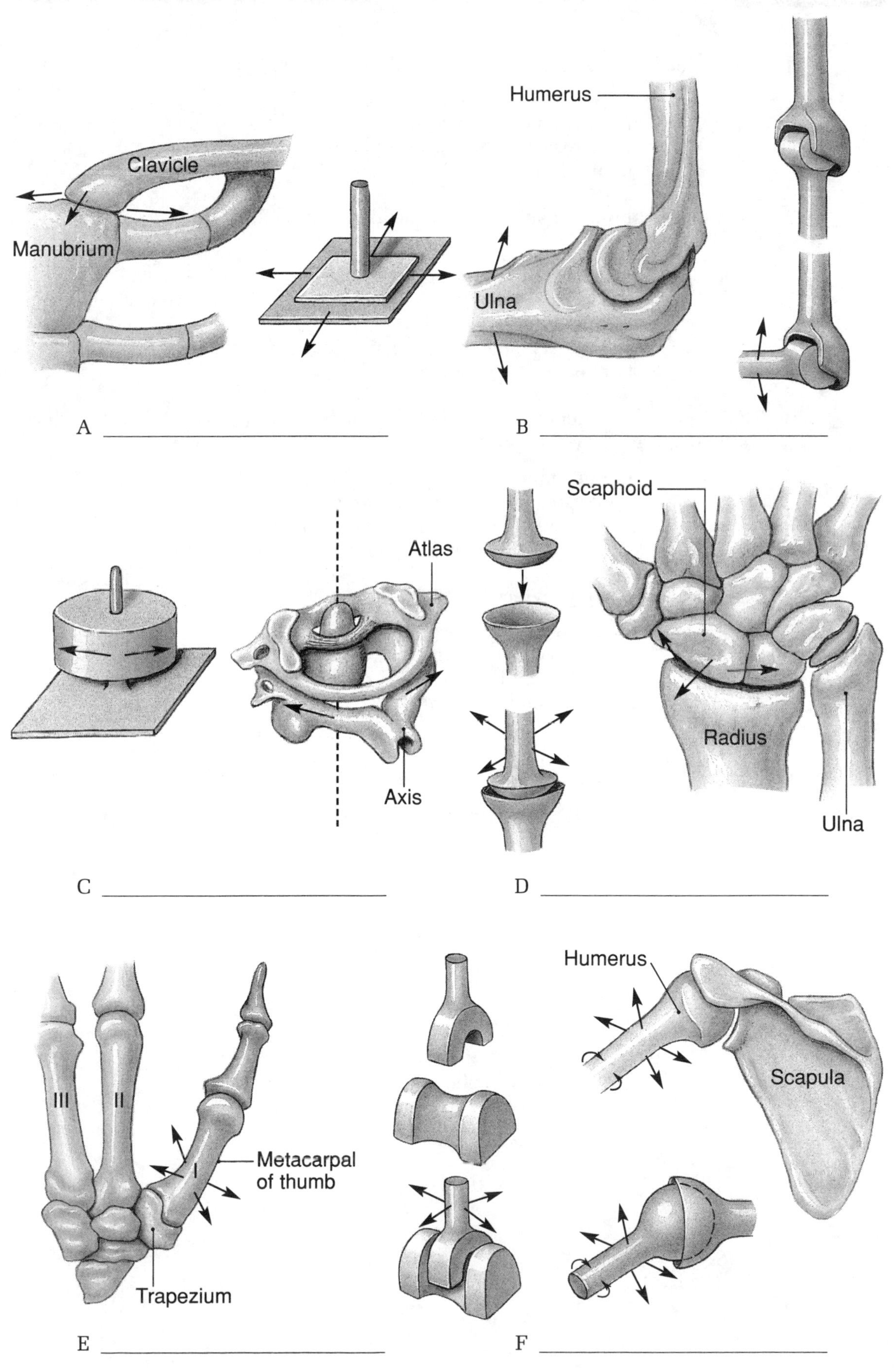

Objective 12 Discuss the functional relationship between the skeletal system and other body systems.

_____ 1. The skeletal system is associated with all the other systems in the body in that it

a. provides mechanical support
b. stores energy reserves
c. stores calcium and phosphate reserves
d. all of the above

_____ 2. The functional relationship of the skeletal system to the nervous system is that the skeletal system

a. stores calcium for neural functions
b. has receptors at joints which provide information about body positions
c. protects the brain and spinal cord
d. a, b, and c are correct

_____ 3. The respiratory system is functionally associated with the skeletal system due to the respiratory system providing

a. nutrients including calcium and phosphate ions
b. oxygen and eliminating carbon dioxide
c. skeletal growth regulated by hormones
d. protection for the lungs and associated structures

_____ 4. The urinary system provides the skeletal system with

a. disposal of waste products
b. conserving calcium and phosphate ions
c. protection for the kidneys and ureters
d. both a and b are correct

Part II: Chapter Comprehensive Exercises

A. Word Elimination

Circle the term that does not belong in each of the following groupings.

1. support protection secretion storage leverage
2. Vitamin D_3 growth hormone calcitrol PTH sex hormones
3. colles ostopenia greenstick Pott's comminuted
4. skull vertebral column pelvis sternum rib cage
5. pelvic pictoral upper limbs lower limbs hyoid bone
6. mandible maxilla occipital zygomatic lacrimal
7. occipital sphenoidal frontal maxilla mastoid
8. cervical scoliosis thoracic lumbar sacral
9. olecranon scaphoid lunate trapezium capitate
10. gliding hinge pronation ellipsoidal saddle

B. Matching

Match the terms in Column "B" with the terms is Column "A." Use letters for answers in the spaces provided.

COLUMN A	COLUMN B
___ 1. menisci	A. intervertebral discs
___ 2. cruciate ligaments	B. sella turcica
___ 3. nucleus pulposus	C. lower leg bones
___ 4. patella	D. articular discs
___ 5. tibia; fibula	E. stylohyoid ligaments
___ 6. ulna; radius	F. infant skull
___ 7. hyoid bone	G. air-filled chambers
___ 8. sphenoid bone	H. knee joint capsule
___ 9. paranasal sinuses	I. bones of the forearm
___ 10. fontanel	J. knee cap

C. Concept Map I - Skeletal System

This concept map is a review of Chapter 6. Use the following terms to complete the map by filling in the boxes identified by circled numbers, 1-10.

axial	**lacunae**	**lumbar**	**osteons**
pectoral girdle	**sternum**	**support**	**tibia**
true	**osteocytes**		

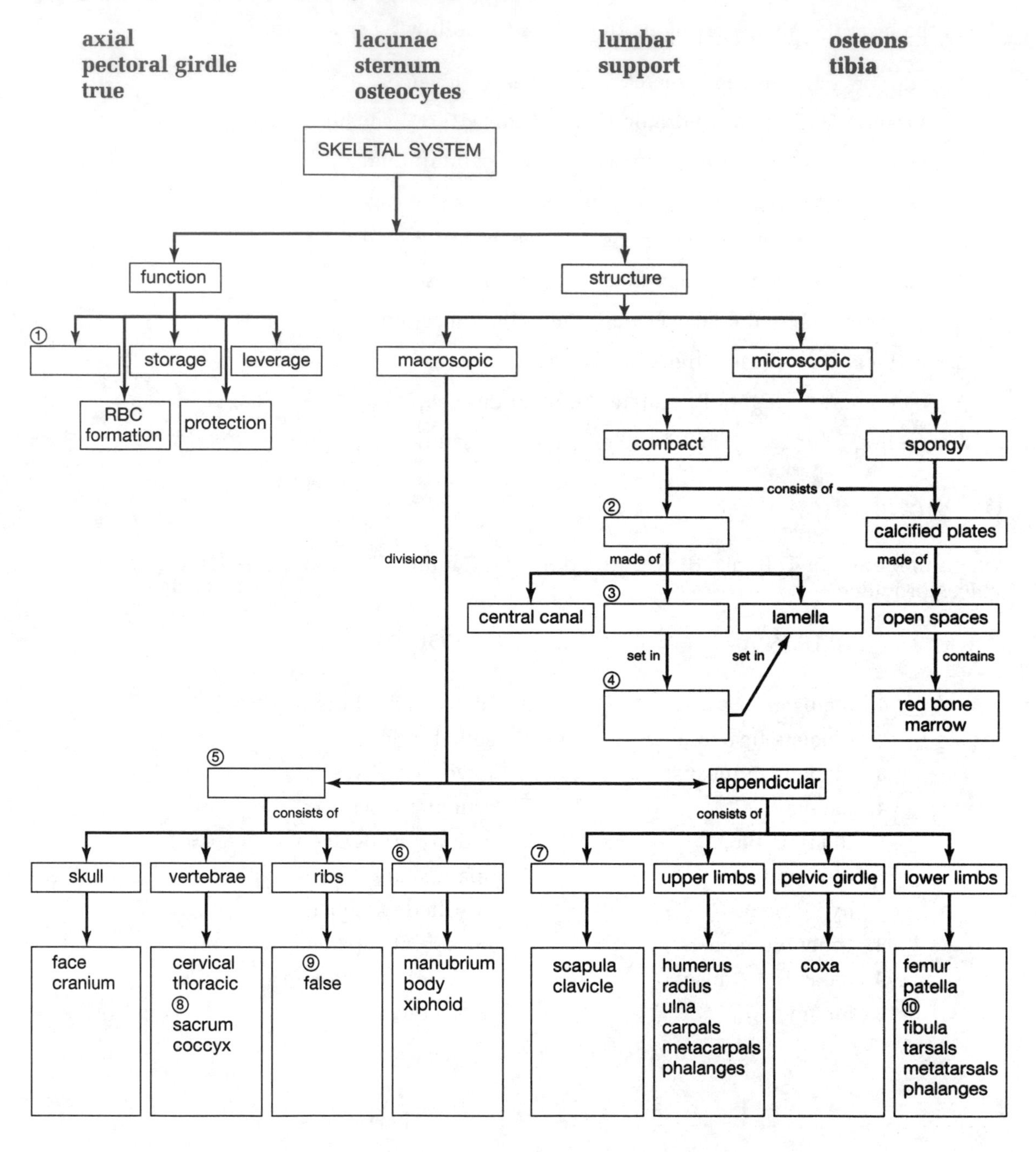

Concept Map II - Joints

Using the following terms, fill in the circled, numbered, blank spaces to complete the concept map. Follow the numbers that comply with the organization of the concept map.

Amphiarthrosis
Symphysis
No movement
Sutures
Cartilaginous
Wrist
Synovial
Monoaxial
Fibrous

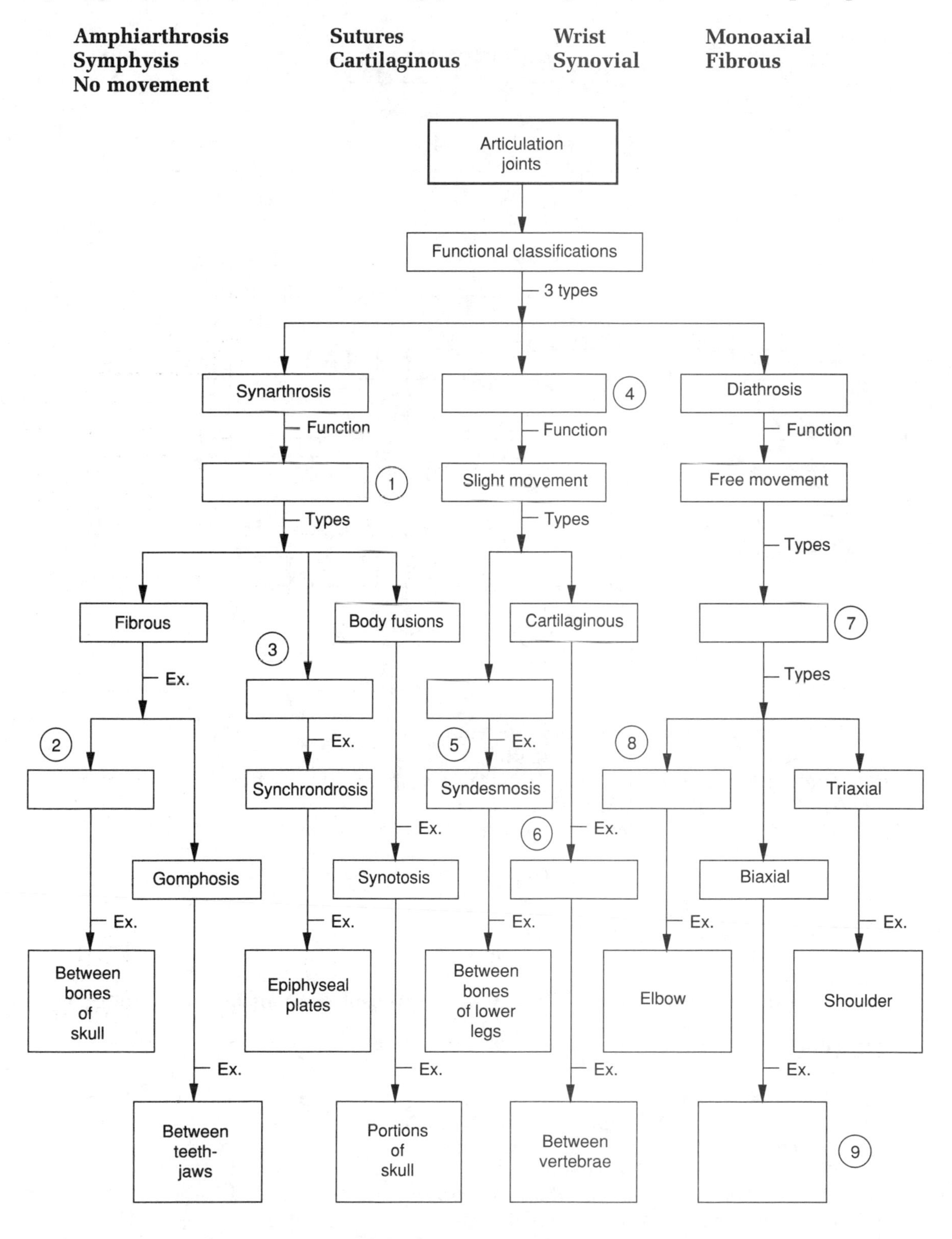

D. Crossword Puzzle

The following puzzle reviews the material in Chapter 6. To complete the puzzle, you must know the answers to the clues given, and you must be able to spell the terms correctly.

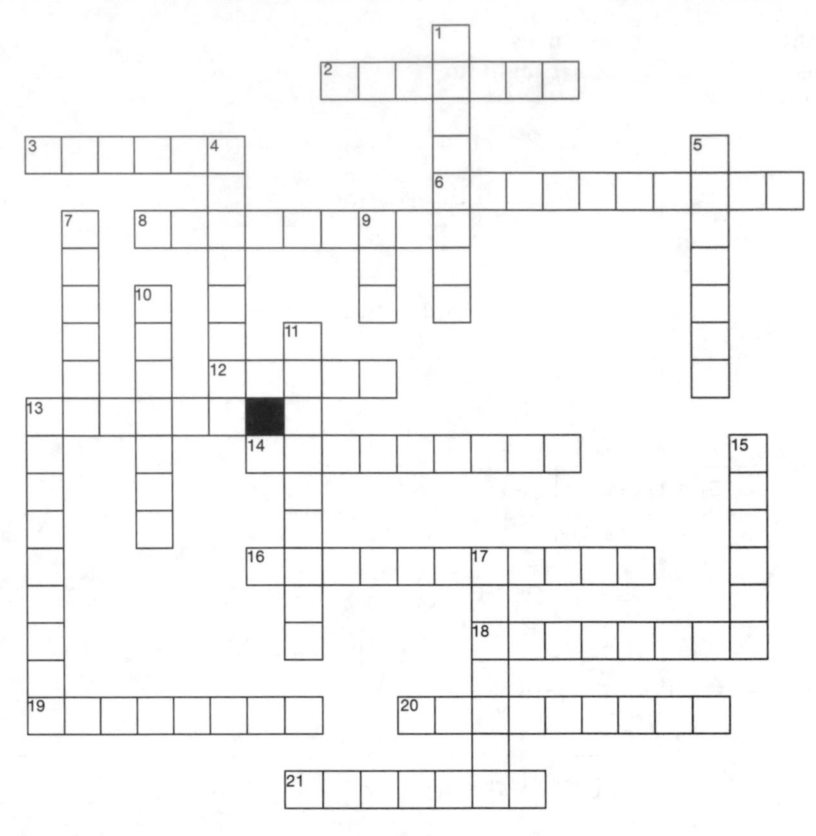

ACROSS

2. The most inferior portion of the sternum.
3. The lateral bone of the antebrachium.
6. The cavity that serves as the "socket" into which the femur fits.
8. The anatomical name for the cheekbone.
12. The division of the skeletal system that includes the skull.
13. The capitate is a _______ bone.
14. This term refers to the ends of the bone.
16. The bone-forming cells.
18. The anatomical name for the collarbone.
19. The fluid that is associated with many joints.
20. A baby's "soft spots."
21. Because osteoclasts remodel bone, they are probably responsible for creating _____.

DOWN

1. The vertebrae to which the ribs attach.
4. The suture that connects the two parietal bones.
5. The cavity that serves as the "socket" of the scapula.
7. The lateral bone of the lower leg.
9. The number of coxae bones that make up the hip.
10. The area of bone that is made of osteons.
11. This term refers to the shaft of the bone.
13. The anatomical name for the heel bone.
15. The fluid-filled sacs that reduce friction at some joint areas.
17. The depressions in the osteons in which the osteocytes set in.

E. Short Answer Questions

Briefly answer the following questions in the spaces provided below.

1. What are the four (4) primary functions of the axial skeleton?

2. Why are the auditory ossicles and hyoid bone referred to as associated bones of the skull?

3. What is craniostenosis and what are the results of this condition?

4. What is the difference between a primary curve and a secondary curve of the spinal column?

5. Distinguish among the abnormal spinal curvature distortions of kyphosis, lordosis, and scoliosis.

6. What is the difference between the true ribs and the false ribs?

7. What are the primary functions of the appendicular skeleton?

8. What bones comprise the pectoral girdle? The pelvic girdle?

9. What is the functional difference between a ligament and a tendon?

10. What are the structural and functional differences between: (a) bursa and (b) a meniscus?

11. What is the functional role of snynovial fluid in a diarthrotic joint?

12. Functionally, what is the relationship between the elbow and the knee joint?

13. Functionally, what is the commonality between the shoulder joint and the hip joint?

14. What regions of the vertebral column do not contain intervertebral discs? Why are they unnecessary in these regions?

15. Identify the unusual types of movements which apply to the following examples:

    (1) twisting motion of the foot that turns the sole inward

    (2) grasping and holding an object with the thumb

    (3) standing on tiptoes

    (4) crossing the arms

    (5) opening the mouth

    (6) shrugging the shoulders

CHAPTER

7

The Muscular System

Overview

This chapter focuses on the three types of muscle tissue with emphasis on the organization of skeletal muscle tissue and the functional organization of the muscular system. Muscles are specialized tissues that support and facilitate body movement and movement of materials within the body. Movement is an important function of life for adjusting to the changing conditions of the external environment in which we live, and the internal environment within the body.

Most of the muscle or "red meat" of the body is skeletal or voluntary muscle. It is called skeletal muscle because it is attached to the bony skeleton by ligaments. If you weigh 150 pounds, approximately 60 pounds or 40% of the body weight consists of skeletal muscle. Cardiac and smooth muscles form the walls of hollow organs and the heart and are involved in transporting materials within the body.

After reviewing and successfully completing the exercises in Chapter 7, you should be able to understand the basic principles of microscopic and gross structure of muscle and muscle physiology, identification of major skeletal muscles, and muscle performance including body movements.

Review of Chapter Objectives

1. Describe the properties and functions of muscle tissue.
2. Describe the organization of muscle at the tissue level.
3. Identify the structural components of a sarcomere.
4. Explain the key steps involved in the contraction of a skeletal muscle fiber.
5. Compare the different types of muscle contractions.
6. Describe the mechanisms by which muscle fibers obtain and use energy to power contractions.
7. Relate the types of muscle fibers to muscle performance.
8. Distinguish between aerobic and anaerobic endurance and explain their implications for muscular performance.
9. Contrast skeletal, cardiac, and smooth muscle in terms of structure and function.
10. Identify the principal axial muscles of the body, together with their origins and insertions.
11. Identify the principal appendicular muscles of the body, together with their origins and insertions.
12. Describe the effects of exercise and aging on muscle tissue.
13. Discuss the functional relationships between the muscular system and other body systems.

Part I: Objective Based Questions

Objective 1 Describe the properties and functions of muscle tissue.

_____ 1. The four basic properties shared by muscle tissue are

a. cardiac, smooth, skeletal, muscle
b. exitability, contractility, extensibility, elasticity
c. movement, support, posture, body temperature
d. epimysium, endomysium, perimysium, sarcomere

_____ 2. The function(s) of skeletal muscle include(s)

a. produce movement
b. support soft tissues
c. maintain body temperature
d. all of the above

_____ 3. When muscles contract, they

a. extend
b. stretch
c. shorten
d. both a and b

_____ 4. The ability to respond to stimuli is a property of muscle called

a. excitability
b. elasticity
c. contractility
d. extensibility

Objective 2 Describe the organization of muscle at the tissue level.

_____ 1. The three layers of connective tissues comprising each muscle are

a. cardiac, smooth, skeletal
b. epimysium, perimysium, endomysium
c. sarcolemma, sarcomers, T-tubules
d. A-band, I-band, Z-lines

_____ 2. The dense layer of collagen-fibers that surround an entire skeletal muscle is the

a. epimysium
b. tendon
c. perimysium
d. fascicle

_____ 3. Nerves and blood vessels that service the muscle fibers are located in the connective tissues of the

a. endomysium
b. sarcolemma
c. sarcomere
d. perimysium

_____ 4. The bundle of collagen fibers at the end of a skeletal muscle that attaches muscle to bone is called a

a. ligament
b. fascicle
c. myofibril
d. tendon

_____ 5. The command to contract is distributed throughout a muscle fiber by the

a. sarcolemma
b. myofibrils
c. transverse tubules
d. sarcomere

_____ 6. Muscle fiber cells differ from "typical cells" in that muscle fibers

a. are multinucleated
b. lack mitochondria
c. are very small
d. all of the above

7. Identify the following structures in Figure 7-1. Place your answers in the spaces provided below the drawing.

Figure 7-1 Organization of Skeletal Muscle

epimysium
blood vessels and nerves
muscle fiber
perimysium
skeletal muscle
tendon
endomysium
muscle fascicle

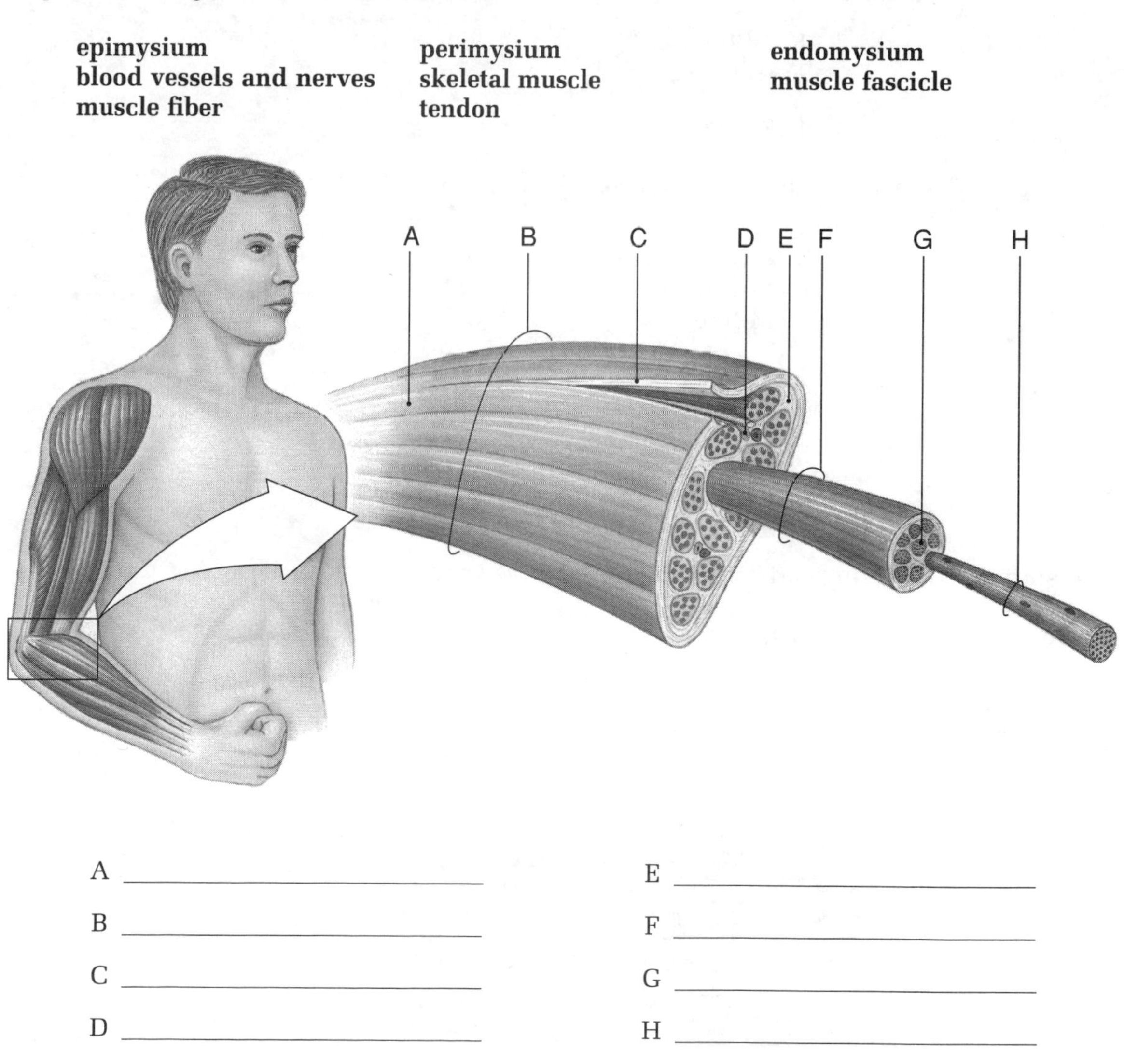

A ______________________________

B ______________________________

C ______________________________

D ______________________________

E ______________________________

F ______________________________

G ______________________________

H ______________________________

Objective 3 Identify the structural components of a sarcomere.

_____ 1. The thin filaments of a sarcomere consist of

a. actin
b. myosin
c. troponin
d. tropomyosin

_____ 2. The thick filaments of a sarcomere consist of

a. actin
b. myosin
c. troponin
d. muscle fibers

_____ 3. Thin filaments at either end of the sarcomere are attached to interconnecting filaments that make up the

a. T-tubules
b. A-band
c. I-band
d. Z lines

_____ 4. The region of the sarcomere containing the thick filaments is the

a. A-band
b. Z-line
c. I-band
d. M-line

_____ 5. The cross-bridges are a part of the __________ molecules.

a. actin
b. myosin
c. troponin
d. tropomyosin

6. Identify the following parts of a sarcomere. Place your answers in the spaces provided.

Figure 7-2 Sarcomere Structure

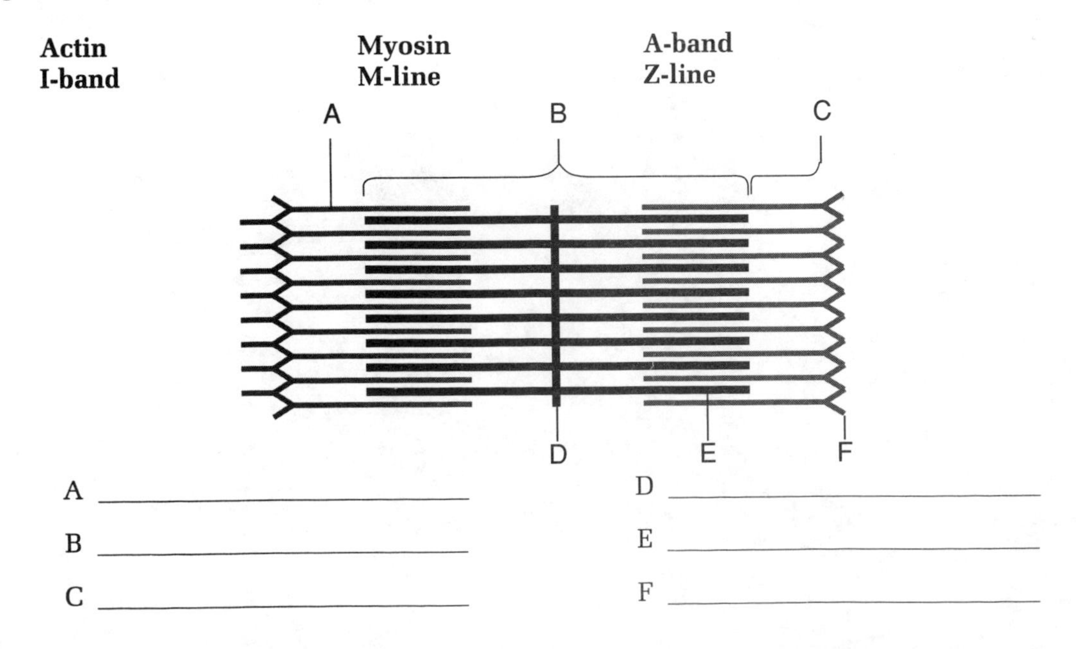

Objective 4 Explain the key steps involved in the contraction of a skeletal muscle fiber.

_____ 1. Skeletal muscle fiber contraction begins when

a. the muscle cell relaxes and lengthens
b. depolarization occurs and an action potential is generated
c. the calcium in concentration at the myofilament increases
d. acetylcholine is released into the neuromuscular junction by the axon terminal

_____ 2. The final step involved in skeletal muscle contraction is

a. the muscle cell relaxes and lengthens
b. the sarcoplasmic reticulum absorbs calcium ions
c. action potential generation ceases as Ach is removed
d. repeated cycles of cross-bridge binding occurs

_____ 3. Active sites on the actin become available for binding when

a. actin binds to troponin
b. myosin binds to troponin
c. calcium binds to troponin
d. troponin binds to tropomysosin

_____ 4. In response to action potentials arriving from the transverse tubules, calcium ions are released from the

a. synaptic cleft
b. sarcoplasmic reticulum
c. neuromuscular junction
d. motor plate

_____ 5. The neurotransmitter released from the synaptic vesicles which initiates an action potential in the sarcolemma is

a. troponin
b. calcium ions
c. acetylcholine
d. actin

6. Identify the steps in the contraction process (A-F). place your answers in the spaces provided below each drawing.

Figure 7-3 Summary of Contraction Process

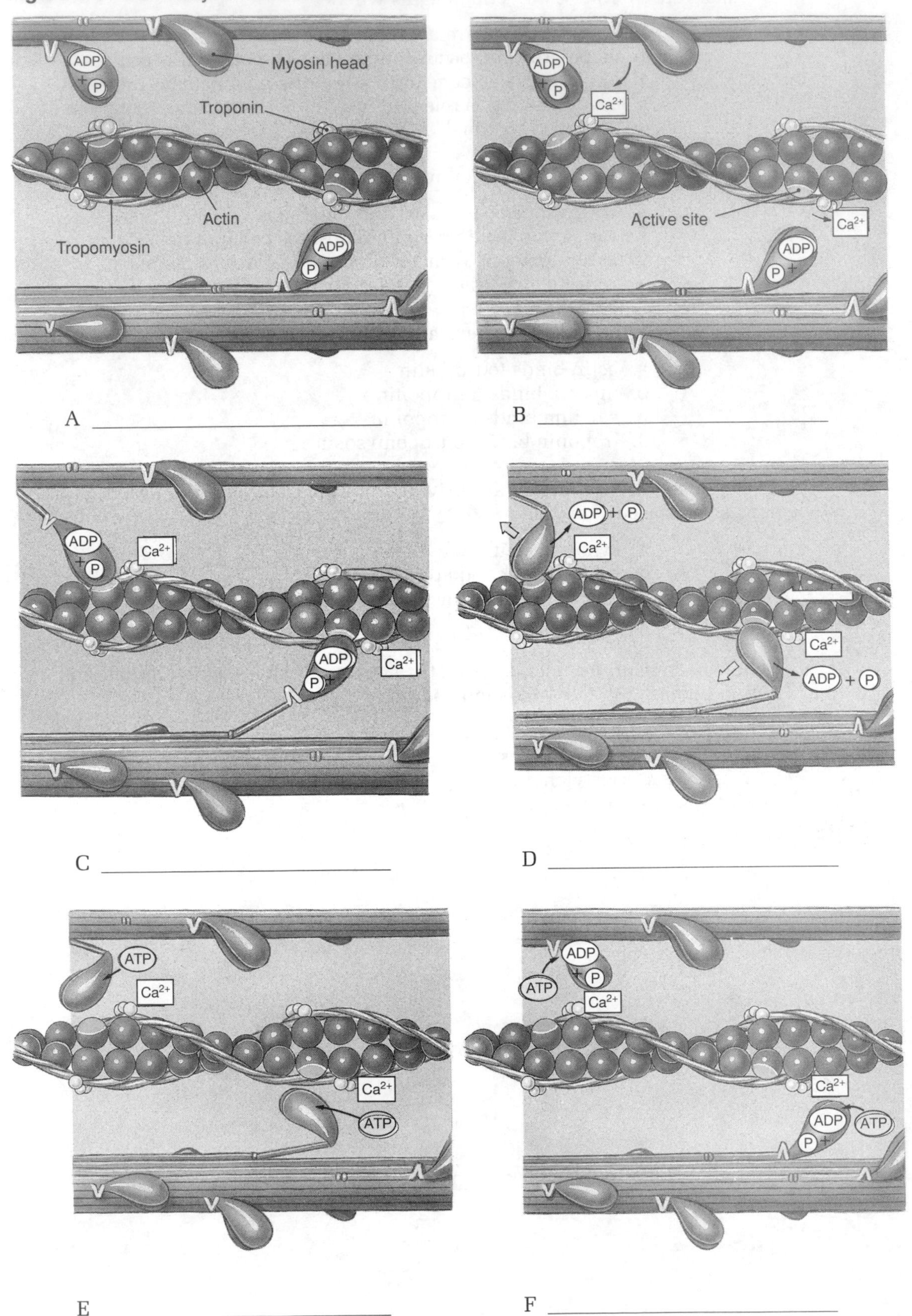

Objective 5 Compare the different types of muscle contractions.

_____ 1. The type of contraction represented by a single stimulus – contraction – relaxation sequence is

a. tetany
b. recovery
c. a twitch
d. recruitment

_____ 2. The smooth but steady increase in muscle tension produced by increasing the number of active motor units is called

a. recruitment
b. tetany
c. recovery
d. isotonic

_____ 3. In an *isotonic* contraction the

a. tension in the muscle varies as it shortens
b. muscle length doesn't change due to resistance
c. cross-bridges must produce enough tension to overcome the resistance
d. tension in the muscle decreases as the resistance increases

_____ 4. In an *isometric* contraction the

a. tension rises but the length of the muscle remains constant
b. tension rises and the muscle shortens
c. tension produced by the muscle is greater than the resistance
d. tension of the muscle increases as the resistance decreases

5. The "staircase" phenomenon during which the peak muscle tension rises in stages is called _________.

6. A muscle producing peak tension during rapid cycles of contraction and relaxation is said to be in _____________ tetany.

Objective 6 Describe the mechanisms by which muscle fibers obtain and use energy to power contractions.

_____ 1. Mitochondrial activities are relatively efficient, but their rate of ATP generation is limited by the

a. presence of enzymes
b. availability of carbon dioxide and water
c. energy demands of other organelles
d. availability of oxygen

_____ 2. When muscles are actively contracting, the process requires large amounts of energy in the form of

a. creatine phosphate (CP)
b. oxygen
c. ATP
d. ADP

_____ 3. The primary energy reserve in muscle tissue is

a. creatine phosphate (CP)
b. adenosine triphosphate (ATP)
c. adenosine diphosphate (ADP)
d. adenosine monophosphate (AMP)

_____ 4. During anaerobic glycolysis

a. a large amount of ATP energy is produced
b. NAD is oxidized via electron transport
c. pyruvic acid is produced
d. all of the above

_____ 5. When energy reserves in a muscle are exhausted or lactic acid levels increase

a. fatigue occurs
b. the muscle contracts
c. tetany is occurring
d. recruitment resumes

_____ 6. A resting muscle generates most of its ATP by

a. anaerobic respiration
b. aerobic respiration
c. glycolysis
d. electron transport

Objective 7 Relate the types of muscle fibers to muscle performance.

_____ 1. The type of muscle fibers that produce powerful contractions but fatigue rapidly are

a. slow fibers
b. fast fibers
c. red muscle fibers
d. pink muscle fibers

_____ 2. The type of muscle fiber that is best adapted for endurance is the

a. fast fiber
b. intermediate fiber
c. white muscle
d. slow fiber

_____ 3. Extensive blood vessels, mitochondria, and myoglobin are found in the greatest concentration in

a. fast fibers
b. slow fibers
c. white muscles
d. smooth muscles

Objective 8 Distinguish between *aerobic* and *anaerobic* endurance and explain their implications for muscular performance.

_____ 1. The length of time a muscle can continue to contract while supported by mitochondrial activities is referred to as

a. anaerobic endurance
b. aerobic endurance
c. hypertrophy
d. recruitment

_____ 2. __________ exercise requires oxygen and is of ______________ duration than ____________.

a. anaerobic, longer, aerobic
b. aerobic, longer, anaerobic
c. anaerobic, shorter, aerobic
d. aerobic, shorter, anaerobic

_____ 3. The amount of oxygen used in the recovery period to restore normal pre-exertion conditions is referred to as the

a. endurance rate
b. citric acid cycle
c. electron transport system
d. oxygen-debt

_____ 4. An example of an activity that requires anaerobic endurance is

a. a 50-yard dash
b. a 3-mile run
c. a 10-mile bicycle ride
d. running a marathon

_____ 5. Athletes training to develop anaerobic endurance perform

a. few, long, relaxing workouts
b. a combination of weight training and running marathons
c. frequent, brief, intensive workouts
d. stretching, flexibility, and relaxation exercises

Objective 9 Contrast skeletal, cardiac, and smooth muscle in terms of structure and function.

_____ 1. An example of striated, involuntary muscle is

a. skeletal muscle
b. cardiac muscle
c. smooth muscle
d. all of the above

_____ 2. The types of muscle tissue in which filament organization consists of sarcomeres along myofibrils are

a. cardiac, skeletal, smooth
b. cardiac and smooth
c. skeletal and smooth
d. skeletal and cardiac

_____ 3. Smooth muscle tissue can be found in

a. walls of blood vessels
b. around hollow organs
c. layers around the respiratory, circulatory, digestive, and reproductive tracts
d. a, b, and c are correct

Objective 10 Identify the principal axial muscles of the body, together with their origins and insertions.

_____ 1. The axial musculature consists of

a. muscles of the head and neck
b. muscles of the spine and pelvic floor
c. oblique and rectus muscles
d. all of the above

_____ 2. From the following selections choose the one that includes only muscles of facial expression

a. lateral rectus, medial rectus, hypoglossus, stylohyoid
b. splenius, masseter, scalenus, platymsa
c. procerus, capitis, cervicis, zygomaticus
d. buccinator, orbicularis oris, risorius, frontalis

_____ 3. The names of the muscles of the tongue are readily identified because their descriptive names end in

a. genio
b. glossus
c. pollicus
d. hallucis

_____ 4. The superficial muscles of the spine are identified by subdivisions that include

a. cervicis, thoracis, lumborum
b. iliocoastalis, longissimus, spinalis
c. longissimus, transversus, longus
d. capitis, splenius, spinalis

_____ 5. The muscluar floor of the pelvic cavity is formed by muscles that make up the

a. urogenital and anal triangle
b. sacrum and coccyx
c. ischium and the pubis
d. ilium and the ischium

_____ 6. The more movable end of a muscle is the

a. origin
b. belly
c. insertion
d. proximal end

_____ 7. A muscle that inserts on the body of the mandible is probably involved in

a. hissing
b. blowing
c. frowning
d. chewing

_____ 8. Muscles that insert on the olecranon process of the ulna act to

a. extend the forearm
b. flex the forearm
c. abduct the forearm
d. adduct the forearm

_____ 9. The origin of the frontalis muscle is the

a. frontal bone
b. galea aponeurotica
c. temporal bone
d. sphenoid bone

_____ 10. The iliac crest is the origin of the

a. longissimus
b. iliocostalis
c. quadratus lumborum
d. supraspinatus

Objective 11 Identify the principal appendicular muscles of the body, together with their origins and insertions.

_____ 1. Of the following selections the one that includes muscles that move the shoulder girdle is the

a. teres major, deltoid, pectoralis major, triceps
b. procerus, capitis, pterygoid, brachialis
c. trapezius, sternocleidomastoid, pectoralis minor, subclavicus
d. internal oblique, thoracis, deltoid, pectoralis minor

_____ 2. From the following selections choose the one that includes the muscles that move the upper arm.

a. deltoid, teres major, latissimus dorsi, pectoralis major
b. trapezius, pectoralis minor, subclavius, triceps
c. rhomboideus, serratus anterior, subclavius, trapezius
d. brachialis, brachioradialis, pronator, supinator

_____ 3. The muscles comprising the rotator cuff are the

a. deltoid and teres major
b. extensor digitorium and palmaris longus
c. deltoid and trapezius
d. infraspinatus and teres major

_____ 4. When the gluteus maximus contracts it

a. extends the thigh anteriorly
b. extends the thigh backward
c. moves the thigh laterally
d. adducts the thigh

_____ 5. The *quadriceps* are a group of anterior thigh muscles which include the

a. semitendinosus, biceps femoris, semimembranosus, sartorius
b. gastrocnemius, soleus, tibialis anterior, gracilis
c. rectus femoris, vastus medialis, vastus lateralis, vastus inter medius
d. sartorius, gracilis, peroneus, popliteus

_____ 6. The *hamstrings* are a group of posterior thigh muscles which include the

a. rectus femoris, vastus medialis, vastus lateralis
b. semimembranosus, rectus femoris, semitendinosus
c. semitendenosus, biceps femoris, semimembranosus
d. rectus femoris, biceps femoris, semitendinosus

_____ 7. The muscles that arise on the humerus and the forearm and rotate the radius without producing flexion or extension of the elbow are the

a. pronator teres and supinator
b. brachialis and brachioradialis
c. triceps and biceps brachii
d. carpi ulnaris and radialis

_____ 8. The muscle which inserts on the acromion process and scapular spine is the

a. pectoralis major
b. trapezius
c. sternocleidomastoid
d. latissionus dorsi

_____ 9. The muscle which inserts on the iliotibial tract and gluteal tuberosity of the femur is the

a. gracilis
b. rectus femoris
c. sartorius
d. gluteus maximus

_____ 10. The muscle which originates along the entire length of the linea aspera of the femur is the

a. biceps femoris
b. vastus medialis
c. vastus lateralis
d. rectus femoris

11. Identify the muscles on the axial and appendicular skeleton, seen in Figure 7–4 on page 135. Select from the list of muscles above the figure.

Figure 7-4 Principal Skeletal Muscles - Anterior View

temporalis	**peroneus longus**	**zygomaticus**
vastus medialis	**sartorius**	**orbicular oculi**
gracilis	**biceps brachii**	**frontalis**
rectus femoris	**adductor muscle**	**tensor fascia lata**
masseter	**orbicularis oris**	**deltoid**
sternocleidmastoid	**external oblique**	**vastus lateralis**
rectus abdominis	**transversus abdominis**	**pectoralis major**
tibialis anterior		

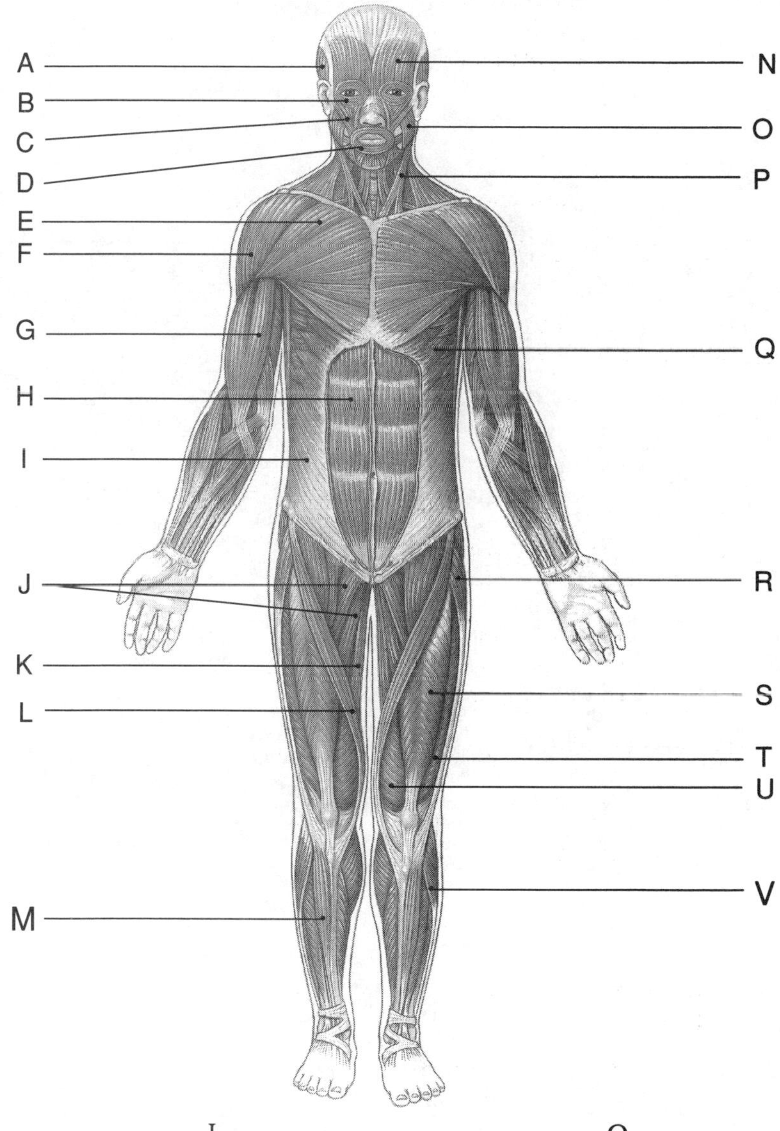

A ____________________ I ____________________ Q ____________________

B ____________________ J ____________________ R ____________________

C ____________________ K ____________________ S ____________________

D ____________________ L ____________________ T ____________________

E ____________________ M ____________________ U ____________________

F ____________________ N ____________________ V ____________________

G ____________________ O ____________________

H ____________________ P ____________________

12. Identify the following muscles on the axial and appendicular skeleton. Select from the following list of muscles.

Figure 7-5 Principal Skeletal Muscles - Posterior View

soleus	**gluteus maximus**	**semimembranosus**
deltoid	**external oblique**	**gastrocnemius**
occipitalis	**biceps femoris**	**semitendinosus**
trapezius	**tricep brachii**	**latissimus dorsi**
gluteus medius		

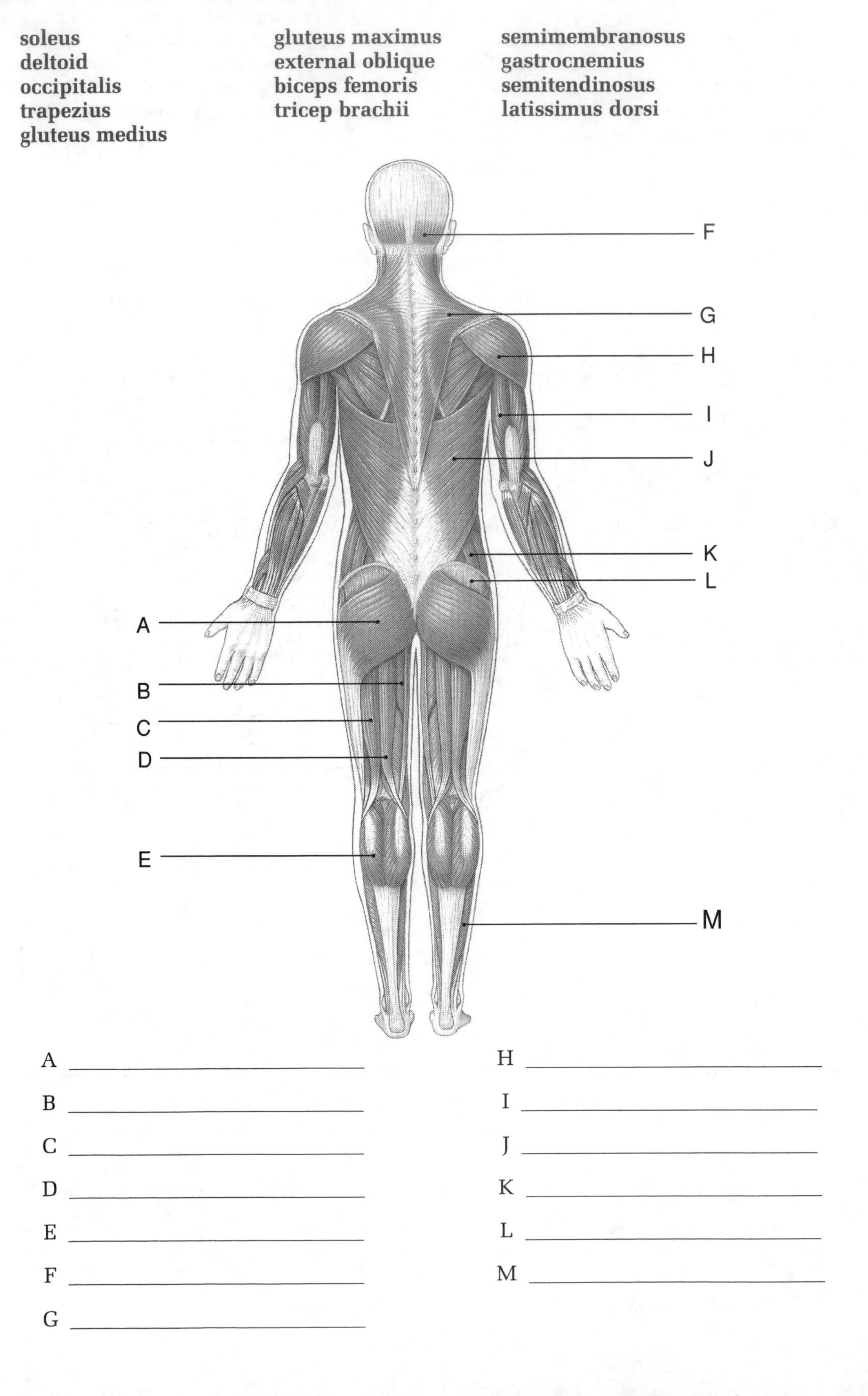

A ______________________

B ______________________

C ______________________

D ______________________

E ______________________

F ______________________

G ______________________

H ______________________

I ______________________

J ______________________

K ______________________

L ______________________

M ______________________

Objective 12 Describe the effects of exercise and aging on muscle tissue.

_____ 1. There is clear evidence that the benefits of regular exercise throughout life include

a. control of body weight
b. increase in bone strength
c. improving the quality of life
d. all of the above

_____ 2. In aging skeletal muscle cells (fibers) the

a. diameter of the fibers decreases
b. fibers begin to disintegrate
c. fibers tend to get shorter
d. fibers become more elastic

_____ 3. The process during which aging skeletal muscles develop increasing amounts of fibrous connective tissue is called

a. atrophy
b. fibrosis
c. tetany
d. hypertrophy

_____ 4. The typical result of repairing aging skeletal muscle tissue is

a. rapid muscular fatigue
b. a decrease in the amount of fibrous connective tissue
c. formation of scar tissue
d. overheating of muscle tissue

Objective 13 Describe the functional relationships between the muscular system and other body systems.

1. The system that provides for muscle attachment in the body is the ____________ system.

2. The system that accelerates oxygen delivery and carbon dioxide removal in muscles is the ______________ system.

3. The system that defends skeletal muscles against infection and assists in tissue repairs after injury is the ______________ system.

4. The system which releases hormones which adjust muscle metabolism and growth is the ______________ system.

5. The system which controls skeletal muscle contractions is the ______________ system.

Part II: Chapter Comprehensive Exercises

A. Word Elimination

Circle the term that does not belong in each of the following groupings.

1. excitability contractility support extensibility elasticity
2. actin thick filament myosin thin filament sarcomere
3. exposure attachment contraction pivoting detachment
4. twitch tetany myogram isometric isotonic
5. ADP CP ATP glucose DNA
6. myoglobin fast fibers white muscles slow fibers red muscles
7. 50-yard dash jogging pole vault weight-lifting 50-yard swim
8. automaticity pacemaker cells uninucleate intercalated discs anaerobic
9. occipitalis parietal temporal sartorius frontalis
10. cardiovascular respiratory heart integumentary nervous

B. Matching

Match the terms in Column "B" with the terms in Column "A". Use letters for answers.

COLUMN A	COLUMN B
___ 1. excitability	A. functional unit of muscle
___ 2. epimysium	B. relaxation phase eliminates
___ 3. sarcomere	C. contraction
___ 4. cross-bridge interaction	D. change in pH
___ 5. complete tetanus	E. response to stimulation
___ 6. lactic acid	F. white muscles
___ 7. fast fibers	G. surrounds muscles
___ 8. skeletal muscle fibers	H. axial muscle
___ 9. diaphragm	I. abducts arm
___ 10. deltoid muscle	J. multinucleated

C. Concept Map - Muscle Tissue

Using the following terms, fill in the numbered, blank spaces to complete the concept map. Follow the numbers that comply with the organization of the map.

Smooth	**Involuntary**	**Striated**	**Heart**
Multinucleated	**Bones**	**Non-striated**	

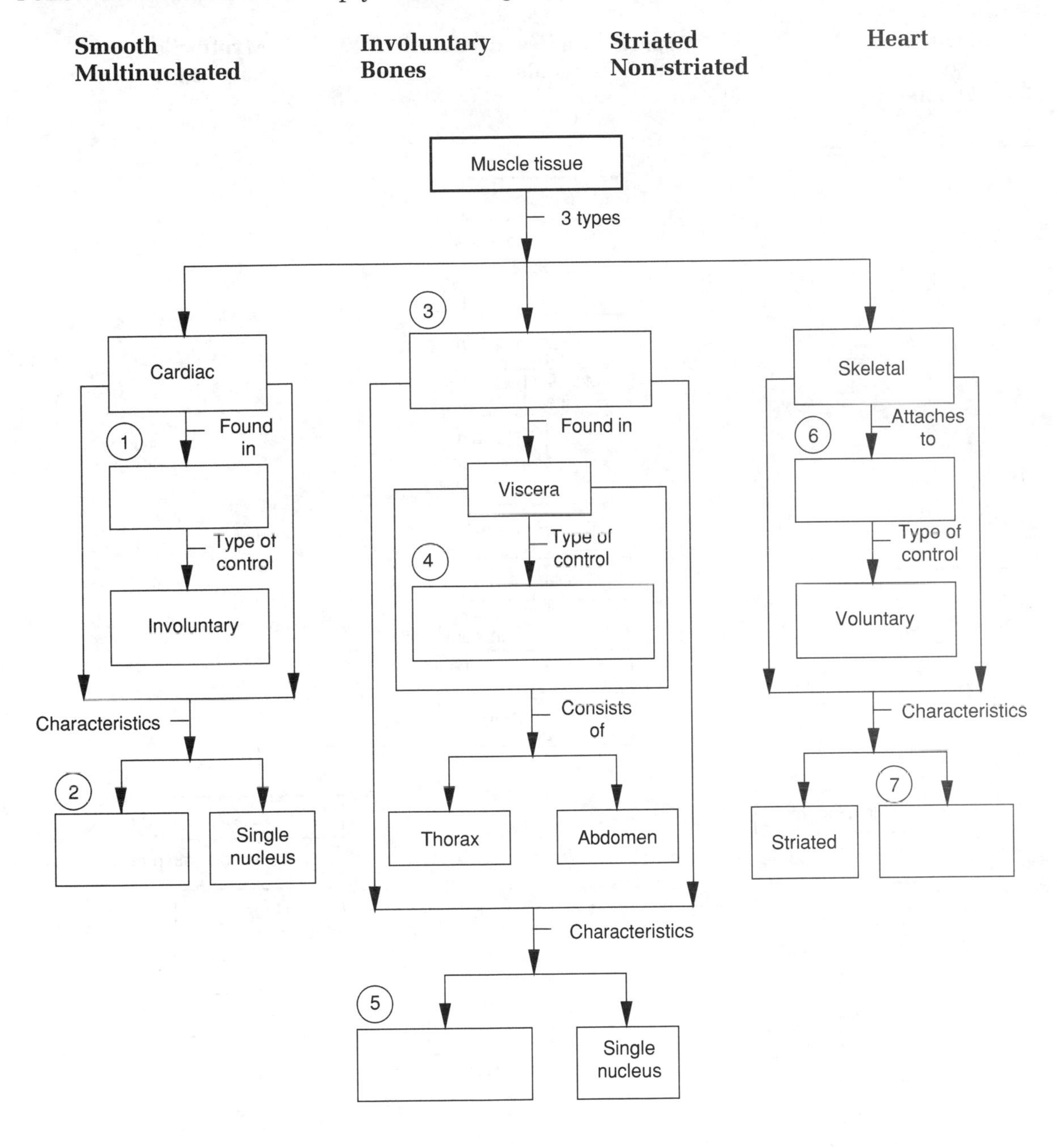

C. Concept Map II - Muscle Structure

Using the following terms, fill in the circled, numbered, blank spaces to complete the concept map. Follow the numbers that comply with the organization of the map.

Z lines	**Muscle bundles (fascicles)**	**Myofibrils**
Actin	**Thick filaments**	**Sarcomeres**
H zone		

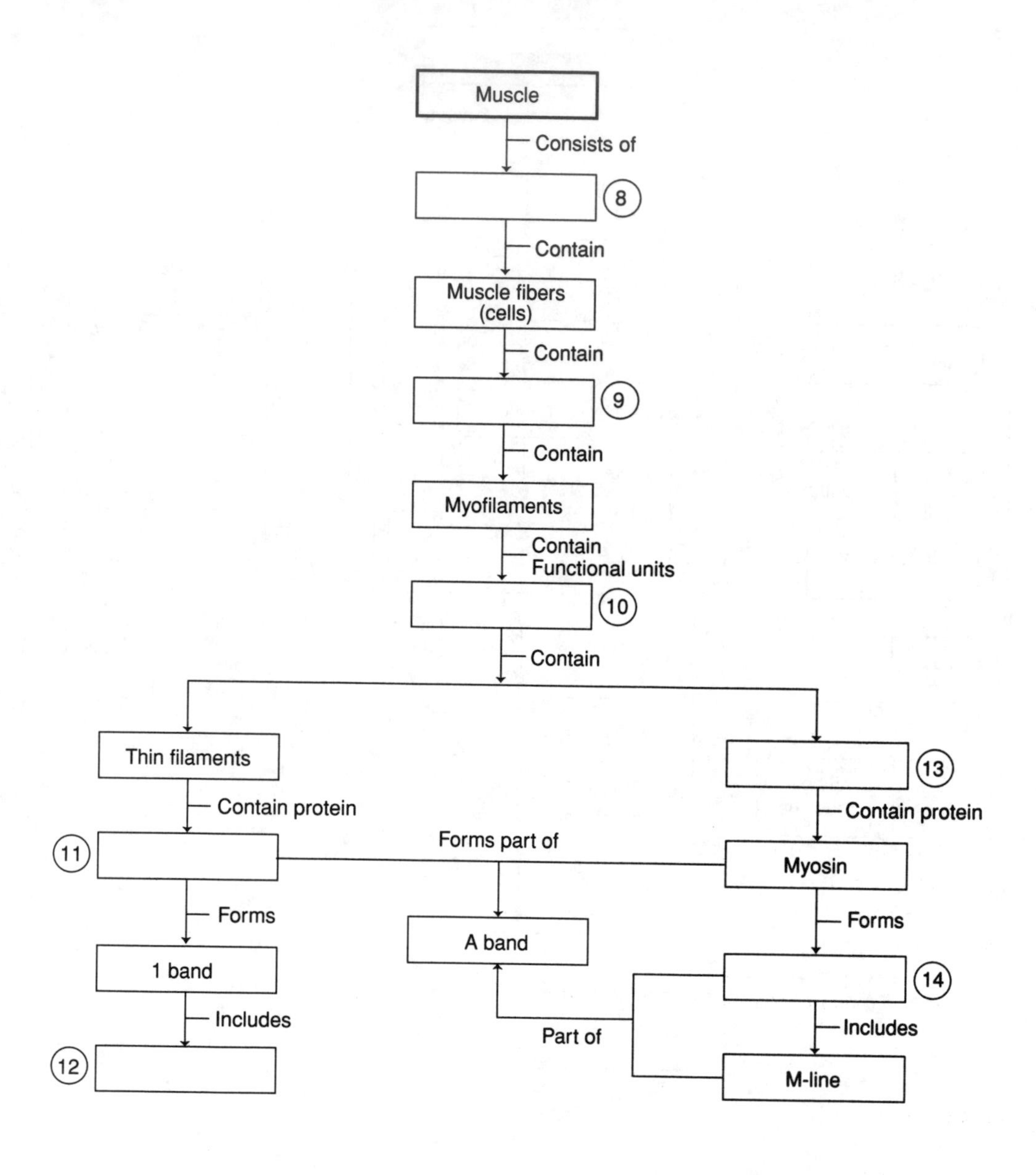

D. Crossword Puzzle

This crossword puzzle reviews the material in Chapter 7. To complete the puzzle, you must know the answers to the clues given, and you must be able to spell the terms correctly.

ACROSS

1. The gracilis will ________ the leg.
3. One of the quadriceps muscles.
4. The biceps brachii will _____ the antebrachium.
5. The filament that slides.
7. The involuntary type of muscle.
10. A unit of muscle that is made of actin and myosin.
12. The muscle associated with the cheeks.
14. The rectus femoris will _________ the lower leg.
15. A major lower back muscle.
16. This type of metabolism requires the use of oxygen.
17. This type of metabolism does not require the use of oxygen.
18. The deltoid will _________ the arm.

DOWN

2. A major upper back muscle.
5. A sheath that connects the frontalis to the occipitalis.
6. The thick filaments that have cross-bridges.
7. The voluntary type of muscle.
8. These connect muscle to bone.
9. A neuromuscular junction is the connection between a motor neuron and a muscle _________.
11. The biceps brachii and the triceps brachii are examples of _________ mucsles.
13. Cardiac muscle has _____ discs. No other muscle has them.
16. A muscle that is not used may undergo ______.

E. Short Answer Questions

Briefly answer the following questions in the spaces provided below.

1. What are the five functions performed by skeletal muscles?

2. What are the three layers of connective tissue that are part of each muscle?

3. Cite the five interlocking steps involved in the contraction process.

4. What are the differences between an isometric and an isotonic contraction?

5. What is the relationship among fatigue, anaerobic glycolysis, and oxygen debt?

6. Why do fast fibers fatigue more rapidly than slow fibers?

7. What is the primary functional difference between an origin and an insertion?

8. What are the three (3) primary actions used to group muscles?

9. What are the four (4) groups of muscles that comprise the axial musculature?

10. What two (2) major groups of muscles comprise the appendicular musculature?

CHAPTER

8

Neural Tissue and the Central Nervous System

Overview

The nervous system includes all the neural tissues in the body, and along with the endocrine system, the nervous system coordinates organ system activities in response to changes in environmental conditions. Due to the complexity and versatility of the nervous system, the structures are described in terms of two major anatomical divisions – the central nervous system (CNS) and the peripheral nervous system (PNS). The CNS consists of the brain and spinal cord and is responsible for integrating and coordinating sensory data and motor commands. The CNS is also the seat of higher functions, such as intelligence, memory and emotion. The PNS, consisting of cranial nerves and spinal nerves and ganglia, provides the communication pathways between the CNS and the muscles, glands, and sensory receptors.

In order to clarify and classify the structure and function of the nervous system, motor activities are identified by the activities occurring in the somatic and autonomic divisions.

The integration and interrelation of the nervous system with all other body systems is an integral part of comprehending how the body functions as a whole and how the nervous system's control mechanisms provide the necessary adjustments to meet changing internal and external environmental conditions.

Beginning with cellular organization in neural tissue, the Chapter 8 exercises are designed to assist you in mastering some principles of neurophysiology, anatomical features of the brain and spinal cord, and functional activities of the CNS.

Review of Chapter Objectives

1. Describe the anatomical organization and general functions of the nervous system.
2. Distinguish between neurons and neuroglia and compare their structure and functions.
3. Discuss the events that generate action potentials in the membranes of nerve cells.
4. Distinguish between continuous and saltatory nerve impulse conduction.
5. Explain the mechanisms of nerve impulse transmission at the synapse.
6. Describe the process of a neural reflex.

7. Describe the three meningeal layers that surround the central nervous system.
8. Discuss the structure and functions of the spinal cord.
9. Name the major regions of the brain and describe their functions.
10. Locate the motor, sensory and association areas of the cerebral cortex and discuss their functions.

Part I: Objective Based Questions

Objective 1 Describe the anatomical organization and general functions of the nervous system.

_____ 1. The two major anatomical divisions of the nervous system are

a. central nervous system (CNS) and peripheral nervous system (PNS)
b. somatic nervous system (SNS) and autonomic nervous system (ANS)
c. neurons and neuroglia
d. afferent division and efferent division

_____ 2. The central nervous system (CNS) consists of

a. afferent and efferent division
b. somatic and visceral division
c. brain and spinal cord
d. autonomic and somatic division

_____ 3. The primary function(s) of the nervous system include

a. providing sensation of the internal and external environments
b. integrating sensory information
c. regulating and controlling peripheral structures and systems
d. all of the above

_____ 4. The peripheral nervous system consists of two divisions, the

a. autonomic and somatic
b. cranial nerves and spinal nerves
c. efferent and afferent
d. sympathetic and parasympathetic

_____ 5. _____ nerves carry impulses from the PNS to the CNS; _____ nerves carry impulses from the CNS to the PNS.

a. Afferent, efferent
b. Autonomic, somatic
c. Somatic, autonomic
d. Efferent, afferent

_____ 6. Voluntary control of skeletal muscles is provided by the

a. somatic nervous system
b. autonomic nervous system
c. sympathetic nervous system
d. parasympathetic nervous system

Objective 2 Distinguish between neurons and neuroglia and compare their structures and functions.

_____ 1. The types of neuroglia (glial cells) in the central nervous system are

a. unipolar, bipolar, multipolar cells
b. astrocytes, oligodendrocytes, microglia and ependymal cells
c. efferent, afferent, association cells
d. motor, sensory, interneuron cells

_____ 2. The white matter of the CNS represents a region dominated by the presence of

a. astrocytes
b. neuroglia
c. oligodendrocytes
d. unmyelinated neurons

_____ 3. Neurons are responsible for

a. creating a three dimensional framework for the CNS
b. performing repairs in damaged neural tissue
c. controlling the interstitial environment
d. information transfer and processing in the nervous system

_____ 4. Neurons are classified on the basis of their structure as

a. motor, sensory, association, interneurons
b. anaxonic, unipolar, bipolar, multipolar
c. astrocytes, oligodendrocytes, microglia, ependymal
d. efferent, afferent, association, interneurons

_____ 5. Neurons are classified on the basis of their function as

a. unipolar, bipolar, multipolar
b. somatic, visceral, autonomic
c. motor, sensory, association
d. central, peripheral, somatic

_____ 6. Small phagocytic cells that are quite obvious in damaged tissue in the CNS are the

a. microglia
b. Schwann cells
c. astrocytes
d. oligodendrocytes

_____ 7. The neurilemma of axons in the PNS is formed by

a. astrocytes
b. microglia
c. ependymal cells
d. Schwann cells

_____ 8. The site of intercellular communication between neurons is the

a. dendrite
b. synapse
c. hillock
d. collateral

_____ 9. Interneurons

a. are found only in the CNS
b. carry only sensory impulses
c. carry only motor impulses
d. are found between neurons and their effectors

_____ 10. Neurons that have several dendrites and a single axon are called

a. unipolar
b. bipolar
c. multipolar
d. polypolar

11. Identify and label the structures in a typical neuron. Place your answers in the spaces provided below the drawing.

Figure 8-1 Neuron Structure

Axon
Dendrites
Axon hillock
Soma
Nucleus
Neurilemma
Axon terminals

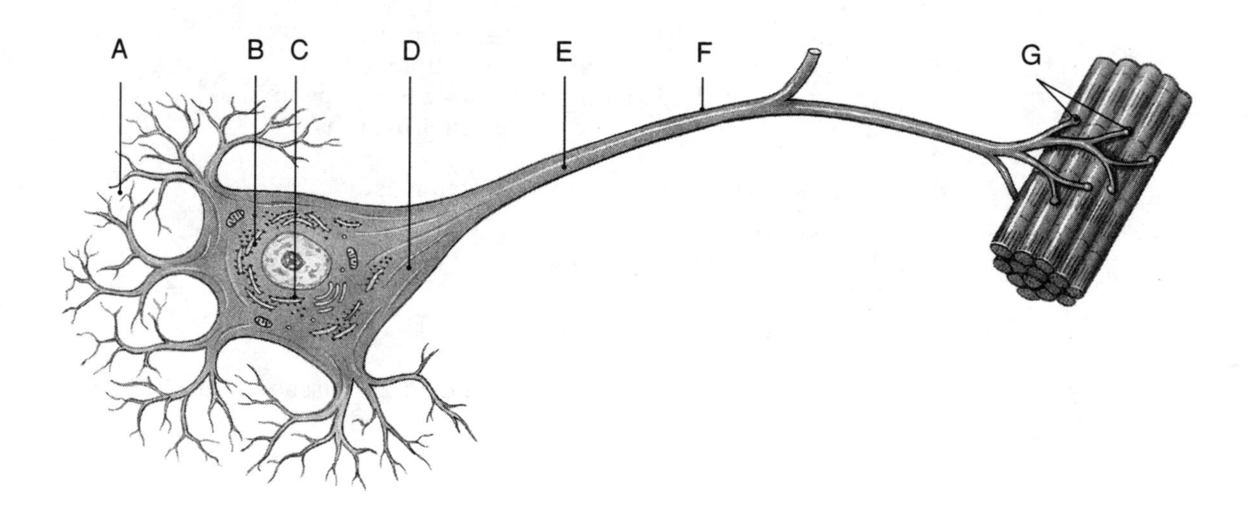

A ______________________

B ______________________

C ______________________

D ______________________

E ______________________

F ______________________

G ______________________

12. On the basis of structure, the neuron illustrated in Figure 8-1 is ___________.

13. Identify and label the following structures in Figure 8-2. Place your answers in the spaces provided below the drawings.

Figure 8-2 Neurons and Neuroglia

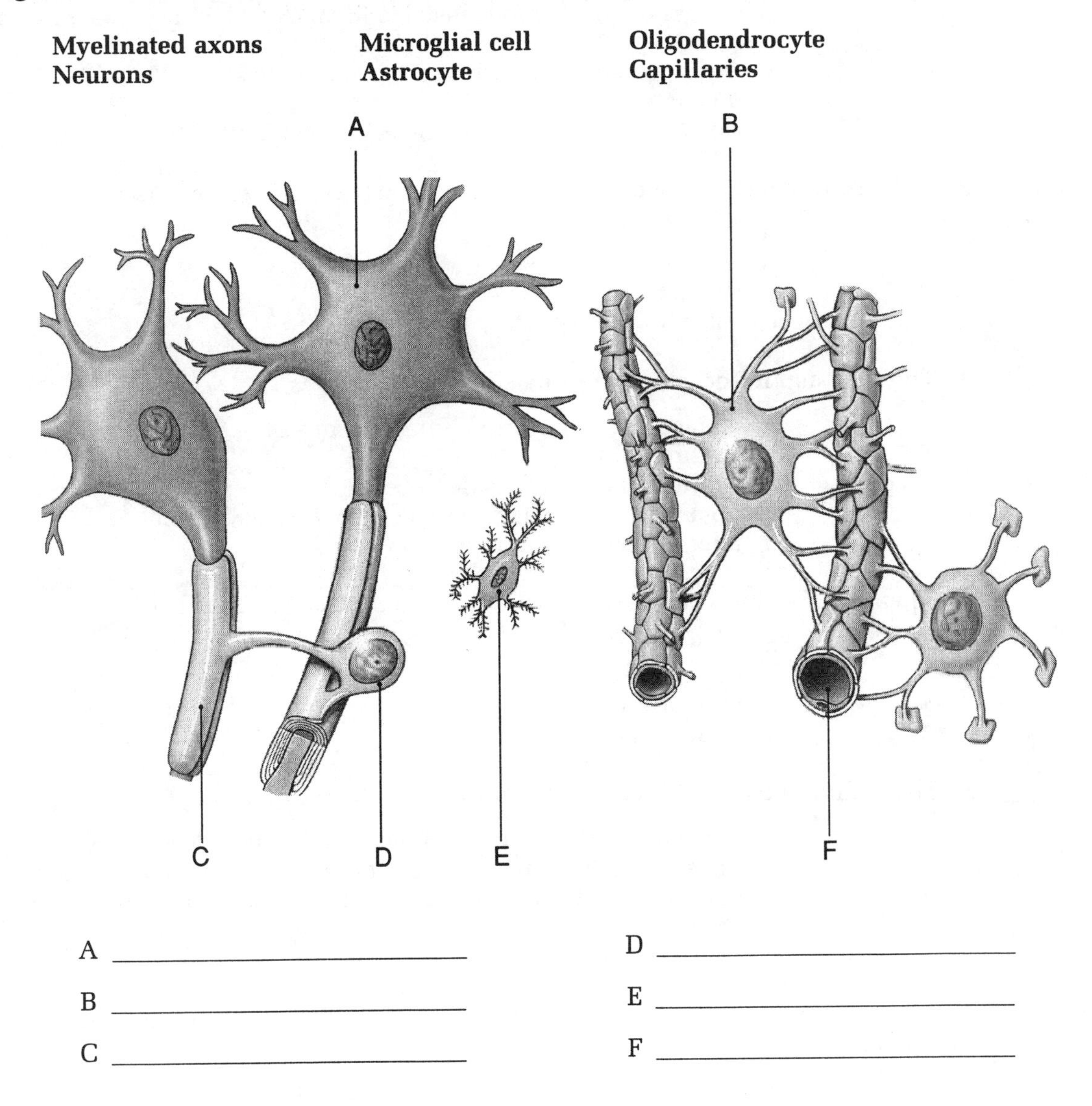

Objective 3 Discuss the events that generate action potentials in the membranes of nerve cells.

_____ 1. At the site of an action potential, the membrane contains an

a. excess of negative ions inside and an excess of negative ions outside
b. excess of positive ions inside and an excess of negative ions outside
c. equal amount of positive and negative ions on either side of the membrane
d. equal amount of positive ions on either side of the membrane

_____ 2. If the resting membrane potential is -70 mV, a hyperpolarized membrane is

a. 0 mV
b. +30 mV
c. -80 mV
d. -65 mV

_____ 3. The first step in the generation of an action potential is

a. a graded depolarization brings an area of excitable membrane to threshold
b. sodium channel activation occurs
c. potassium channels open and potassium moves out
d. a temporary hyperpolarization occurs

_____ 4. Opening of sodium channels in the membrane of a neuron results in

a. hyperpolarization
b. depolarization
c. repolarization
d. none of the above

_____ 5. The sodium-potassium exchange pump

a. transports sodium ions into the cell during depolarization
b. transports potassium ions out of the cell during repolarization
c. moves sodium and potassium in the direction of their chemical gradients
d. requires ATP energy to function

_____ 6. At the normal resting potential of a typical neuron, the ion exchange pump exchanges

a. 1 intracellular Na ion for 2 extracellular K ions
b. 2 intracellular Na ions for 1 extracellular K ion
c. 3 intracellular Na ions for 1 extracellular K ion
d. 3 intracellular Na ions for 2 extracellular K ions

_____ 7. The all-or-none-principle states that

a. all stimuli will produce identical action potentials
b. stimuli that are strong enough to bring the membrane to threshold will produce identical action potentials.
c. the greater the magnitude of the stimuli, the greater the intensity of the action potential
d. only motor stimuli can activate action potentials

Objective 4 Distinguish between continuous and saltatory nerve impulse conduction.

1. The form of an action potential in which the process occurs like a chain reaction along the membrane is __________ conduction.

2. The form of an action potential which occurs at successive nodes along the length of the stimulated axons called __________ conduction.

Objective 5 Explain the mechanism of nerve impulse transmission at the synapse.

_____ 1. The most common type of synapse found in the nervous system is

a. chemical
b. electrical
c. mechanical
d. time released

_____ 2. The release of acetylcholine into the synaptic cleft is initiated by

a. sodium ions
b. potassium ions
c. calcium ions
d. chloride ions

_____ 3. Cholinergic synapses release the neurotransmitter

a. cholinesterase
b. acetylcholine
c. norepinephrine
d. serotonin

_____ 4. Adrenergic synapses release the neurotransmitter

a. norepinepherine
b. adrenalin
c. dopamine
d. acetylcholine

_____ 5. The processing of the same information at the same time by several neuronal pools is called

a. serial processing
b. parallel processing
c. convergent processing
d. divergent processing

_____ 6. The last event to occur at a typical cholinergic synapse is

a. calcium ions enter the cytoplasm of the synaptic cleft
b. ACh release ceases because Ca^{++} are removed from the cytoplasm
c. ACh is broken down into acetate and choline by AChE
d. The synaptic knob reabsorbs choline from the synaptic cleft and uses it to resynthesize ACh.

Objective 6 Describe the process of a neural reflex.

_____ 1. In a reflex arc, a stimulus initiates a nerve impulse which travels along a

a. sensory neuron to the CNS which sends an impulse on a motor neuron to the effector
b. motor neuron to the CNS which sends an impulse to the effector on a sensory neuron
c. sensory neuron to the PNS which sends an impulse to the effector via of a motor neuron
d. sensory neuron to the effector which simulates the CNS to send an impulse to the PNS

_____ 2. The sensory neuron associated with a reflex arc transmits the impulse

a. away from the interneuron
b. away from the spinal cord
c. towards the effector
d. toward the CNS

_____ 3. The motor neuron associated with a reflex arc transmits the impulse

a. towards the effector
b. away from the CNS
c. toward the CNS
d. a and b are correct

_____ 4. In a typical reflex arc, the correct pathway of an action potential beginning with the receptor is

a. interneuron → motor neuron → CNS → sensory neuron→ effector
b. sensory neuron → interneuron → CNS → motor neuron → effector
c. motor neuron → interneuron → PNS → sensory neuron → effector
d. sensory neuron → motor neuron → CNS → interneuron → effector

_____ 5. In a reflex arc, neurotransmitter activity occurs at

a. the site of stimulation
b. the effector
c. synapses
d. the receptor

6. In the drawing below, identify and label the parts of a reflex arc. Place your answers in the spaces provided below the drawings.

Figure 8-3 The Reflex Arc

Effector	**Interneuron**	**Gray matter**
Receptor	**Synapse**	**Stimulus**
Sensory neuron	**Motor neuron**	**White matter**

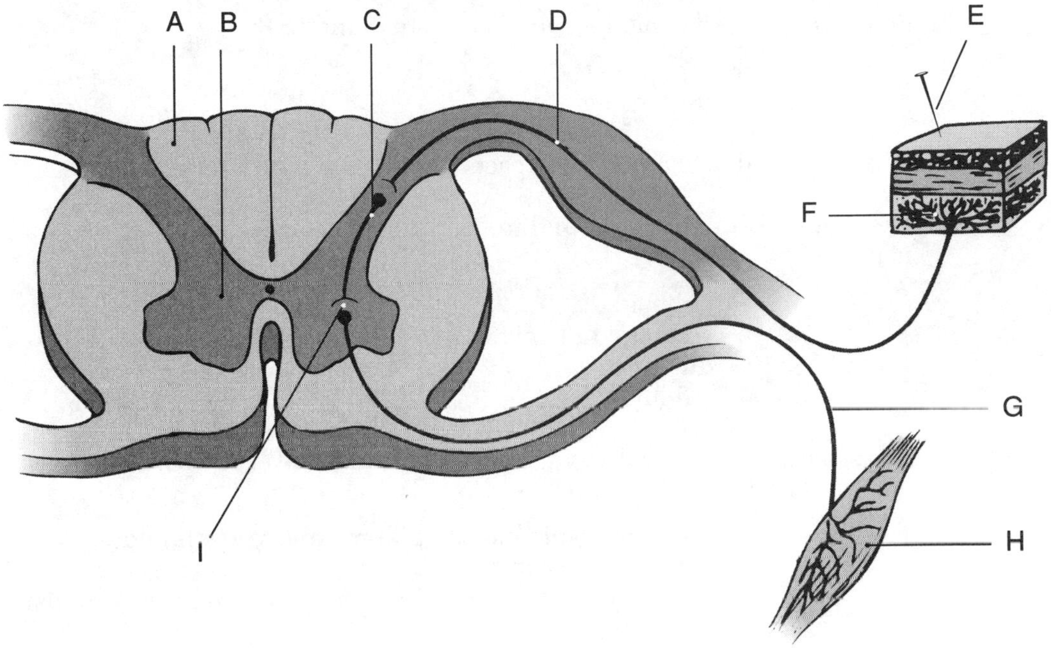

A ______________________

B ______________________

C ______________________

D ______________________

E ______________________

F ______________________

G ______________________

H ______________________

I ______________________

Objective 7 - Describe the three layers that surround the central nervous system.

_____ 1. Delicate neural tissues in the CNS are protected and defended by a series of specialized membranes called

a. pleura
b. peritoneum
c. meninges
d. pericardium

_____ 2. Blood vessels servicing the spinal cord are found in the

a. pia mater
b. dura mater
c. epidural space
d. subarachnoid space

_____ 3. The menix that is firmly bound to neural tissue and deep to the other meninges is the

a. pia mater
b. arachnoid membrane
c. dura mater
d. epidural space

_____ 4. Progressing from the outward layer to the inward layer the correct sequence of meningeal layers of the spinal cord is

a. dura mater, pia mater, epidural space, arachnoid, subarachnoid space
b. arachnoid, subarachnoid space, epidural space, dura mater, pia mater
c. epidural space, dura mater, arachnoid, subarachnoid space, pia mater
d. pia mater, subarachnoid space, arachnoid, dura mater, epidural space

_____ 5. If cerebrospinal fluid was drawn during a spinal tap, a needle would be inserted into the

a. subdural space
b. subarachnoid space
c. epidural space
d. dura mater

_____ 6. In the spinal cord the cerebrospinal fluid is found within the

a. subarachnoid and epidural space
b. central canal and subarachnoid space
c. central canal and epidural space
d. subdural and epidural spaces

Objective 8 Discuss the structure and functions of the spinal cord.

_____ 1. The spinal cord is part of the

a. peripheral nervous system
b. somatic nervous system
c. autonomic nervous system
d. central nervous system

_____ 2. The identifiable areas of the spinal cord that are based on the regions they serve include

a. axillary, radial, median, ulnar
b. pia mater, dura mater, arachnoid mater
c. cervical, thoracic, lumbar, sacral
d. cranial, visceral, autonomic spinal

_____ 3. The white matter of the spinal cord contains

a. cell bodies of neurons and glial cells
b. somatic and visceral sensory nuclei
c. large numbers of myelinated and unmyelinated axons
d. sensory and motor nuclei

_____ 4. The area of the spinal cord that surrounds the central canal and is dominated by the cell bodies of neurons and glial cells is the

a. white matter
b. gray matter
c. ascending tracts
d. descending tracts

_____ 5. The posterior gray horns of the spinal cord contain

a. somatic and visceral sensory nuclei
b. somatic and visceral motor nuclei
c. ascending and descending tracts
d. anterior and posterior columns

_____ 6. The axons in the white matter of the spinal cord that carry sensory information up toward the brain are organized into

a. anterior white column
b. descending rami
c. descending tracts
d. ascending tracts

7. Identify and label the following structures of the spinal cord. Place your answers in the spaces provided below the drawing.

Figure 8-4 Organization of the Spinal Cord

White matter	**Anterior median fissure**	**Posterior median sulcus**
Gray commissure	**Subarachnoid space**	**Central canal**
Dura mater	**Anterior horn**	**Dorsal root**
Ventral root	**Spinal nerve**	**Posterior horn**
Pia mater		

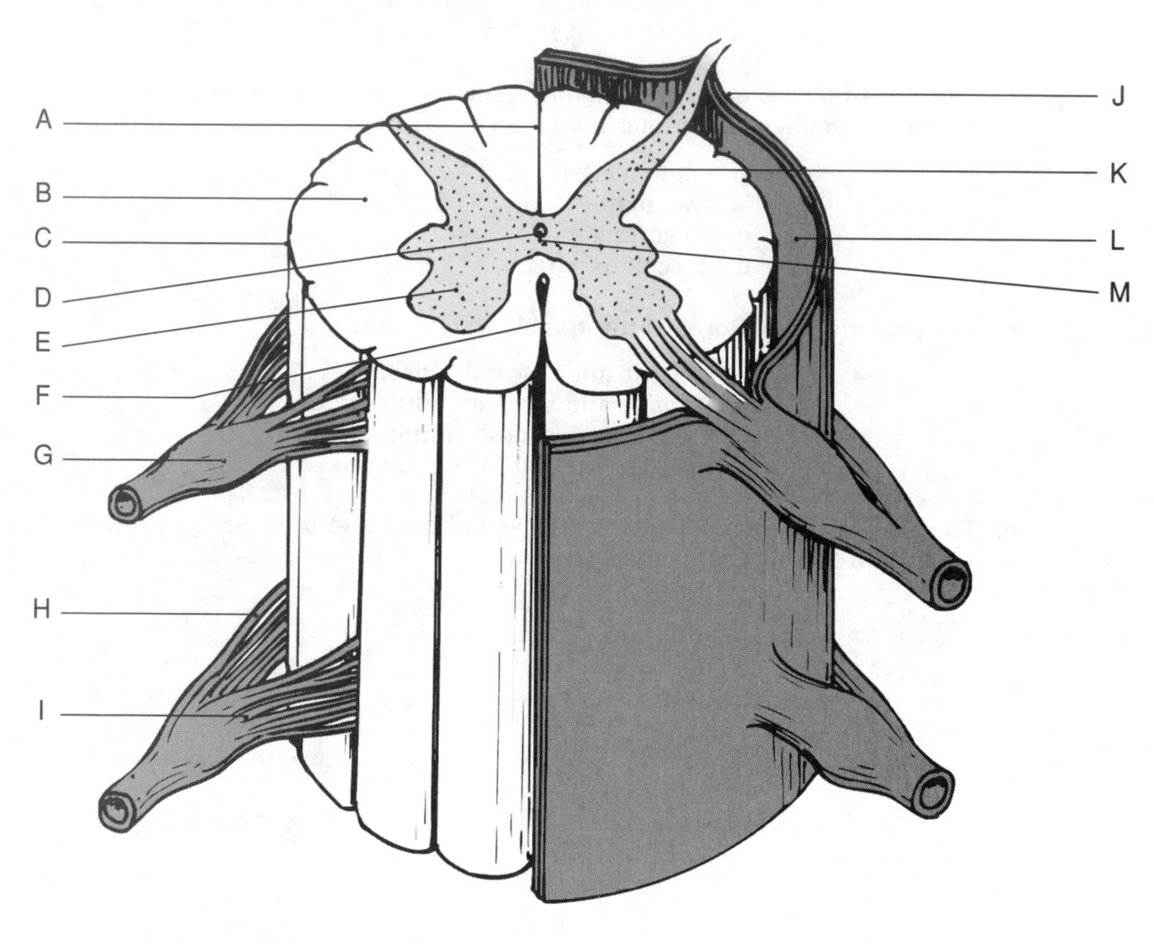

A ______________________

B ______________________

C ______________________

D ______________________

E ______________________

F ______________________

G ______________________

H ______________________

I ______________________

J ______________________

K ______________________

L ______________________

M ______________________

Objective 9 Name the major regions of the brain and describe their functions.

_____ 1. The major region of the brain responsible for conscious thought processes, sensations, intellectual functions, memory, and complex motor patterns is the

a. cerebellum
b. medulla
c. pons
d. cerebrum

_____ 2. The region of the brain that adjusts voluntary and involuntary motor activities on the basis of sensory information and stored memories of previous movements is the

a. cerebrum
b. cerebellum
c. medulla
d. diencephalon

_____ 3. The brain stem consists of the

a. midbrain, pons, medulla oblongata
b. cerebellum, medulla, pons
c. thalamus, hypothalamus, medulla
d. spinal cord, cerebellum, medulla

_____ 4. The thalamus contains

a. centers involved with emotions and hormone production
b. relay and processing centers for serious information
c. centers for involuntary somatic motor responses
d. nuclei involved with visceral motor control

_____ 5. The hypothalamus contains centers involved with

a. voluntary somatic motor responses
b. somatic and visceral motor control
c. maintenance of consciousness
d. emotions, autonomic function, and hormone production

_____ 6. The walls of the diencephalon form the

a. hypothalamus
b. thalamus
c. midbrain
d. cerebellum

_____ 7. Major centers concerned with autonomic control of breathing, blood pressure heart rate, and digestive activities are located in the

a. medulla oblongata
b. pons
c. midbrain
d. diencephalon

_____ 8. The diencephalic chamber is called the

a. fourth ventricle
b. lateral ventricle
c. second ventricle
d. third ventricle

_____ 9. Alzheimer's disease is associated with damage to the

a. hypothalamus in the diencephalon
b. cerebellar peduncles in the cerebellum
c. hippocampus in the limbic system
d. choroid plexus in the roof of the fourth ventricle

_____ 10. Cerebrospinal fluid which circulates between the different ventricles forms at the

a. mesencephalic aqueducts
b. choroid plexus
c. central canal
d. corpus callosum

11. Identify and label the structures in Figure 8-5. Place your answers in the spaces provided below the drawing.

Figure 8-5 Lateral View of the Human Brain

Frontal lobe	**Parietal lobe**	**Temporal lobe**
Occipital lobe	**Central sulcus**	**Lateral fissure**
Precentral gyrus	**Medulla oblongata**	**Parieto-occipital fissure**
Cerebellum	**Post-central gyrus**	**Pons**

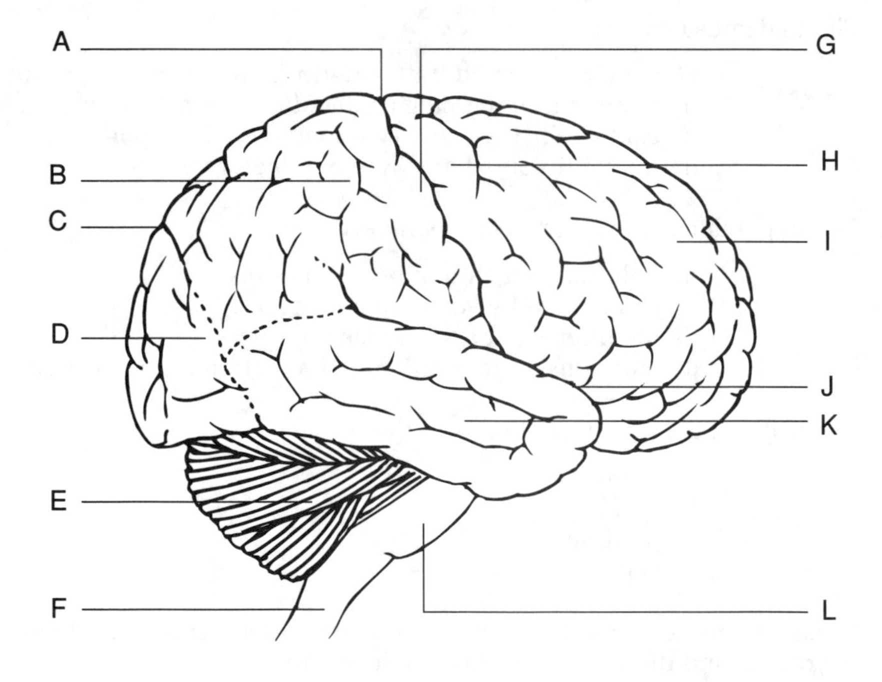

A _______________ G _______________

B _______________ H _______________

C _______________ I _______________

D _______________ J _______________

E _______________ K _______________

F _______________ L _______________

12. Identify and label the structures in Figure 8-6. Place your answers in the spaces provided below the drawing.

Figure 8-6 Sagittal View of the Human Brain

Cerebral hemispheres	**Corpus callosum**	**Pineal gland**
Cerebellum	**Thalamus**	**Fornix**
Third ventricle	**Pituitary gland**	**Pons**
Medulla oblongata	**Mammilary gland**	**Choroid plexus**
Cerebral peduncle	**Cerebral aqueduct**	**Fourth ventricle**
Corpora quadrigemina	**Optic chiasma**	

A ____________________

B ____________________

C ____________________

D ____________________

E ____________________

F ____________________

G ____________________

H ____________________

I ____________________

J ____________________

K ____________________

L ____________________

M ____________________

N ____________________

O ____________________

P ____________________

Q ____________________

Objective 10 Locate the motor, sensory, and association areas of the cerebral cortex and discuss their functions.

_____ 1. The neurons in the primary sensory cortex receive somatic information from

a. commisural fibers
b. touch, pressure, pain, taste, and temperature receptors
c. visual and auditory receptors in the eyes and ears
d. receptors in muscle spindles and Golgi tendon organs

_____ 2. The neurons of the primary motor cortex are responsible for directing

a. visual and auditory responses
b. responses to taste and temperature
c. voluntary movements
d. involuntary movements

_____ 3. The *somatic motor association* area is responsible for the

a. ability to hear, see, and smell
b. coordination of learned motor responses
c. coordination of all reflex activity throughout the body
d. assimilation of neural responses to tactile stimulation

_____ 4. The primary *motor* areas are located in the

a. precentral gyrus area
b. postcentral gyrus area
c. corpus callosum
d. limbic system

_____ 5. The primary *sensory* areas are located in the

a. choroid plexus
b. amygdaloid bodies
c. postcentral gyrus area
d. cerebral peduncles

_____ 6. The interpretive association area for *vision* is the

a. occipital lobe
b. parietal lobe
c. frontal lobe
d. temporal lobe

_____ 7. The inability to interpret what is read or heard indicates that there is damage to the

a. left hemisphere
b. right hemisphere
c. cerebellum
d. medulla oblongata

_____ 8. The auditory cortex is located in the

a. occipital lobe
b. parietal lobe
c. temporal lobe
d. frontal lobe

_____ 9. Interconnecting neurons and communication between cerebral hemispheres occurs through the

a. hypothalamus
b. medulla oblongata
c. pons
d. corpus callosum

_____ 10. The series of elevated ridges that increase the surface area of the cerebral hemispheres and the number of neurons in the cortical areas are called

a. sulci
b. gyri
c. fissures
d. all of the above

11. Identify and label the primary association areas in figure 8-7. Place the answers in the spaces provided below the drawing.

Figure 8-7 Left Cerebral Hemisphere - Association Areas

Occipital lobe	**Parietal lobe**	**Central sulcus**
Premotor cortex	**Frontal lobe**	**Temporal lobe**
Precentral gyrus	**Postcentral gyrus**	

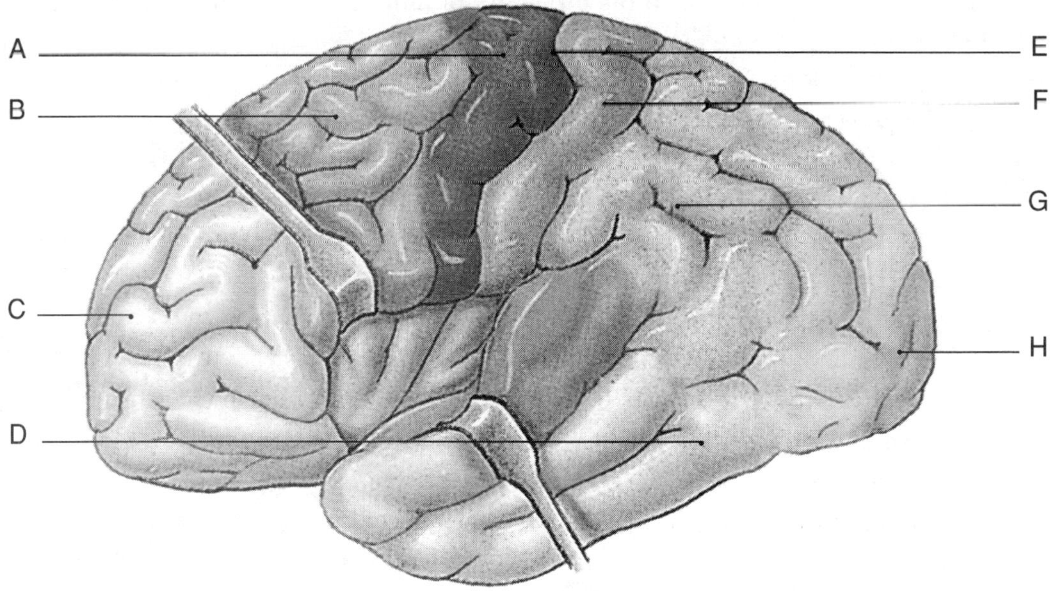

A
B
C
D
E
F
G
H

A ____________________ E ____________________

B ____________________ F ____________________

C ____________________ G ____________________

D ____________________ H ____________________

Part II: Chapter Comprehensive Exercises

A. Word Elimination

Circle the term that does not belong in each of the following groupings.

1. afferent efferent somatic autonomic CNS
2. neuron exteroceptors proprioceptors receptors interoceptors
3. soma dendrites neuroglia axon synaptic knob
4. astrocyte neuron oligodendrocyte microglia ependymal cells
5. resting threshold conduction depolarization repolarization
6. norepinephrine dopamine serotonin GABA adrenergic
7. nucleus neural cortex tracts ganglia pathways
8. sensory cervical thoracic lumbar sacral
9. frontal parietal occipital gyrus temporal
10. cerebrum meninges diencephalon cerebellum medulla oblongata

B. Matching

Match the terms in Column "B" with the terms in Column "A". Use letters for answers in the spaces provided.

COLUMN A	COLUMN B
___ 1. somatic nervous system	A. glial cells in PNS
___ 2. sensory neuron	B. involuntary control smooth, cardiac muscle
___ 3. K^+; Pr^-	C. release of ACh
___ 4. cholinergic synapse	D. ECF ions
___ 5. autonomic nervous system	E. maintained by astrocytes
___ 6. Schwann cells	F. voluntary control skeletal muscle
___ 7. Na^+; Cl^-	G. ICF ions
___ 8. reflex arc	H. proprioceptors
___ 9. blood-brain barrier	I. speech center
___ 10. Broca's area	J. negative feedback control mechanism

C. Concept Map I - Neural Tissue

Using the following terms, fill in the circled, numbered, blank spaces to complete the concept map. Follow the numbers that comply with the organization of the map.

Glial cells
Surround peripheral ganglia
Central Nervous System
Transmit Nerve Impulses
Schwann cells

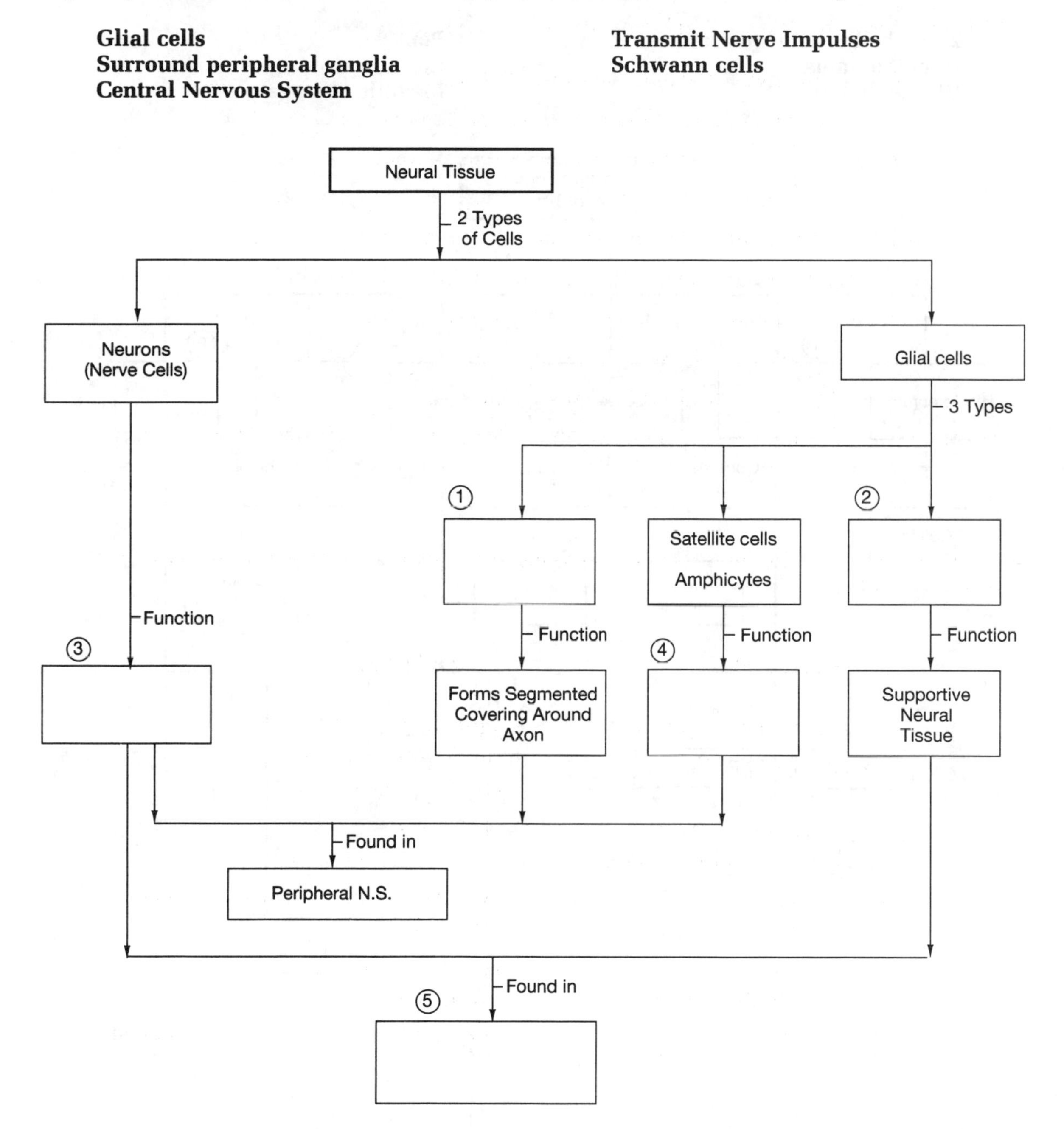

Concept Map II - Major Regions of the Brain

Using the terms below, fill in the circled, numbered, blank spaces to complete the concept map. Follow the numbers as they apply to the organization of the map.

2 Cerebellar hemispheres
Hypothalamus
Diencephalon
Pons
Corpora quadrigemina
Medulla oblongata

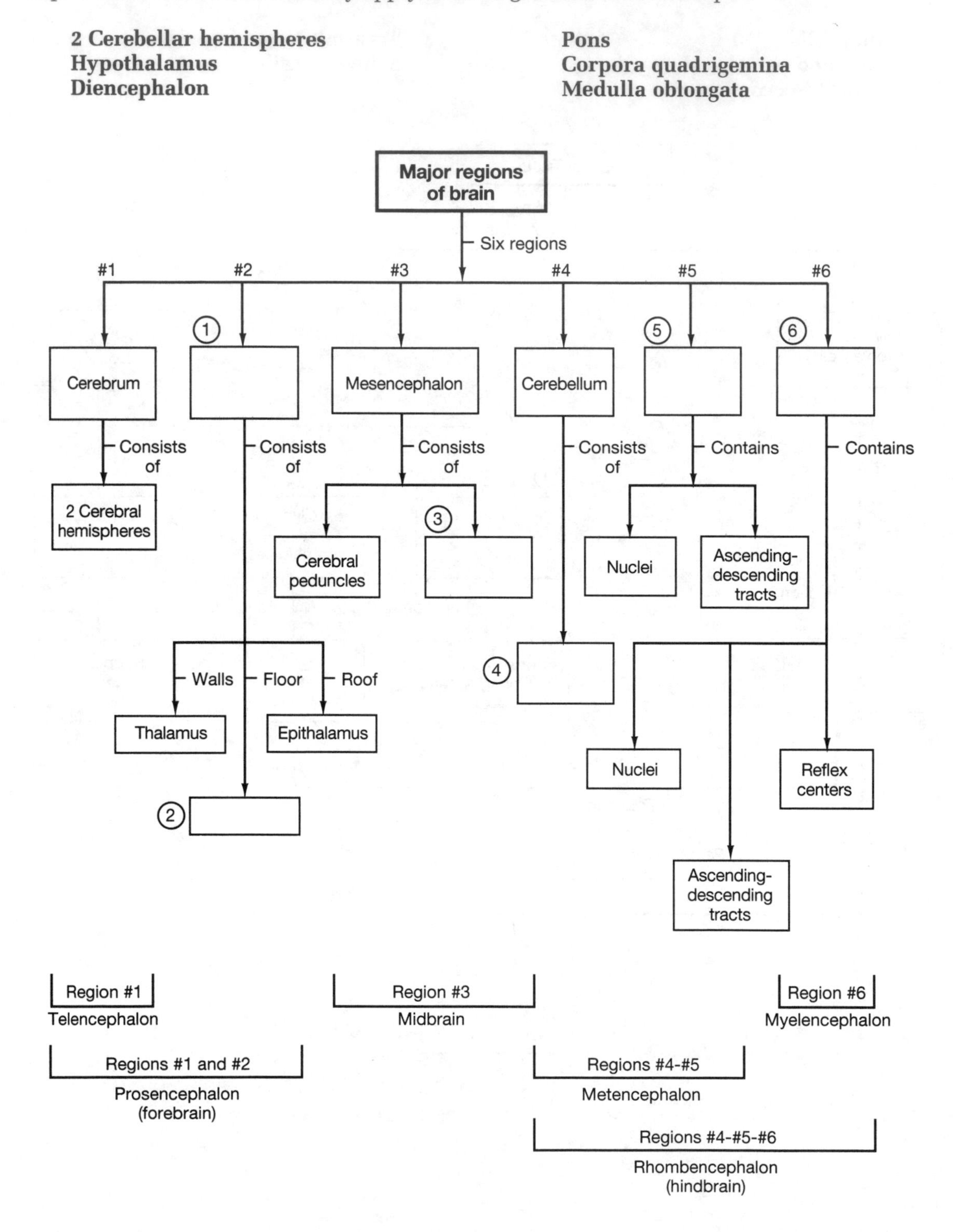

D. Crossword Puzzle

The following crossword puzzle reviews the material in Chapter 8. To complete the puzzle, you must know the answers to the clues given, and you must be able to spell the terms correctly.

ACROSS

1. One of the many types of neurotransmitters.
5. Nerves antagonistic to the parasympathetic nervous system.
6. An integral part of the autonomic nervous system.
7. The same as a sensory nerve.
9. The abbreviation for acetylcholine.
10. There are seven cervical vertebrae but ______ pairs of cervical nerves.
11. A nerve emerging from the cervical plexus involved in breathing.
12. The descending tracts in the spinal cord are _____ nerves.

DOWN

2. One of the two components of the CNS.
3. A major nerve of the lumbosacral plexus.
4. The portion of the neuron that contains organelles.
8. The same as a motor nerve.

E. Short Answer Questions

Briefly answer the following questions in the spaces provided below.

1. What are the four (4) major functions of the nervous system?

2. What are the major components of the central nervous system and the peripheral nervous system?

3. What four types of glial cells are found in the central nervous system?

4. Functionally, what is the major difference between neurons and neuroglia?

5. What is the difference between divergence and convergence?

6. Why is the brain more versatile than the spinal cord when responding to stimuli?

7. List the six (6) major regions in the adult brain.

8. What does the term "higher centers" refer to in the brain?

9. Where is the limbic system and what are its functions?

10. What are the hypothalamic contributions to the endocrine system?

11. What are the two (2) primary functions of the cerebellum?

12. What are the four (4) primary structural components of the pons?

CHAPTER

9

The Peripheral Nervous System and Integrated Neural Function

Overview

The peripheral nervous system (PNS) consists of all the neuron cell bodies and processes located outside the brain and spinal cord. All sensory information and motor commands are carried by axons of the PNS, which facilitate communication processes via pathways, nerve tracts, and nuclei that relay sensory and motor information from the spinal cord to the higher centers in the brain. The peripheral nerves of the PNS connected to the brain are called cranial nerves and those attached to the spinal cord are called spinal nerves. Figure 8-1 in your Essentials textbook shows a functional overview of the nervous system. Reference to this flow diagram will help you to visualize the functional role the afferent division of the PNS plays in relaying sensory information to the CNS from receptors in peripheral tissues and organs and how the efferent division of the PNS carries motor commands from the CNS to muscles and glands. You will readily notice that the efferent division has a somatic component, somatic N.S. (SNS), that provides voluntary control over skeletal muscle contraction. The efferent division has a visceral component, or autonomic N.S. (ANS), which provides automatic involuntary regulation of smooth muscle, cardiac muscle, and glandular activity. Sound complicated? You may want to draw your own diagram to help you organize and visualize this information and make it more meaningful to you.

The exercises in Chapter 9 will facilitate your understanding of the complex physical and mental activities which require integration and coordination of nervous system activity if homeostasis is to be maintained within the human body.

Review of Chapter Objectives

1. Identify the cranial nerves and relate each pair of cranial nerves to its principal functions.
2. Relate the distribution pattern of spinal nerves to the regions they innervate.

3. Distinguish between the motor responses produced by simple and complex reflexes.
4. Explain how higher centers control and modify reflex responses.
5. Identify the principal sensory and motor pathways.
6. Explain how we can distinguish among sensations that originate in different areas of the body.
7. Compare the autonomic nervous system with the other divisions of the nervous system.
8. Explain the functions and structures of the sympathetic and parasympathetic divisions.
9. Discuss the relationship between the sympathetic and parasympathetic divisions and explain the implications of dual innervation.
10. Summarize the effects of aging on the nervous system.
11. Discuss the interrelationships among the nervous system and other organ systems.

Part I: Objective Based Questions

Objective 1 Identify the cranial nerves and relate each pair of cranial nerves to its principal functions.

_____ 1. The special sensory cranial nerves include the

a. oculomotor, trochlear, abducens
b. spinal accessory, hypoglossal, glossopharyngeal
c. olfactory, optic, vestibulocochlear
d. vagus, trigeminal and facial

_____ 2. Cranial nerves III, IV, and XI, which provide motor control, are

a. trigeminal, facial, glossopharyngeal, vagus
b. oculomotor, trochlear, abducens, spinal accessory
c. olfactory, optic, vestibulocochlear, hypoglossal
d. oculomotor, hypoglossal, optic, olfactory

_____ 3. The cranial nerves that carry sensory information and involuntary motor commands are

a. I, II, III, IV
b. II, IV, VI, VIII
c. V, VI, VIII, XII
d. V, VII, IX, X

4. The cranial nerve responsible for the sense of smell is the _____________.

5. The cranial nerve responsible for vision is the _____________.

6. The pair of cranial nerves that controls the pupil of the eye is the _____________.

7. The cranial nerve involved when a person feels a sinus headache is the _____________.

8. The number of the pair of cranial nerves involved in taste is _____________.

9. The cranial nerve which controls the diaphragm is the _____________.

10. The hypoglossal cranial nerves control movement of the _____________.

11. Identify and label the cranial nerves and their number in Figure 9-1. (Ex. NI - OLFACTORY). Place your answers in the spaces provided below the drawing

Figure 9-1 The 12 Pairs of Cranial Nerves

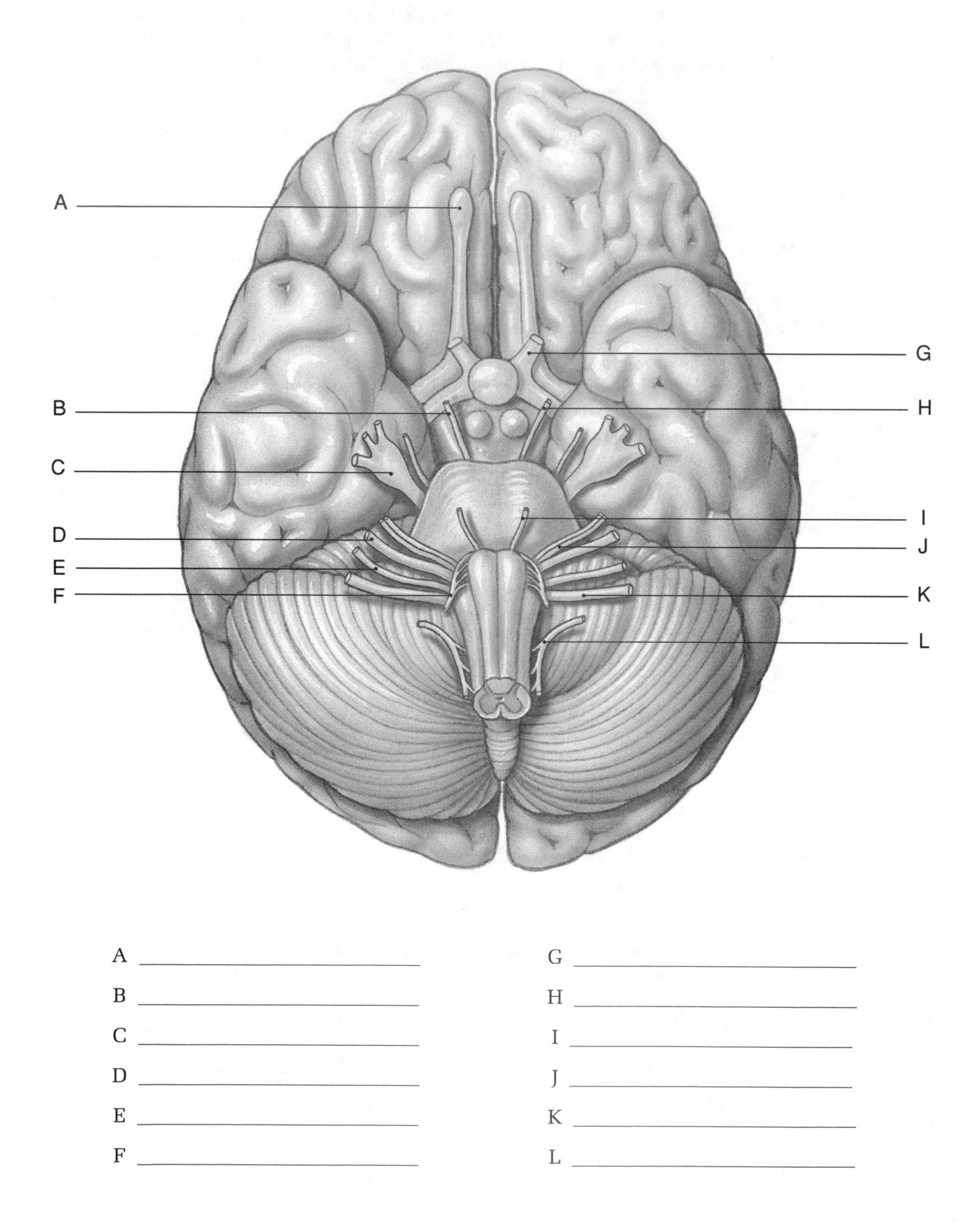

A ____________________ G ____________________

B ____________________ H ____________________

C ____________________ I ____________________

D ____________________ J ____________________

E ____________________ K ____________________

F ____________________ L ____________________

Objective 2 Relate the distribution pattern of spinal nerves to the regions they innervate.

_____ 1. The 31 pairs of spinal nerves include

a. 12 pr. cervical; 8 pr. thoracic; 5 pr. lumbar; 5 pr. sacral; 1 pr. coccygeal
b. 8 pr. cervical; 12 pr. thoracic; 5 pr. lumbar; 5 pr. sacral; 1 pr. coccygeal
c. 5 pr. cervical; 12 pr. thoracic; 8 pr. lumbar; 5 pr. sacral; 1 pr. coccygeal
d. 8 pr. cervical; 10 pr. thoracic; 7 pr. lumbar; 5 pr. sacral; 1 pr. coccygeal

_____ 2. Muscles of the neck and shoulder are innervated by spinal nerves from the

a. cervical region
b. lumbar region
c. thoracic region
d. sacral region

_____ 3. Spinal nerves from the sacral region of the cord innervate the

a. shoulder muscles
b. arm and leg muscles
c. muscles of the chest
d. abdominal and leg muscles

_____ 4. The joining of adjacent spinal nerves is termed a(n)

a. pair of cranial nerves
b. nerve tract
c. nerve plexus
d. conjugated axon terminal

Objective 3 Distinguish between the motor responses produced by simple and complex reflexes.

_____ 1. Stretch reflexes are important in the

a. delay between a stimulus and response
b. effect they have on the muscles of a limb
c. maintenance of posture and balance
d. all of the above

_____ 2. The receptors which are activated in the patellar reflex are the

a. Golgi tendon organs
b. muscle spindles
c. motor neurons
d. interneurons

_____ 3. The reflexes which are capable of producing more complicated responses are __________ reflexes.

a. monosynaptic
b. stretch
c. withdrawal
d. polysynaptic

_____ 4. The *flexor* reflex

a. prevents a muscle from overstretching
b. prevents a muscle from generating excessive tension
c. moves a limb away from a painful stimulus
d. is usually initiated due to application of excessive pressure

_____ 5. When one set of motor neurons is stimulated, those controlling antagonistic muscles are inhibited, describes the principle of

a. dual innervation
b. referred pain
c. reciprocal inhibition
d. accommodation

Objective 4 Explain how higher centers control and modify reflex responses.

_____ 1. Control of spinal reflexes by the brain involves

a. conscious control
b. activation of the precentral gyrus
c. descending tracts of the spinal cord
d. all of the above

_____ 2. In an adult, the curling of the toes in response to a stimulus is called the

a. Babinski reflex
b. plantar reflex
c. patellar reflex
d. flexor reflex

_____ 3. Motor patterns for walking, running and jumping are primarily directed by neuronal pools in the

a. spinal cord
b. cerebrum
c. cerebellum
d. precentral gyrus

_____ 4. The simplest reflexes are mediated at the level of the

a. cerebrum
b. medulla
c. spinal cord
d. cerebellum

_____ 5. The highest levels of information processing occur in the

a. cerebrum
b. corpus callosum
c. cerebellum
d. thalamus

Objective 5 Identify the principal sensory and motor pathways.

_____ 1. The three major somatic sensory pathways include the

a. first-order, second-order, and third-order pathways
b. nuclear, cerebellar, and thalamic pathways
c. posterior column, spinothalamic, and spinocerebellar pathways
d. anterior, posterior, and lateral pathways

_____ 2. The motor (descending) pathways include the

a. spinothalamic and cerebellar pathways
b. pyramidal and extrapyramidal systems
c. posterior and anterior column pathway
d. cerebral and cerebellar systems

_____ 3. The axons of the posterior column ascend within the

a. posterior and lateral spinothalamic tracts
b. fasciculus gracilis and fasciculus cuneatus
c. posterior and anterior spinocerebellar tracts
d. all of the above

_____ 4. The spinothalamic pathway consists of

a. lateral and anterior tracts
b. anterior and posterior tracts
c. gracilis and cuneatus nuclei
d. proprioceptors and interneurons

_____ 5. The somatic nervous system (SNS) issues somatic motor commands that direct the

a. activities of the autonomic nervous system
b. contractions of smooth and cardiac muscles
c. activities of glands and fat cells
d. contractions of skeletal muscles

_____ 6. The autonomic nervous system (ANS) issues motor commands that control

a. the somatic nervous system
b. smooth and cardiac muscles, glands, and fat cells
c. contractions of skeletal muscles
d. voluntary activities

_____ 7. The pyramidal system consists of

a. rubrospinal and reticulospinal tracts
b. vestibulospinal and tectospinal tracts
c. corticobulbar and corticospinal tracts
d. sensory and motor neurons

_____ 8. The extrapyramidal system consists of

a. gracilis and cuneatus nuclei
b. pyramidal, corticobulbar, and corticospinal tracts
c. sensory and motor fibers
d. vestibulospinal, tectospinal, rubrospinal , reticulospinal tracts

_____ 9. The motor commands carried by the tectospinal tracts are triggered by

a. visual and auditory stimuli
b. stimulation of the red nucleus
c. the vestibular nuclei
d. all of the above

10. Identify and label the following pathways in Figure 9-2. Place your answers in the spaces below the drawing.

Figure 9-2 Sensory and Motor Pathways of the Spinal Cord

Posterior spinocerebellar	**Tectospinal**	**Lateral corticospinal**
Anterior spinothalamic	**Fasciculus gracilis**	**Reticulospinal**
Anterior corticospinal	**Anterior spinocerebellar**	**Vestibulospinal**
Rubrospinal	**Lateral spinothalamic**	**Fasciculus cuneatus**

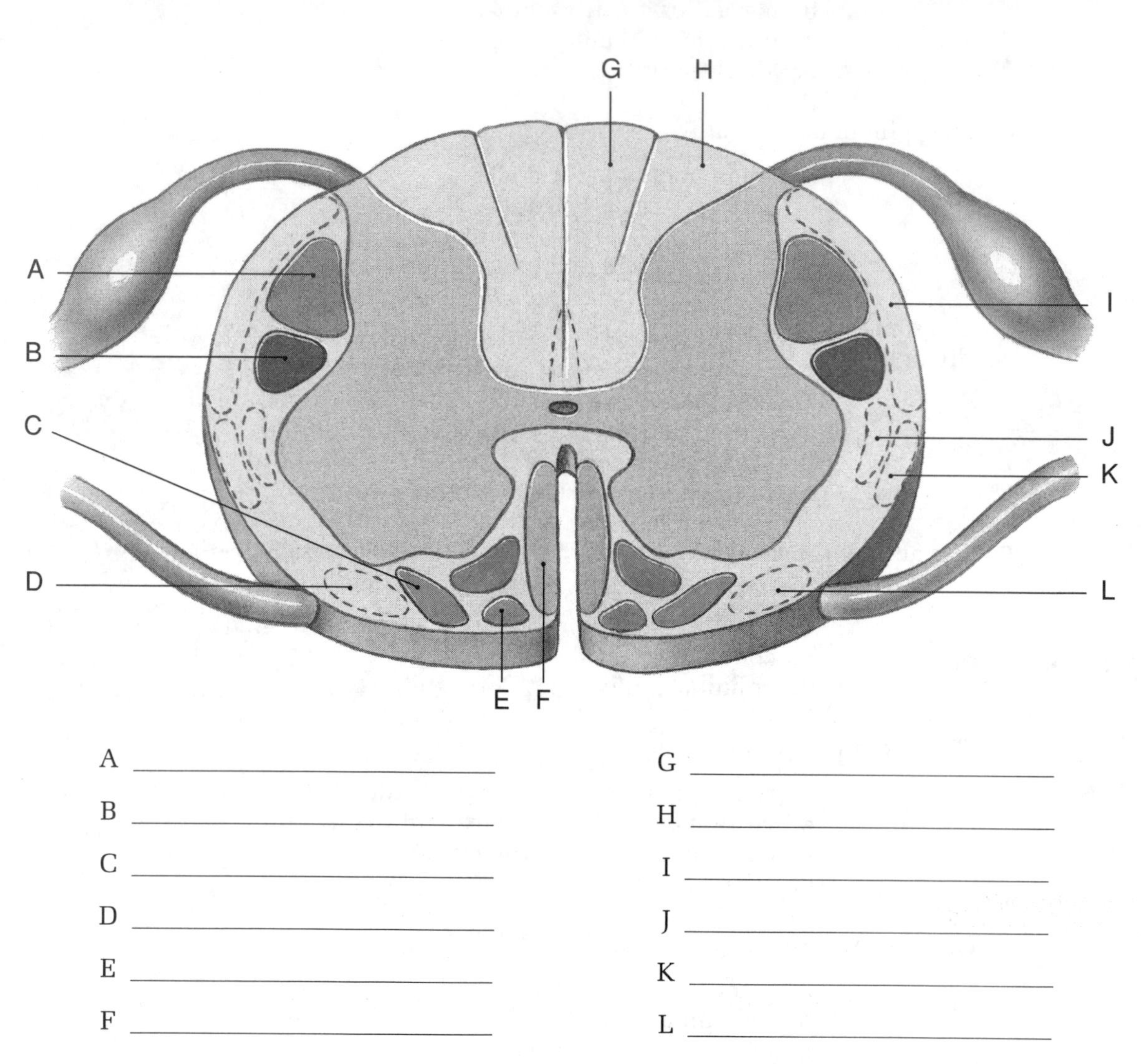

A ______________________ G ______________________

B ______________________ H ______________________

C ______________________ I ______________________

D ______________________ J ______________________

E ______________________ K ______________________

F ______________________ L ______________________

Objective 6 Explain how we can distinguish among sensations that originate in different areas of the body.

_____ 1. Information received by the brain concerning internal or external environmental conditions is called a(n)

a. stimuli
b. sensation
c. response
d. action potential

_____ 2. The area of sensory cortex devoted to a body region is relative to the

a. size of the body area
b. distance of the body area from the brain
c. number of motor units in the area of the body
d. number of sensory receptors in the area of the body

_____ 3. We can distinguish between sensations that originate in different areas of the body because

a. sensory neurons transmit only one type of information
b. sensory neurons from each body region synapse in specific brain regions
c. incoming sensory information is assessed by the thalamus
d. sensory neurons in different parts of the body vary in function

_____ 4. The cerebellum adjusts voluntary and involuntary motor activity in response to

a. visual information
b. proprioceptive data
c. visual information
d. all of the above

_____ 5. Complex motor activities like riding a bicycle or eating

a. require neural processing in the cerebrum
b. do not involve activity from the brain
c. require coordinated activity of several regions of the brain
d. are controlled by the cerebellum

Objective 7 Compare the autonomic nervous system with other divisions of the nervous system.

_____ 1. The autonomic nervous system (ANS) is a subdivision of the

a. central nervous system
b. peripheral nervous system
c. somatic nervous system
d. visceral nervous system

_____ 2. In the autonomic nervous system there is

a. voluntary and involuntary control of skeletal muscles
b. voluntary control of skeletal muscles
c. indirect voluntary control of skeletal muscles
d. a synapse interposed between the CNS and the peripheral effector

_____ 3. The two divisions of the ANS are the

a. somatic and peripheral
b. sympathetic and parasympathetic
c. central and somatic
d. peripheral and parasympathetic

4. In the ANS, the point at which information is transmitted from one excitable cell to another is the ________.

5. In the ANS, the control of smooth and cardiac muscle is __________.

Objective 8 Explain the function and structures of the sympathetic and parasympathetic divisions.

_____ 1. Sympathetic (preganglionic) neurons originate between

a. T1 and T2 of the spinal cord
b. T1 and L2 of the spinal cord
c. L1 and L4 of the spinal cord
d. S2 and S4 of the spinal cord

_____ 2. The important function(s) of the postganglionic fibers that enter the thoracic cavity in autonomic nerves include

a. accelerating the heart rate
b. increasing the force of cardiac contractions
c. dilating the respiratory passageways
d. a, b, and c are correct

_____ 3. Preganglionic neurons in the parasympathetic division of the ANS originate in the

a. peripheral ganglia adjacent to the target organ
b. thoracolumbar area of the spinal cord
c. walls of the target organ
d. brain stem and sacral segments of the spinal cord

_____ 4. Since second-order neurons in the parasympathetic division are all located in the same ganglion, the effects of parasympathetic stimulation are

a. more diversified and less localized than those of the sympathetic division
b. less diversified but less localized than those of the sympathetic division
c. more specific and localized than those of the sympathetic division
d. more diversified and more localized than those of the sympathetic division

_____ 5. The parasympathetic division of the ANS includes visceral motor nuclei associated with cranial nerves

a. I, II, III, IV
b. III, VII, IX, X
c. IV, V, VI, VIII
d. V, VI, VII, XII

_____ 6. At their synapses with ganglionic neurons, all ganglionic neurons in the sympathetic division release

a. epinephrine
b. norepinephrine
c. acetylcholine
d. a, b, and c are correct

_____ 7. At neuroeffector junctions, typical sympathetic postganglionic fibers release

a. epinephrine
b. norepinephrine
c. acetylcholine
d. dopamine

_____ 8. At synapse and neuroeffector junctions, all preganglionic and postganglionic fibers in the parasympathetic division release

a. epinephrine
b. norepinephrine
c. acetylcholine
d. a, b, and c are correct

_____ 9. The functions of the parasympathetic division center on

a. accelerating the heart rate and the force of contraction
b. dilation of the respiratory passageways
c. relaxation, food processing, and energy absorption
d. a, b, and c are correct

_____ 10. During a crisis, the event necessary for the individual to cope with stressful and potentially dangerous situations is called

a. the effector response
b. sympathetic activation
c. parasympathetic activation
d. a, b, and c are correct

_____ 11. Parasympathetic preganglionic fibers of the vagus nerve entering the thoracic cavity join the

a. cardiac plexus
b. pulmonary plexus
c. hypogastric plexus
d. a and b are correct

_____ 12. The major structural difference between sympathetic and pre- and postganglionic fibers is that

a. preganglionic fibers are short and postganglionic fibers are long
b. preganglionic fibers are long and postganglionic fibers are short
c. preganglionic fibers are close to target organs and postganglionic fibers are close to the spinal cord
d. preganglionic fibers innervate target organs while postganglionic fibers originate from cranial nerves

_____ 13. The effects of parasympathetic stimulation are usually

a. brief in duration and restricted to specific organs and sites
b. long in duration and diverse in distribution
c. brief in duration and diverse in distribution
d. long in duration and restricted to specific organs and sites

14. Because the *sympathetic* division of the PNS stimulates tissue metabolism and increases alertness, it is called the ______________ or ______________ division.

15. Because the *parasympathetic* division of the ANS conserves energy and promotes sedentary activity, it is known as the ___________ or _____________ division.

Objective 9 Discuss the relationship between the sympathetic and parasympathetic divisions and explain the implications of dual innervation.

_____ 1. The functional relationship between sympathetic fibers and parasympathetic fibers in the ANS is usually

a. complimentary
b. antagonistic
c. supportive
d. unpredictable

_____ 2. Dual innervation refers to an organ or tissue receiving

a. two nerves from the spinal cord
b. both autonomic and somatic motor nerves
c. both sympathetic and parasympathetic nerves
d. nerves from the brain and the spinal cord

_____ 3. When vital organs receive dual innervation, the result is usually

a. a stimulatory effect within the organ
b. an inhibitory effect within the organ
c. sympathetic - parasympathetic opposition
d. sympathetic - parasympathetic inhibitory effect

_____ 4. Proper control of the respiratory passages depends upon

a. sympathetic and parasympathetic stimulation
b. sympathetic stimulation
c. parasympathetic stimulation
d. somatic motor stimulation

_____ 5. The release of small amounts of acetylcholine and norepinephrine on a continual basis in an organ innervated by sympathetic and parasympathetic fibers is referred to as

a. reinforcement
b. autonomic tone
c. autonomy
d. complimentation

_____ 6. Autonomic tone exists in the heart because

a. ACh (acetylcholine) released by the parasympathetic division decreases the heart rate, and norepinephrine (NE) released by the sympathetic division accelerates the heart rate
b. ACh released by the parasympathetic division accelerates the heart rate, and NE released by the sympathetic division decreases the heart rate
c. NE released by the parasympathetic division accelerates the heart rate, and ACh released by the sympathetic division decreases the heart rate
d. NE released by the sympathetic division and ACh released by the parasympathetic division accelerate the heart rate

_____ 7. A decrease in the autonomic tone of the smooth muscle in blood vessels would result in

a. no change in vessel diameter
b. decreased blood flow in the blood vessels
c. a decrease in vessel diameter
d. an increase in vessel diameter

Objective 10 Summarize the effects of aging on the nervous system.

_____ 1. Differences in the brains of the elderly and younger individuals include

a. narrower gyri, wider sulci, and large subarachnoid space in the young
b. narrower gyri, wider sulci, and large subarachnoid space in the elderly
c. decreased numbers of cortical neurons in the young
d. increased blood flow to the brain in the elderly

_____ 2. Changes that occur in the CNS during the aging process include

a. increased numbers of synaptic connections
b. increased numbers of cortical neurons
c. reduction in brain size and weight
d. reduction of abnormal intracellular deposits in neurons

_____ 3. Alzheimer's disease is characterized by

a. decreased blood flow to the brain
b. reduced neuronal loss due to dementia
c. increased rate of neurotransmitter production
d. a gradual deterioration of mental organization

Objective 11 Discuss the interrelationships among the nervous system and other organ systems.

1. Sexual hormones which affect CNS development and sexual behaviors result from the activity of the _________ system.

2. Storage of calcium for neural function and protection of the brain and spinal cord results from the activity of the ____________ system.

3. Hormonal effect on CNS neural metabolism and CNS development results from the activity of the _________ system.

4. The system that provides nutrients for energy production and neurotransmitter synthesis for the nervous system is the ______________ system.

5. The system that provides endothelial cells to maintain the blood brain is the ______________ system.

Part II: Chapter Comprehensive Exercises

A. Word Elimination

Circle the term that does not belong in each of the following groupings.

1. axons peripheral nerves cell bodies ganglia CNS
2. N I N III N VII N IX N X
3. trigeminal opthalamic N IV maxillary mandibular
4. cervical peripheral brachial lumbar sacral
5. sensory excitatory motor muscle inhibitory
6. posterior column spinothalamic spinocerebellar pyramidal
7. cardiac muscles ganglia smooth muscles glands adipose tissues
8. splanchnic chain celiac superior mesenteric inferior mesenteric
9. epinephrine adrenalin acetylcholine norepinephrine lymph
10. ciliary submandibular sphenopalatine cardiac otic

B. Matching

Match the terms in Column "B" with the terms in Column "A". Use letters for answers in the spaces provided.

COLUMN A	COLUMN B
___ 1. opposition	A. splanchnic nerves
___ 2. vagus	B. cranial nerve VII
___ 3. facial	C. releases ACh
___ 4. adrenergic fibers	D. involuntary control
___ 5. collateral ganglia	E. dual innervation
___ 6. cholinergic fibers	F. stretch reflex
___ 7. pyramidal system	G. tongue movements
___ 8. muscle spindles	H. cranial nerve X
___ 9. extrapyramidal system	I. voluntary control
___ 10. hypoglossal	J. releases NE

C. Concept Map I - Autonomic Nervous System

Using the following terms, fill in the numbered, blank spaces to complete the concept map. Follow the numbers that comply with the organization of the map.

Ganglia outside CNS	**Motor neurons**	**Sympathetic**
Smooth muscle	**Postganglionic**	**First-order neurons**

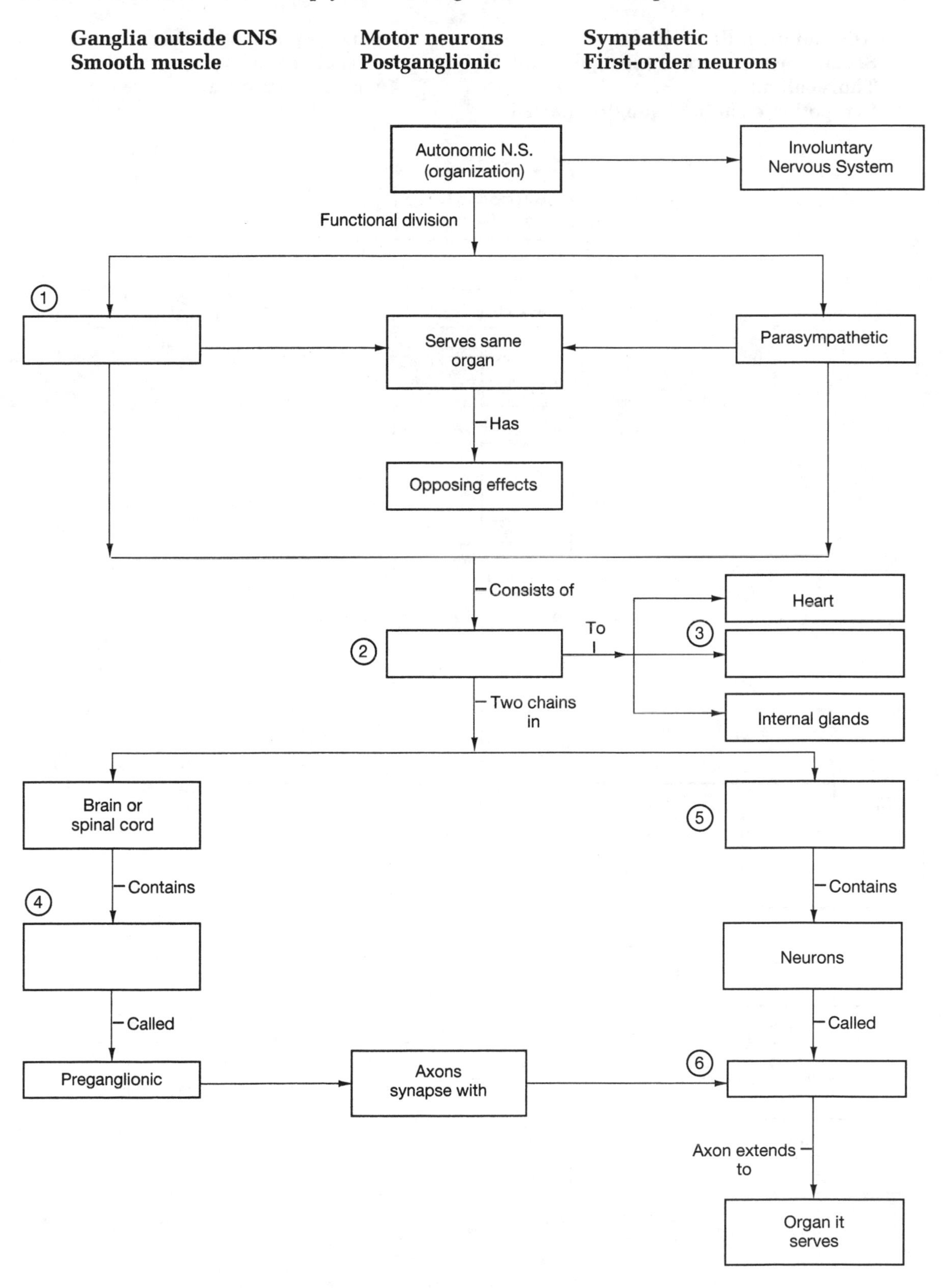

Concept Map II - Sympathetic Division of ANS

Using the following terms, fill in the circled, numbered, blank spaces to complete the concept map. Follow the numbers that comply with the organization of the map.

Adrenal medulla (paired)
Second-order neurons (postganglionic)
Thoracolumbar
Sympathetic chain of ganglia (paired)
Spinal segments T1 - L2
Visceral effectors
General circulation

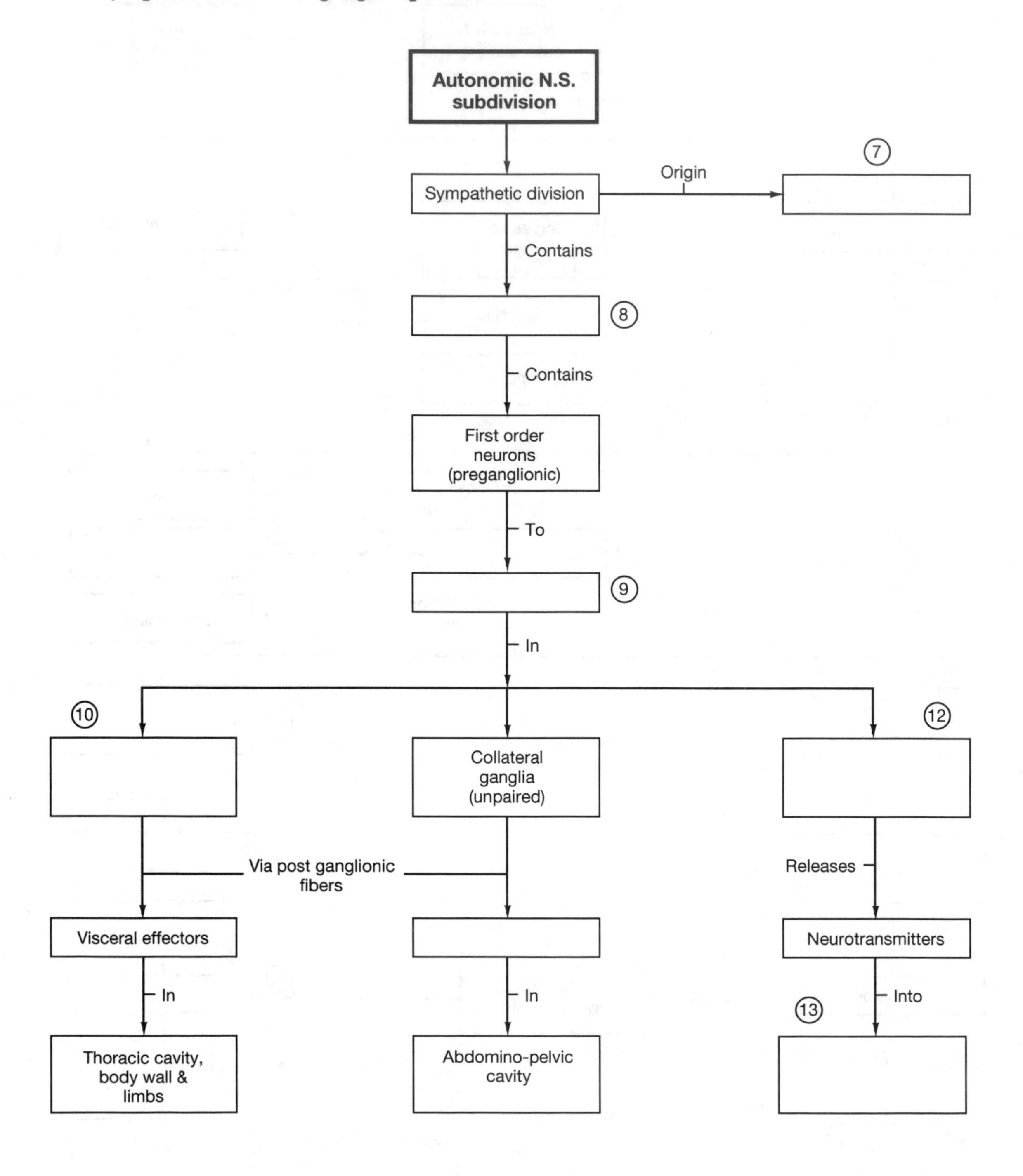

Concept Map III - Parasympathetic Division of ANS

Using the following terms, fill in the circled, numbered, blank spaces to complete the concept map. Follow the numbers that comply with the organization of the map.

Lower abdominopelvic cavity	**Otic ganglia**	**N X**
Nasal, tear, salivary glands	**Intramural ganglia**	**N VII**
Craniosacral	**Ciliary ganglion**	**Brain stem**
Segments S2 - S4		

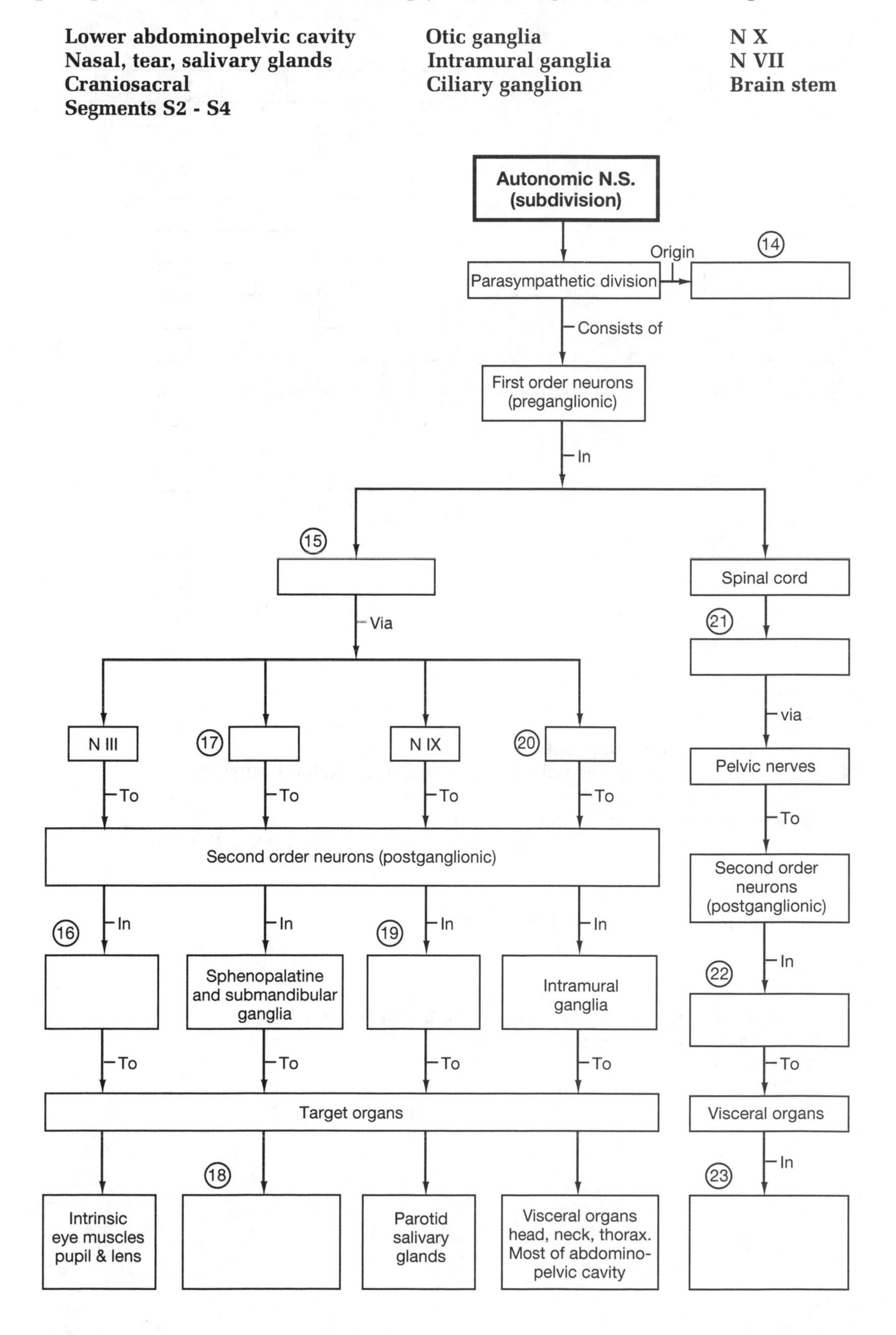

D. Crossword Puzzle

The following crossword puzzle reviews the material in Chapter 9. To complete the puzzle, you must know the answers to the clues given, and you must be able to spell the terms correctly.

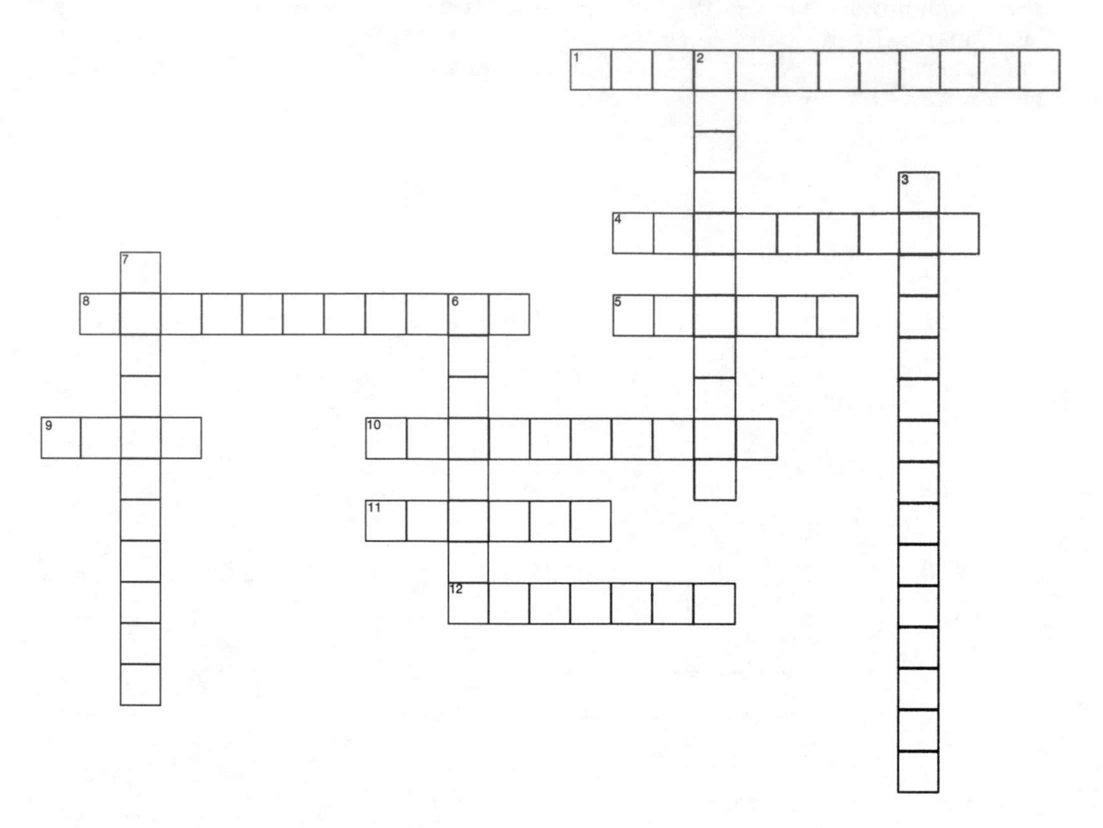

ACROSS

1. Fanning of the toes in infants.
4. The system that provides voluntary control over skeletal muscles.
5. A type of withdrawal reflex.
8. CN responsible for tongue movements.
9. Type of innervation where instructions come from both ANS divisions.
10. CN III is the _________ nerve.
11. A network of a nerve trunk.
12. An example of a monosynaptic reflex.

DOWN

2. Found between a sensory neuron and a motor neuron.
3. Rest and repose division of the ANS.
6. CN VI is the __________ nerve.
7. Fight or flight division of the ANS.

E. Short Answer Questions

Briefly answer the following questions in the space below each question.

1. What are the three major somatic sensory pathways in the body?

2. What motor pathways make up the pyramidal system?

3. What motor pathways make up the extrapyramidal system?

4. What is the primary functional difference between the pyramidal and extrapyramidal system?

5. What are the three major components of the sympathetic division?

6. What are the two major components of the parasympathetic division of the ANS?

7. What effect does dual innervation have on autonomic control throughout the body?

8. Why is the parasympathetic division sometimes referred to as the anabolic system?

9. Why is the sympathetic division of the ANS called the "fight or flight" system?

10. Why is the parasympathetic division of the ANS known as the "rest and repose" system?

11. What three patterns of neurotransmission are of primary importance in the autonomic nervous system?

12. List four (4) of the anatomical changes in the nervous system that are commonly associated with the aging process.

CHAPTER

10

Sensory Functions

Overview

Seeing, hearing, smelling, tasting, and balance; the *special senses* that allow us to be aware of the world within and around us by reacting to stimuli are usually taken for granted; in fact, it is not unusual to ignore their importance until we don't have them. Man's chances for survival would be severely limited if the general senses of temperature, pain, touch, pressure, vibration, and proprioception would cease to allow us to respond in positive or negative ways so that we are consciously aware of the sensory information we receive. All these specialized parts of the nervous system allow the body to assess and make adjustments to changing conditions and maintain homeostasis within the body. Sensory receptors in the specialized sensory structures receive stimuli, initiating production of action potentials which are transmitted on neurons to the CNS where the sensations are processed, resulting in motor responses that preserve the integrity of the body's steady state and ability to survive.

The activities and exercises in Chapter 10 are designed to reinforce your understanding of the structure and function of general sensory receptors and specialized receptor cells which are structurally more complex than those of the general senses. Mastery of this material may require more than memorization, therefore, you may want to synthesize and organize the material in Sensory Functions by referring to the concept maps at the Prentice Hall web site on the Internet.

Review of Chapter Objectives

1. Distinguish between the general and special senses.
2. Identify the receptors for the general senses and describe how they function.
3. Describe the receptors and processes involved in the sense of smell.
4. Discuss the receptors and processes involved in the sense of taste.
5. Identify the parts of the eye and their functions.
6. Explain how we are able to see objects and distinguish colors.
7. Discuss how the central nervous system processes information related to vision.
8. Discuss the receptors and processes involved in the sense of equilibrium.
9. Describe the parts of the ear and their roles in the process of hearing.

Part I: Objective Based Questions

Objective 1 Distinguish between the general and special senses.

_____ 1. The term *general senses* refers to sensations of

a. smell, taste, balance, hearing, and vision
b. pain, smell, pressure, balance, and vision
c. temperature, pain, touch, pressure, vibration, and proprioception
d. touch, taste, balance, vibration, and hearing

_____ 2. The *special senses* refer to

a. balance, taste, smell, hearing, and vision
b. temperature, pain, taste, touch, and hearing
c. touch, pressure, vibration, and proprioception
d. proprioception, smell, touch, and taste

_____ 3. Receptors for special senses are located

a. in specific sense organs
b. throughout the body
c. in the cerebral area
d. in the integument

_____ 4. The general senses

a. are localized in specific areas of the body
b. involve receptors that are relatively simple
c. are located in the sense organs
d. do not require action potentials for effector responses

Objective 2 Identify the receptors for the general senses and describe how they function.

_____ 1. The receptors for the general senses are the

a. axons of the sensory neurons
b. cell bodies of sensory neurons
c. dendrites of sensory neurons
d. all of the above

_____ 2. An injection or deep cut causes sensations of fast or prickling pain carried on

a. unmyelinated Type A fibers
b. myelinated Type C fibers
c. myelinated Type B fibers
d. myelinated Type A fibers

_____ 3. The three classes of mechanoreceptors are

a. Meissner's corpuscles, Pacinian corpuscles, Merkel's discs
b. fine touch, crude touch, and pressure receptors
c. tactile, baroreceptors, proprioceptors
d. slow-adapting, fast-adapting, and central-adapting receptors

_____ 4. Nociceptors are sensitive to

a. pain
b. light touch
c. pressure
d. light

_____ 5. Sensory receptors that monitor the position of joints are called

a. nociceptors
b. baroreceptors
c. chemoreceptors
d. proprioceptors

_____ 6. Baroreceptors are sensory receptors that

a. monitor the position of joints
b. respond to changes in blood pressure
c. respond to light touch
d. respond to temperature changes

_____ 7. Tactile receptors respond to all of the following except

a. vibration
b. touch
c. pain
d. pressure

_____ 8. Chemoreceptors are located in the

a. carotid and aortic bodies
b. special senses of taste and smell
c. respiratory control center of the medulla
d. all of the above

Objective 3 Describe the receptors and processes involved in the sense of smell.

_____ 1. The first step in olfactory reception occurs on the surface of the

a. olfactory cilia
b. columnar cells
c. basal cells
d. olfactory glands

_____ 2. The CNS interprets smell on the basis of the particular pattern of

a. cortical arrangement
b. neuronal replacement
c. receptor activity
d. sensory impressions

_____ 3. The only type of sensory information that reaches the cerebral cortex without synapsing in the thalamus is _____ stimuli.

a. visual
b. olfactory
c. gustation
d. all of the above

_____ 4. The human's power of olfaction is the most acute of all mammals.

a. true
b. false

_____ 5. The cerebral interpretation of smell occurs in the

a. olfactory cortex
b. hypothalamus
c. portions of the limbic system
d. all of the above

6. In the structures below, identify and label the following parts. Place the labels in the spaces provided below the drawing.

Figure 10-1 **(a)** Structure of Olfactory Organ **(b)** Olfactory Receptor

Olfactory tract	**Cilia**	**Afferent nerve fiber**
Cribiform plate	**Mucus layer**	**Basal cell**
Olfactory bulb	**Olfactor mucus gland**	**Olfactory receptor cell**

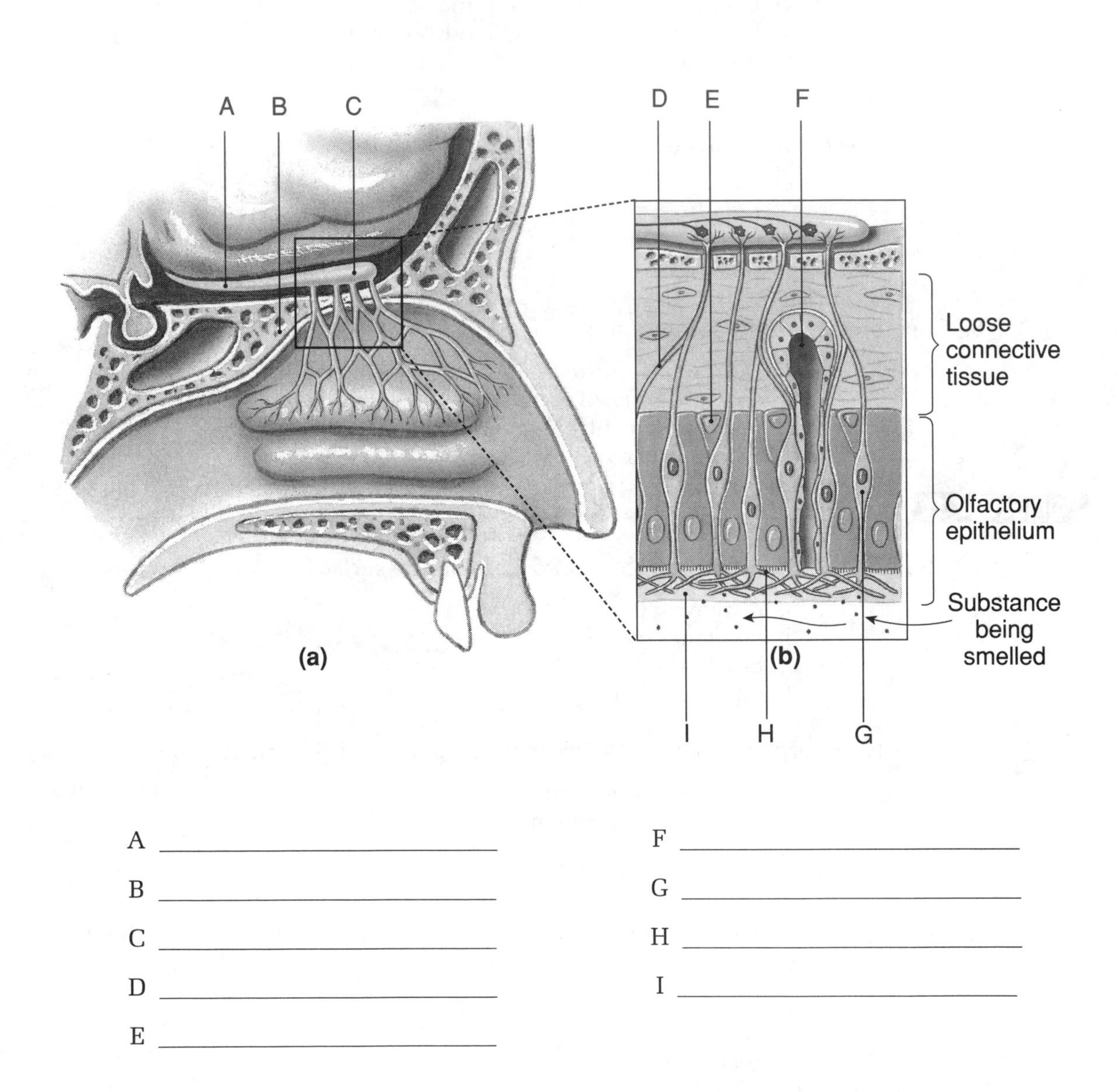

A ______________________________

B ______________________________

C ______________________________

D ______________________________

E ______________________________

F ______________________________

G ______________________________

H ______________________________

I ______________________________

Objective 4 Discuss the receptors and processes involved in the sense of taste.

_____ 1. The three different types of papillae on the human tongue are

a. kinocilia, cupula, and maculae
b. sweet, sour, and bitter
c. cuniculate, pennate, and circumvate
d. filiform, fungiform, and circumvallate

_____ 2. Taste buds are monitored by cranial nerves

a. VIII, IX, X
b. IV, V, VI
c. I, II, III
d. VIII, XI, XII

_____ 3. After synapsing in the thalamus, gustatory information is transmitted to the appropriate portion of the

a. medulla
b. medial lemniscus
c. primary sensory cortex
d. VII, IX, and X cranial nerves

_____ 4. The sensation of taste is due to receptors called

a. papillae
b. gustatory cells
c. taste buds
d. taste pores

5. Identify and label the following structures in Figure 10-2. Place the answers in the spaces provided next to the letters.

Figure 10-2 Gustatory Reception

Sweet taste	**Salt taste**	**Sour taste**
Bitter taste	**Taste buds**	**Taste pore**
Stratified squamous epithelium	**Cranial nerve fibers**	**Connective tissue**
	Gustatory cell	**Supporting cell**

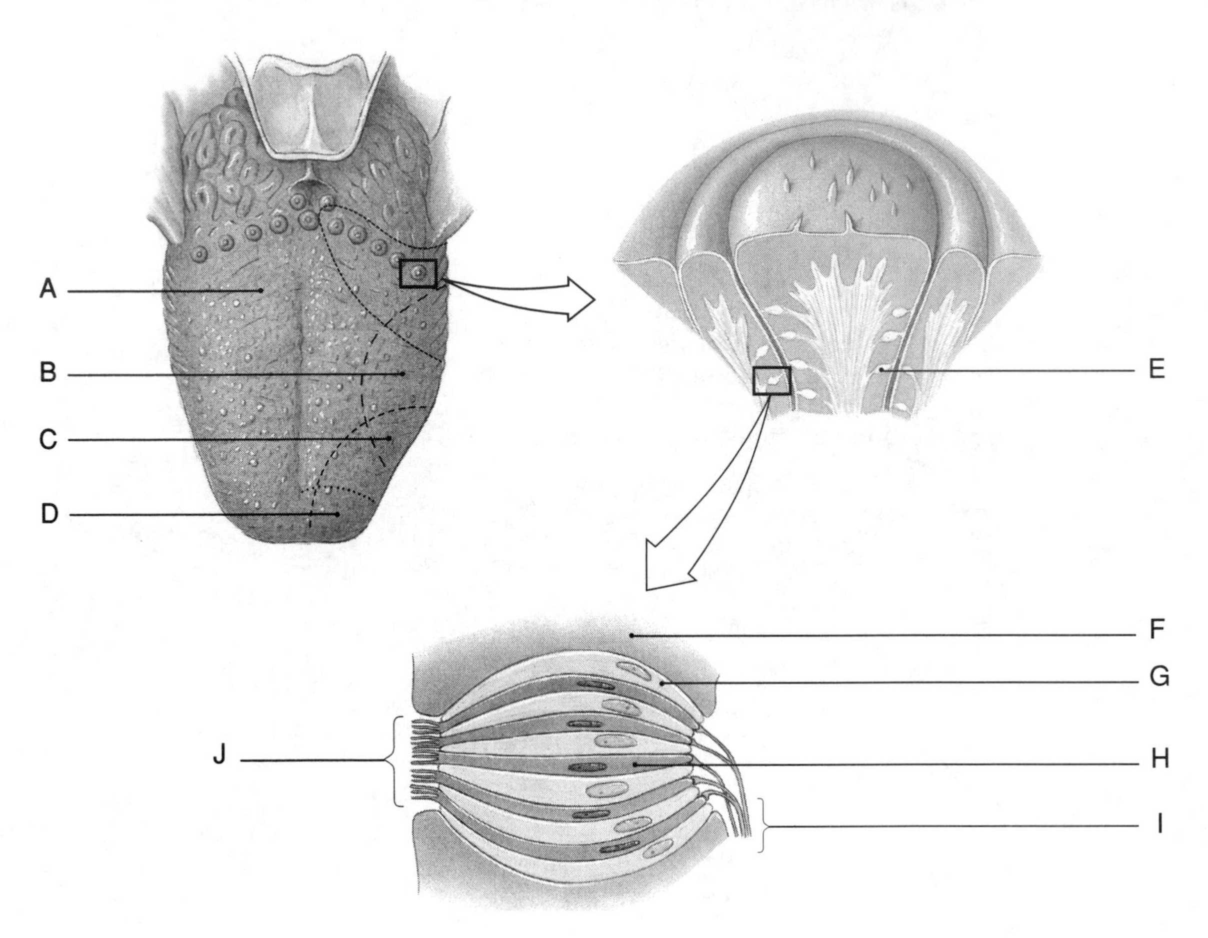

A ______________________

B ______________________

C ______________________

D ______________________

E ______________________

F ______________________

G ______________________

H ______________________

I ______________________

J ______________________

Objective 5 Identify the parts of the eye and their functions.

_____ 1. A lipid-rich product that helps to keep the eyelids from sticking together is produced by the

a. lacrimal glands
b. meibomian glands
c. conjuctiva
d. gland of Zeis

_____ 2. The fibrous tunic, the outermost layer covering the eye, consists of the

a. iris and choroid
b. pupil and ciliary body
c. sclera and cornea
d. lacrimal sac and orbital fat

_____ 3. The vascular tunic consists of three distinct structures that include the

a. iris, ciliary body, choroid
b. scelera, cornea, iris
c. choroid, pupil, lacrimal sac
d. retina, cornea, iris

_____ 4. The function of the vitreous body in the eye is to

a. provide a fluid cushion for protection of the eye
b. serve as a route for nutrient and waste transport
c. stabilize the shape of the eye and give physical support to the retina
d. serve as a medium for cleansing the inner eye

_____ 5. The primary function of the lens of the eye is to

a. absorb light as it passes through the retina
b. biochemically interact with the photoreceptors of the retina
c. focus the visual image on retinal receptors
d. integrate visual information for the retina

_____ 6. When looking directly at an object, its image falls upon the portion of the retina called the

a. fovea centralis
b. choroid layer
c. sclera
d. focal point

_____ 7. When photons of all wavelengths stimulate both rods and cones, the eye perceives

a. "black" objects
b. all the colors of the visible light spectrum
c. either "red" or "blue" light
d. "white" light

_____ 8. The most detailed information about the visual image is provided by the

a. cones
b. rods
c. optic disc
d. rods and cones

9. Most of the ocular surface of the eye is covered by the ___________.

10. The opening surrounded by the iris is called the ___________.

11. The photoreceptors that enable us to see in dimly lit rooms, at twilight, or in pale moon light are the ___________.

12. The photoreceptors that account for the perception of color are the ___________.

13. Visual information is integrated in the cortical area of the ___________ lobe.

14. Identify and label the following structures in Figure 10-3. Place the answers in the spaces below the drawing.

Figure 10-3 Anatomy of the Eye

Vitreous humor	**Aqueous humor**	**Choroid coat**
Fovea centralis	**Cornea**	**Iris**
Lens	**Retina**	**Ciliary body**
Optic disk	**Optic nerve**	**Sclera**
Suspensory ligament		

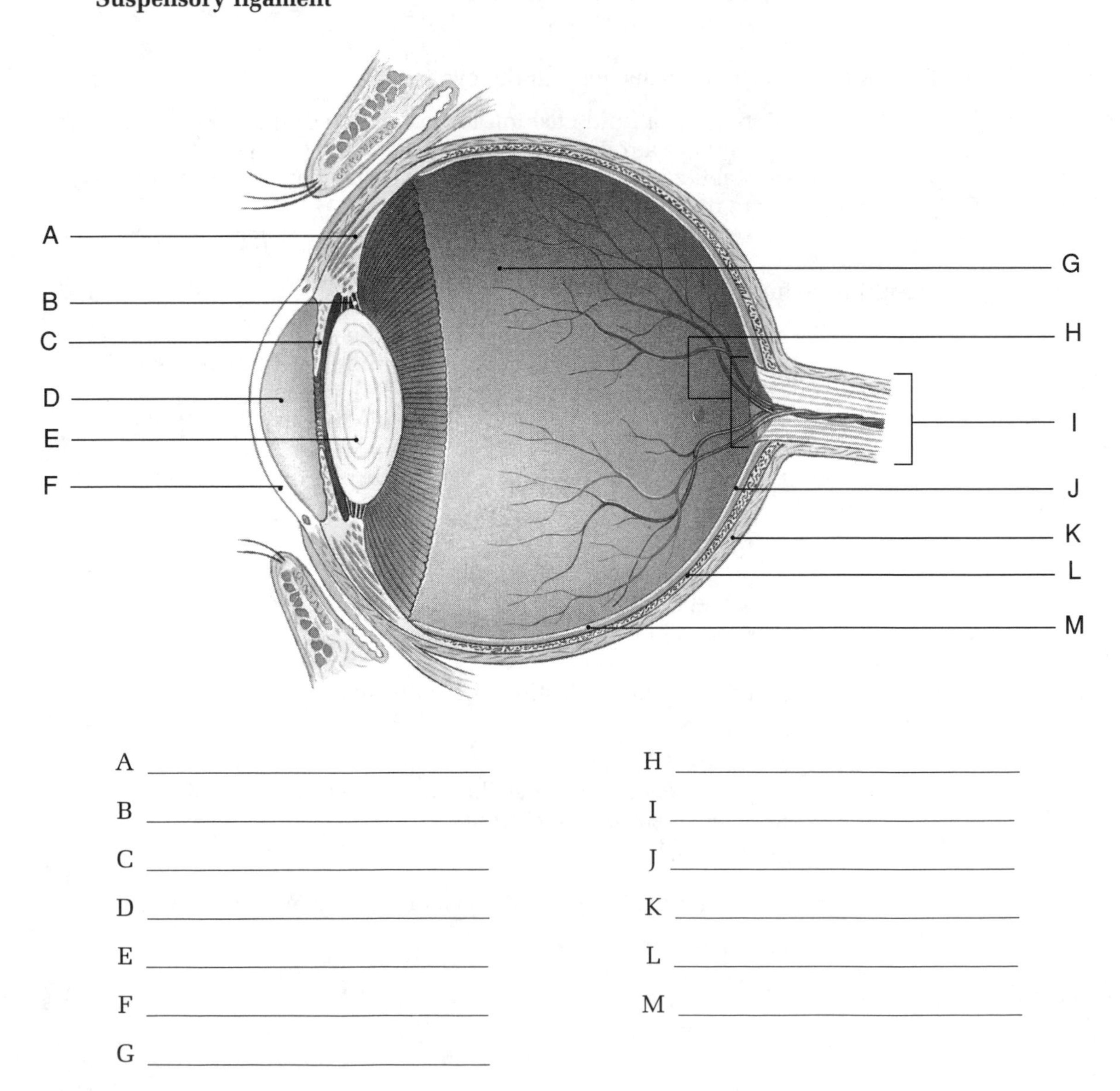

A ___________________

B ___________________

C ___________________

D ___________________

E ___________________

F ___________________

G ___________________

H ___________________

I ___________________

J ___________________

K ___________________

L ___________________

M ___________________

15. Identify and label the following structures in Figure 10-4. Place the answers in the spaces below the drawing.

Figure 10-4 Extrinsic Eye Muscles

Superior oblique	**Inferior oblique**	**Superior rectus**
Lateral rectus	**Inferior rectus**	

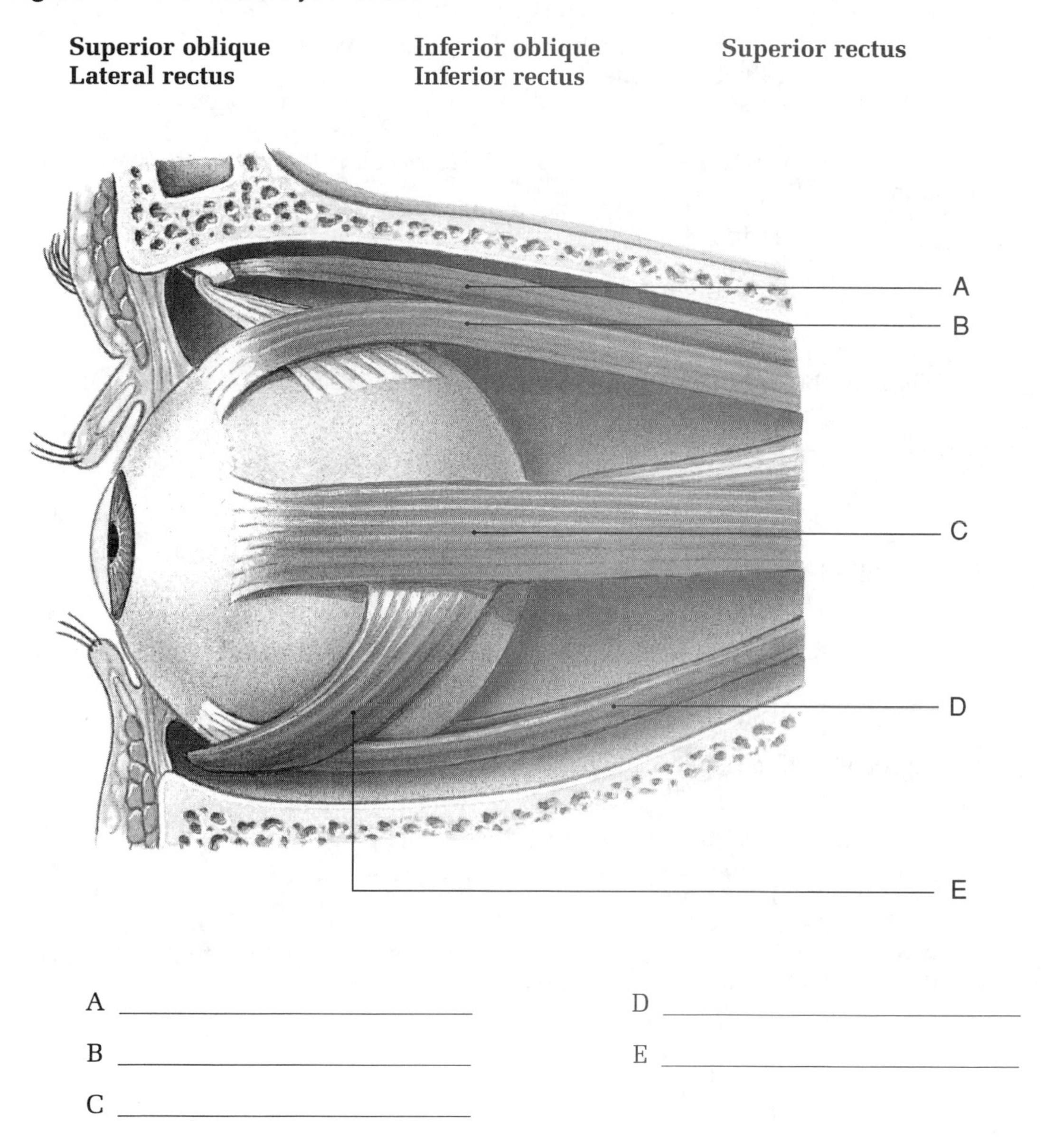

A ______________________

B ______________________

C ______________________

D ______________________

E ______________________

Objective 6 Explain how we are able to see objects and distinguish colors.

_____ 1. When the *cones* in the retina are stimulated the result is

a. the ability to see in dim light
b. the ability to distinguish patterned images
c. color vision
d. accommodation

_____ 2. The area of the retina that contains only cones and is the site of sharpest vision is the

a. iris
b. fovea
c. optic disk
d. neural tunic

_____ 3. When all three cone populations are stimulated we see

a. white
b. black
c. red, blue, green
d. all of the above

_____ 4. The "blind spot" in the retina occurs where

a. ganglion cells synapse with bipolar cells
b. rod cells cluster to form the macula
c. the optic nerve attaches to the retina
d. there is an accumulation of amacrine cells

_____ 5. Light absorption requires the presence of

a. neurotransmitters
b. visual pigments
c. rods and cones
d. vitamin A

_____ 6. The _______ changes in diameter in response to the intensity of light entering the eye. Light is necessary to activate the cells of the ______.

a. lens, retina
b. pupil, optic disc
c. retina, pupil
d. pupil, retina

7. The muscles of the eye responsible for changing the diameter of the pupil are contained in the ___________.

8. Nearsightedness, or being able to see "up close" is called ___________.

9. The absence of rods in the retina is apparent in a region called the ____________.

10. The process of focusing an image on the retina by changing the shape of the lens is called ____________.

11. Identify and label the following structures in Figure 10-5. Place the answers in the spaces provided below the drawing.

Figure 10-5 Cellular Organization of the Retina

Amacrine cell	**Cone**	**Pigment layer of retina**
Ganglion cells	**Horizontal cell**	**Rod**
Bipolar cells		

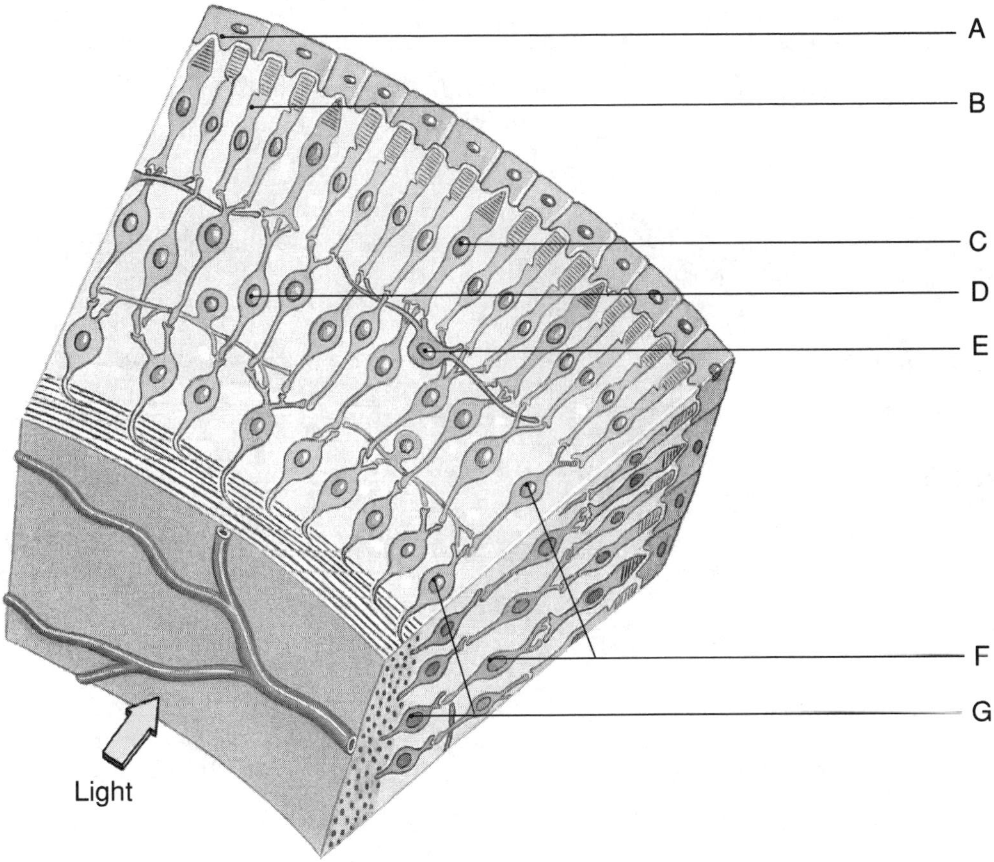

A ______________________

B ______________________

C ______________________

D ______________________

E ______________________

F ______________________

G ______________________

Objective 7 Discuss how the central nervous system processes information related to vision.

_____ 1. Axons converge on the optic disc, penetrate the wall of the eye, and proceed toward the

a. retina at the posterior part of the eye
b. diencephalon as the optic nerve (N II)
c. retinal processing areas below the choroid coat
d. cerebral cortex area of the parietal lobes

_____ 2. The sensation of vision arises from the integration of information arriving at the

a. lateral geniculate of the left side
b. lateral geniculate of the right side
c. visual cortex of the cerebrum
d. reflex centers in the brain

_____ 3. Visual input from the hypothalamus and pineal gland establish a daily pattern of activity tied to the day-night cycle called

a. biological time clock
b. visual photoreception
c. crossover charisma
d. circadian rhythm

Objective 8 Discuss the receptors and processes involved in the sense of equilibrium.

_____ 1. The branch of cranial nerves responsible for monitoring changes in equilibrium is the ________ branch.

a. cochlear
b. vestibular
c. auditory
d. trigeminal

_____ 2. The sense of equilibrium and hearing are provided by receptors in the

a. external ear
b. middle ear
c. inner ear
d. all of the above

_____ 3. The receptors involved in equilibrium are found in the

a. cochlea
b. semicircular canals
c. organ of corti
d. auditory ossicles

_____ 4. The parts of the vestibular complex which provide information about your position with respect to gravity are the

a. malleus and stapes
b. vestibular and tympanic ducts
c. oval and round windows
d. saccule and utricle

_____ 5. Equilibrium is achieved when the fluid in the

a. ear moves and a signal is sent to the brain via the cochlear branch of CN VIII
b. cochlea moves and a signal is sent to the brain via CN VIII
c. cochlea moves and a signal is sent to the brain via the vestibular portion of CN VIII
d. semicircular canals move, and a signal is sent via CN VIII

_____ 6. The name of the fluid in the cochlear duct of the inner ear is

a. perilymph
b. cerumen
c. endolymph
d. exudate

Objective 9 Describe the parts of the ear and their roles in the process of hearing.

_____ 1. The process of hearing begins when sound waves are received by the

a. outer ear
b. middle ear
c. inner ear
d. temporal lobe

_____ 2. In the middle ear, sound waves vibrate the _________ which converts sound energy into mechanical movements of the ossicles, which consist sequentially of the ___________.

a. oval window; malleus incus, stapes
b. round window; malleus, incus, stapes
c. tympanum; stapes, incus, malleus
d. tympanum; malleus, incus, stapes

_____ 3. The receptors that provide the sensation of hearing are located in the

a. vestibular complex
b. cochlea
c. ampulla
d. tympanic membrane

_____ 4. The long labyrinth of the ear is subdivided into the

a. auditory meatus, auditory canal, ceruminous glands
b. saccule, utricle, vestibule
c. vestibule, semicircular canals, and cochlea
d. ampulla, crista, cupula

_____ 5. The dividing line between the external ear and middle ear is the

a. pharyngotympanic tube
b. tympanic membrane
c. round window
d. oval window

_____ 6. The structure in the cochlea of the inner ear that provides information to the CNS is the

a. scala tympani
b. tectorial membrane
c. organ of corti
d. basilar membrane

_____ 7. The fluid in the vestibular and tympanic ducts which is affected by sound vibrations is

a. endolymph
b. ceruminal fluid
c. perilymph
d. lymphoid fluid

8. Identify and label the following structures of the ear. Place the answers in the spaces provided below the drawing.

Figure 10-6 Anatomy of the Ear

Middle ear	**Outer ear**	**Inner ear**
Malleus	**Pinna**	**Incus**
Vestibular complex	**Stapes**	**Temporal bone**
Vestibulocochlear nerve	**Cochlea**	**Bony labyrinth**
Auditory tube	**Tympanum**	**External auditory canal**

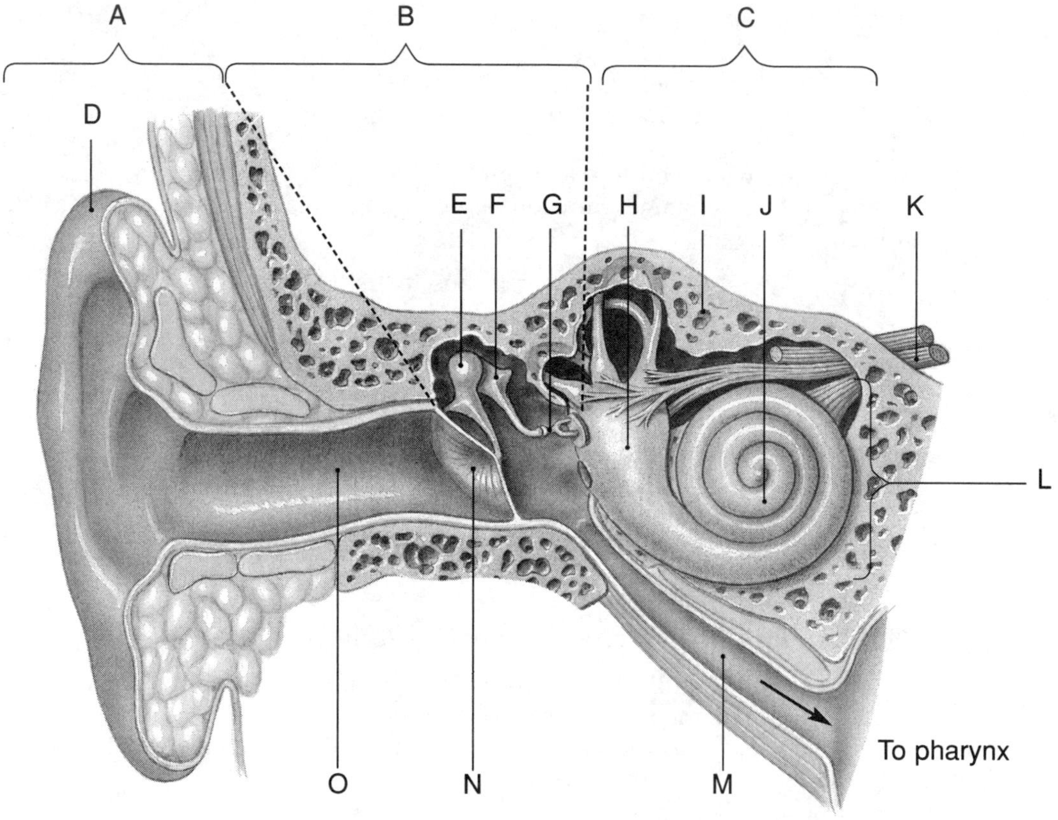

A ______________________

B ______________________

C ______________________

D ______________________

E ______________________

F ______________________

G ______________________

H ______________________

I ______________________

J ______________________

K ______________________

L ______________________

M ______________________

N ______________________

O ______________________

Part II: Chapter Comprehensive Exercises

A. Word Elimination

Circle the term that does not belong in each of the following groupings.

1. smell taste touch hearing vision
2. temperature pain pressure balance vibration
3. nociceptors stereoreceptors thermoreceptors chemoreceptors mechanoreceptors
4. cornea eyelids conjunctiva lacrimal gland extrinsic eye muscles
5. rods cones bipolar cells choroid ganglion cells
6. emmetropia propriopia myopia hyperopia presbyopia
7. tympanum malleus incus stapes
8. tectorial membrane basilar membrane hair cells nerve fibers otolith
9. optic nerve auditory tube optic tract projection fibers occipital lobes
10. rhodopsin retinal opsin cerumen vitamin A

B. Matching

Match the terms in Column B with the terms in Column A. Use letters for answers.

COLUMN A	COLUMN B
___ 1. nociceptor	A. sense of smell
___ 2. baroreceptor	B. far sighted
___ 3. olfaction	C. sclera
___ 4. gustatory receptors	D. ear wax
___ 5. "white of the eye"	E. utricle and saccule
___ 6. hyperopia	F. pain receptor
___ 7. myopia	G. frequency of sound
___ 8. vestibule	H. monitors pressure changes
___ 9. cerumen	I. near sighted
___ 10. hertz	J. sense taste

C. Concept Map I - Special Senses

Using the following terms, fill in the numbered, blank spaces to complete the concept map. Follow the numbers that comply with the organization of the map.

Retina	**Hearing**	**Rods and cones**
Olfaction	**Balance and hearing**	**Smell**
Ears	**Taste buds**	**Tongue**

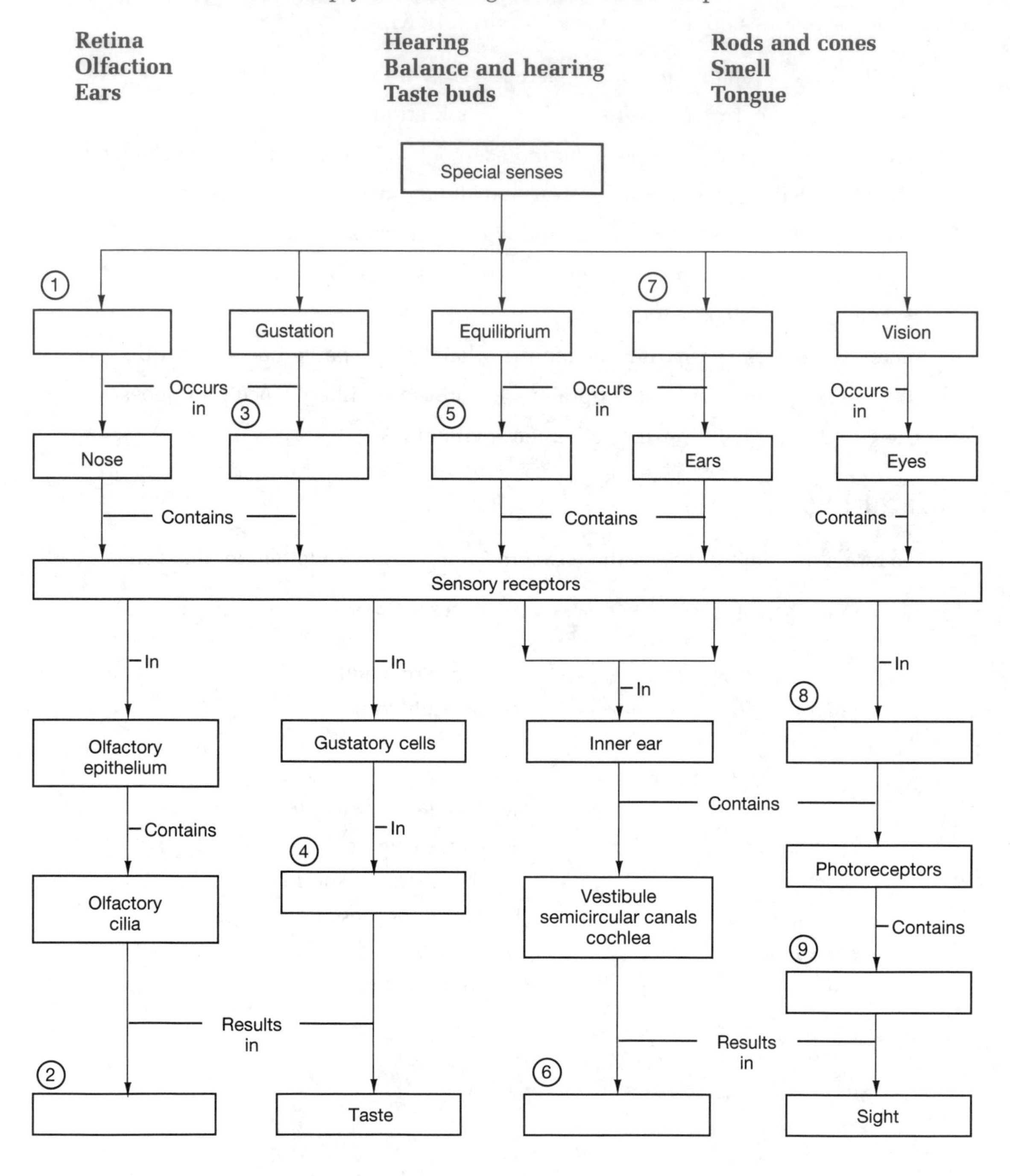

Concept Map II - General Senses

Using the following terms, fill in the circled, numbered, blank spaces to complete the concept map. Follow the numbers that comply with the organization of the map.

Pacinian corpuscles
Proprioception
Thermoreceptors
Aortic sinus
Merkel's discs
Tactile
Muscle spindles
Pain
Baroreceptors
Pressure
Dendritic processes

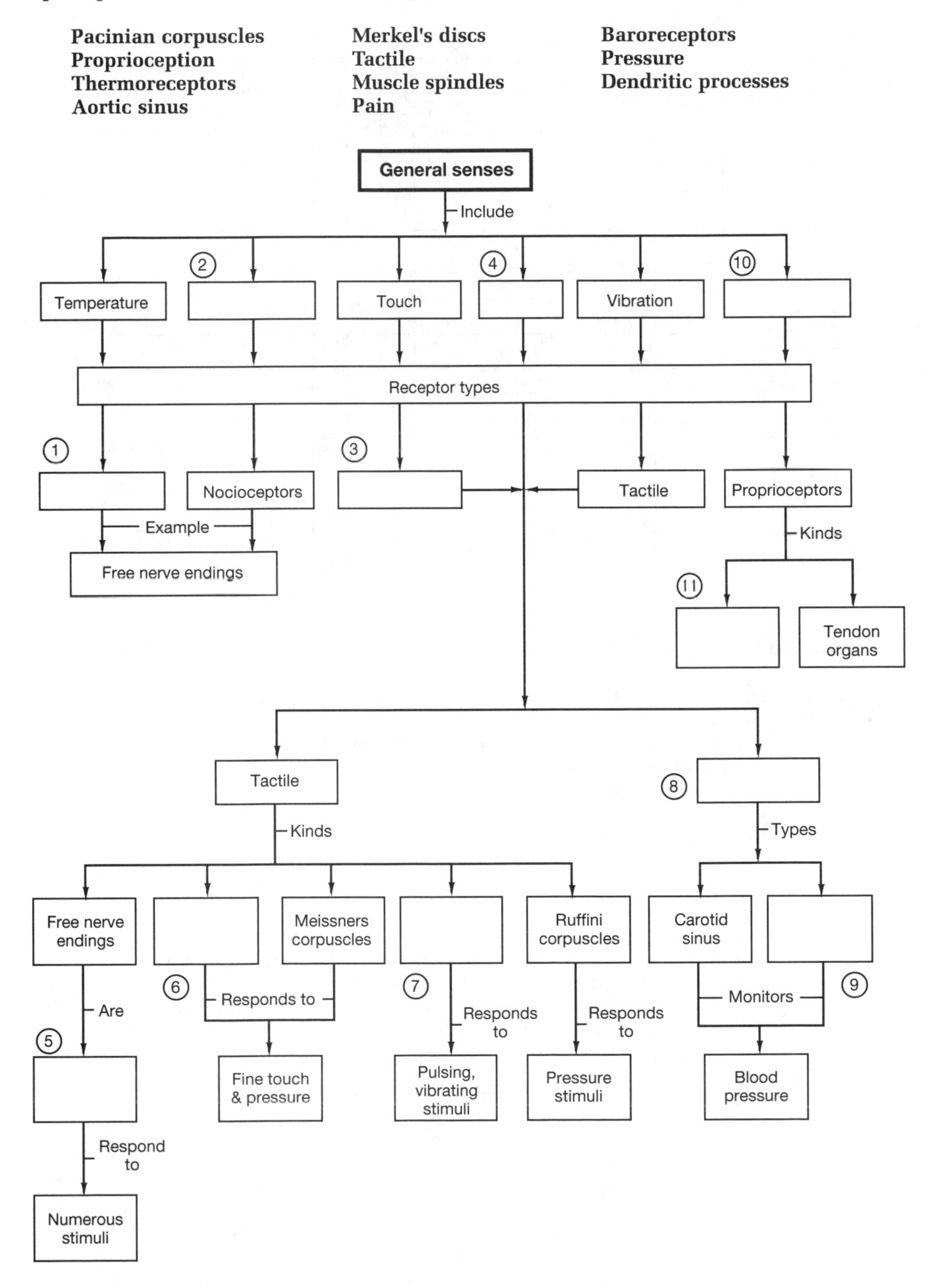

D. Crossword Puzzle

The following crossword puzzle reviews the material in Chapter 10. In order to solve the puzzle, you must know the answers to the clues given, and you must be able to spell the terms correctly.

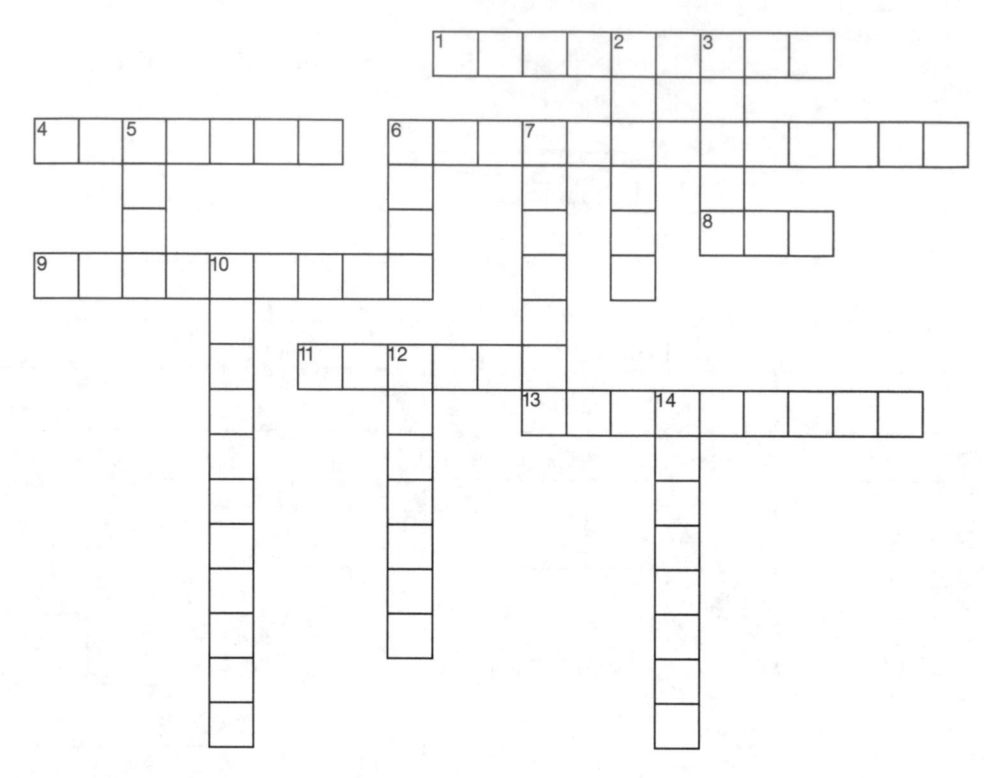

ACROSS

1. The portion of the retina where the optic nerve exits the eye.
4. The name of earwax.
6. Rods and cones are a type of sensory receptor called a ________.
8. An infection of the lacrimal gland.
9. The process of tasting food.
11. A cranial nerve involved in taste.
13. The fluid in the semicircular canals.

DOWN

2. The transparent portion of the sclera.
3. The second ossicle involved in the sequence of transmitting sound waves.
5. The cells of the retina that cannot detect color.
6. ________ receptors are sensory nerve cells called free nerve endings.
7. Sensory receptors involved with touch.
10. Irregularities in the shape of the lens.
12. The location of the organ of Corti.
14. The three smallest bones of the body, which are located in the middle ear.

E. Short Answer Questions

Briefly answer the following questions in the spaces provided below.

1. What sensations are included as "general senses?"

2. What sensations are included as "special senses?"

3. What four (4) kinds of "general sense" receptors are found throughout the body and to what kind(s) of stimuli do they respond?

4. What is the functional difference between a baroreceptor and a proprioceptor?

5. Trace an olfactory sensation from the time it leaves the olfactory bulb until it reaches its final destinations in the higher centers of the brain.

6. What are the four (4) primary taste sensations?

7. What sensations are provided by the *saccule* and *utricle* in the vestibule of the inner ear?

8. What are the three (3) primary functions of the *fibrous tunic*, which consists of the sclera and the cornea?

9. What are the three (3) primary functions of the *vascular tunic*, which consists of the iris, the ciliary body, and the choroid?

10. What are the primary functions of the *neural tunic*, which consists of an outer pigment layer and an inner retina that contains the visual receptors and associated neurons?

11. When referring to the eye, what is the purpose of *accommodation* and how does it work?

12. Why are cataracts common in the elderly?

CHAPTER

11

The Endocrine System

Overview

Along with the nervous system, the endocrine system plays an important role in monitoring and regulating the activity of cells throughout the body to maintain homeostasis. Chemical messengers, called hormones, orchestrate ongoing and long term cellular changes to target tissues throughout the body. The influence of these chemical messengers results in facilitating processes which include growth and development, sexual maturation and reproduction, and the maintenance of homeostasis of many body systems.

To simplify the study of the endocrine system, ask yourself and answer the following questions:

- Where is the endocrine gland located in the body?
- What hormone(s) does the gland secrete?
- What is the hormone's target organ in the body?
- What effect does the hormone have on the body?
- What are the consequences of the hormone's hyper/hypo secretion?

Chapter 11 tests your knowledge of the endocrine organs and their functions and presents opportunities for you to affirm your understanding of the effects of endocrine activity within the body.

Review of Chapter Objectives

1. Compare the similarities between the endocrine and nervous systems.
2. Compare the major chemical classes of hormones.
3. Explain the general mechanisms of hormonal action.
4. Describe how endocrine organs are controlled.
5. Discuss the location, hormones, and functions of the following endocrine glands and tissues: pituitary, thyroid, parathyroids, thymus, adrenals, kidneys, heart, pancreas, testes, ovaries, and pineal gland.
6. Explain how hormones interact to produce coordinated physiological responses.
7. Identify the hormones that are especially important to normal growth and discuss their roles.
8. Explain how the endocrine system responds to stress.
9. Discuss the results of abnormal hormone production.
10. Discuss the functional relationships between the endocrine system and other body systems.

Part I: Objective Based Questions

Objective 1 Compare the similarities between the endocrine and nervous systems.

_____ 1. Response patterns in the endocrine system are particularly effective in

a. regulating ongoing metabolic processes
b. rapid short-term specific responses
c. the release of chemical neurotransmitters
d. crisis management

_____ 2. An example of a functional similarity between the nervous system and the endocrine system is

a. both systems secrete hormones into the bloodstream
b. the cells of the endocrine system and nervous system are functionally the same
c. compounds used as hormones by the endocrine system may also function as neurotransmitters inside the CNS
d. both produce very specific responses to environmental stimuli

_____ 3. The endocrine system

a. produces rapid and specific responses to environmental stimuli
b. produces effects that can last for hours, days or even longer
c. continues to produce a response long after neural output ceases
d. a and c only

_____ 4. Endocrine cells

a. are similar in structure to nerve cells
b. release their secretions onto epithelial surfaces
c. release their secretions into the bloodstream
d. contain very few vesicles

Objective 2 Compare the major chemical classes of hormones.

_____ 1. Peptide hormones consist of chains of

a. glucose molecules
b. fatty acids
c. polysaccharides
d. amino acids

_____ 2. Steroid hormones are derived from

a. prostaglandins
b. cholesterol
c. arachidonic acid
d. muscle tissue

_____ 3. Protein hormones are produced by the

a. adrenal medulla
b. ovaries
c. pituitary gland
d. testes

_____ 4. Steroid hormones are produced by the

a. adrenal cortex and reproductive organs
b. pituitary gland and thymes
c. thyroid and parathyroid glands
d. all of the above

Objective 3 Explain the general mechanisms of hormonal action.

_____ 1. A cell's hormonal sensitivities are determined by the

a. chemical nature of the hormone
b. quantity of circulating hormone
c. shape of the hormone molecule
d. presence or absence of appropriate receptors

_____ 2. Hormones alter cellular operations by changing the

a. cell membrane permeability properties
b. identities, activities, quantities, or properties of important enzymes
c. arrangement of the molecular complex of the cell membrane
d. rate at which hormones affect the target oran cells

_____ 3. Catecholamines and peptide hormones affect target organ cells by

a. binding to receptors in the cytoplasm
b. binding to target receptors in the nucleus
c. enzymatic reactions that occur in the ribosomes
d. second messengers released when receptor binding occurs at the membrane surface

_____ 4. Steroid hormones affect target organ cells by

a. target receptors in peripheral tissues
b. releasing second messengers at cell membrane receptors
c. binding to receptors in the cell membrane
d. binding to target receptors in the nucleus

_____ 5. When adenyl cyclase is activated

a. cyclic AMP is formed
b. cyclic AMP is broken down
c. steroids are produced
d. protein kinases are metabolized

_____ 6. The target organ of a hormone is the

a. origin of the hormone
b. stimulus that has caused the release of the hormone
c. site of the hormone's destination
d. production site of the hormone

Objective 4 Describe how endocrine organs are controlled.

_____ 1. The basis of control for endocrine activity is provided by

a. increased metabolic activity
b. negative feedback mechanisms
c. the nervous system
d. all of the above

_____ 2. The coordinating centers that regulate the activities of the nervous and endocrine systems are in the

a. hypothalamus
b. pituitary gland
c. medulla oblongata
d. cerebrum

_____ 3. Regulatory hormones that regulate the activities of endocrine cells in the anterior pituitary gland include

a. oxygen and ADH
b. parathyroid hormone and calcitonin
c. releasing hormones and inhibiting hormones
d. thyroxine and thymosin

_____ 4. Of the following selections, the pair of hormones which would produce opposite effects is

a. oxytocin, ADH
b. parathyroid hormone, calcitonin
c. testosterone, estrogen
d. thyroid hormone, parathyroid hormone

_____ 5. The gland which is activated when blood calcium ion levels are too high is the

a. hypothalamus
b. thyroid
c. pituitary
d. parathyroid

_____ 6. If blood calcium ion levels become too low, the hormone which will appear in higher concentrations in the blood is

a. calcitonin
b. thyroxin
c. antidiuretic hormone
d. parathyroid hormone

Objective 5 Discuss the location, hormones, and function of the following endocrine glands and tissues: pituitary, thyroid, parathyroids, thymus, adrenals, kidneys, heart, pancreas, testes, ovaries, and pineal gland.

1. ADH and oxytocin are secreted by the _____________ gland.
2. The hormone that initiates uterine contractions is ___________.
3. The pituitary hormone that controls the release of glucocorticoids from the adrenal cortex is _________.
4. The pituitary hormone that promotes egg development in ovaries and sperm develop ment in testes is _______________.
5. The pituitary hormone that stimulates melanocytes to produce melanin is _____________.
6. The C cells of the thyroid gland produce ____________.
7. Thyroid hormone contains the mineral ____________.
8. The glands responsible for producing a hormone that increases the level of calcium ions in the blood are the _____________ glands.
9. The gland responsible for the calorigenic effect is the _____________.
10. The adrenal medulla produces _______________.
11. Cells of the adrenal cortex produce _______________.
12. Calcitrol, erythropoietin, and renin are hormones released by the _____________.

13. The hormone ANP which lowers blood volume is released by the ___________.
14. When blood glucose levels rise, insulin is secreted and released by the ___________.
15. The alpha cells of the pancreas produce the hormone ___________.
16. Testosterone and inhibin are hormones produced by cells in the ___________.
17. When stimulated by FSH, follicle cells in the ovary produce large quantities of ___________.
18. The gland which is believed to be involved with the establishment and maintenance of circadian rhythms in the ___________ gland.

19. Identify and label the endocrine organs and tissues in Figure 11-1. Place your answers in the spaces provided below the drawing.

Figure 11-1 The Endocrine System

Testis
Adrenals
Pancreas
Pituitary
Thyroid
Hypothalamus
Ovaries
Pineal
Thymus
Parathyroid
Atria (heart)

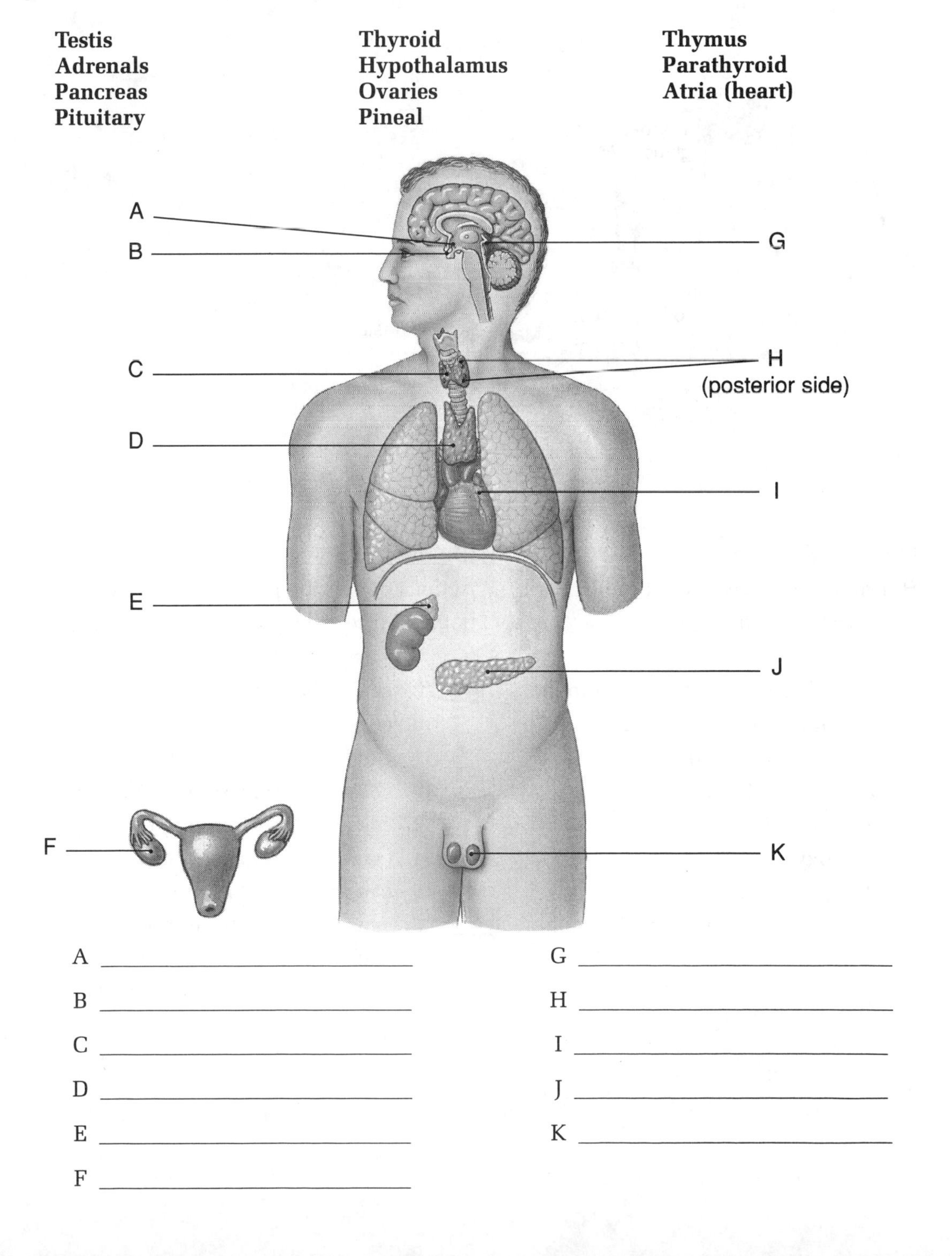

A ___________

B ___________

C ___________

D ___________

E ___________

F ___________

G ___________

H ___________

I ___________

J ___________

K ___________

20. In Figure 11-2, identify the pituitary secretions in column A, and identify the actions of those pituitary secretions in column B.

Figure 11-2 Pituitary Secretions, Hormones, Their Targets, and Actions

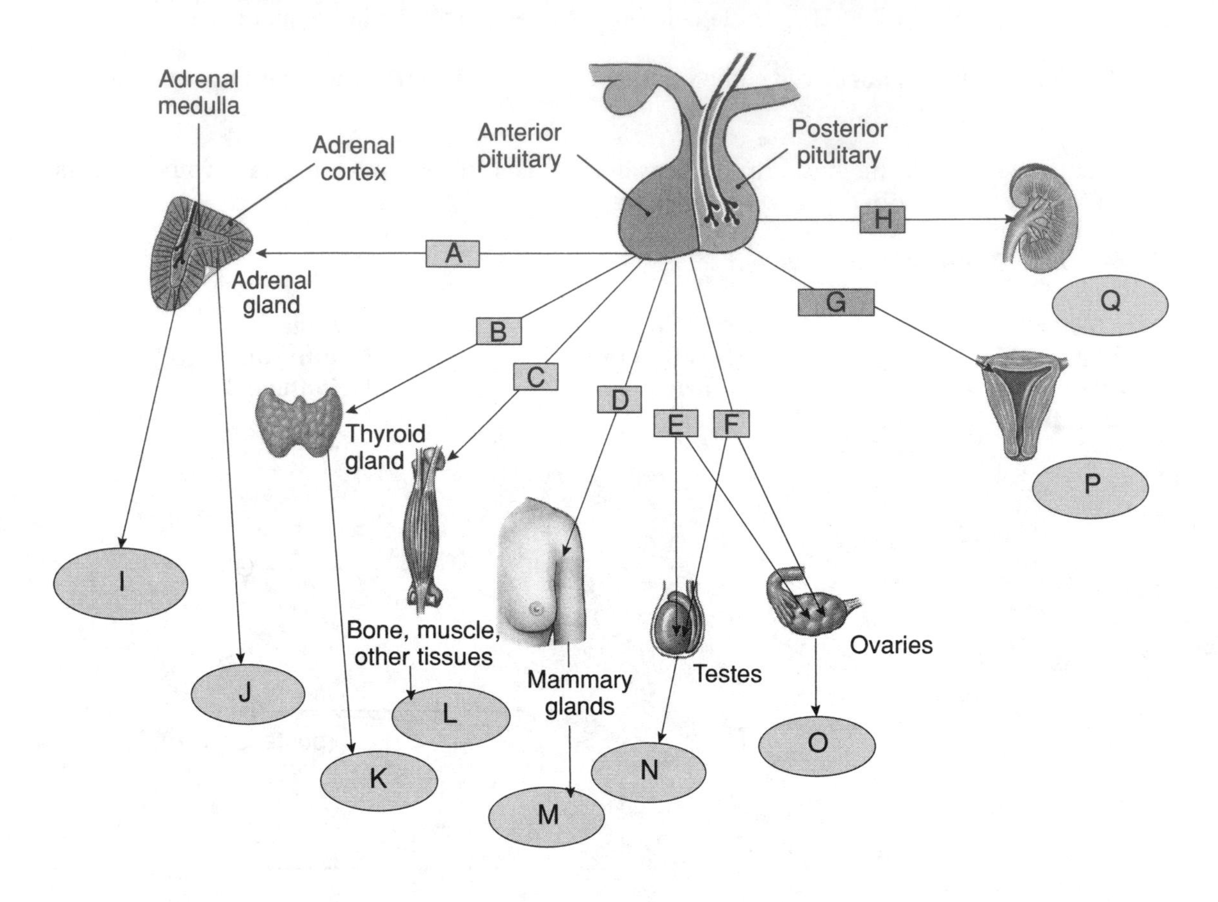

(A)	(B)
Hormones released from the pituitary gland (letters A through H)	Actions due to the pituitary secretions (letters I through Q)

___ 21 adrenocoritcotropic hormone
___ 22 antidiuretic hormone
___ 23 follicle stimulating hormone
___ 24 growth hormone
___ 25 luteinizing hormone
___ 26 oxytocin
___ 27 prolactin
___ 28 thyroid stimulating hormone

___ 29 causes bone growth
___ 30 causes milk production
___ 31 causes the release of thyroid hormones
___ 32 causes the release of epinephrine
___ 33 causes the release of glucocorticoids
___ 34 causes the release of testosterone
___ 35 causes the retention of water
___ 36 causes the release of progesterone
___ 37 causes uterine contractions

Objective 6 Explain how hormones interact to produce coordinated physiological responses.

_____ 1. When a cell receives instructions from two different hormones at the same time the results may be

a. antagonistic or synergistic
b. permissive
c. integrative
d. all of the above

_____ 2. If one hormone has a permissive effect on another the

a. first hormone is needed for the second to produce its effect
b. hormones produce different but complimentary results in specific tissues
c. two hormones have opposing effects
d. two hormones have additive effects

_____ 3. The hormone that is the antagonist of insulin is

a. calcitonin
b. glucagon
c. adrenalin
d. aldosterone

_____ 4. Stimulation of mammary gland development by prolactin, estrogens and GH is an example of the __________ effect.

a. antagonistic
b. synergistic
c. permissive
d. integrative

Objective 7 Identify the hormones that are especially important to normal growth and discuss their roles.

_____ 1. The important hormone for normal growth include

a. growth hormone, LH, ADH, and parathormone
b. insulin, prolactin, ADH, and MSH
c. thyroid hormones, TSH, ACTH, insulin, and LH
d. growth hormone, thyroid hormones, insulin, parathyroid and gonadal hormones

_____ 2. The reason insulin is important to normal growth is that it promotes

a. changes in skeletal proportions and calcium deposition in the body
b. passage of glucose and amino acids across cell membranes
c. muscular development via protein synthesis
d. all of the above

_____ 3. Undersecretion of growth hormone can lead to

a. gigantism
b. pituitary dwarfism
c. cretinism
d. all of the above

Objective 8 Explain how the endocrine system responds to stress.

_____ 1. All stressors produce the same pattern of hormonal and physiological adjustments.

a. true
b. false

_____ 2. The alarm phase of the *general adaptation syndrome* (GAS) is under the direction of the

a. somatic nervous system (SNS)
b. parasympathetic division of the ANS
c. sympathetic division of the ANS
d. both b and c

_____ 3. The dominant hormone during the alarm phase is

a. epinephrine
b. aldosterone
c. insulin
d. growth hormone

_____ 4. The dominant hormones of the *resistance* phase during the GAS are

a. mineralocorticoids
b. glucocorticoids
c. gonadal hormones
d. pancreatic hormones

_____ 5. The *exhaustive* phase of the GAS begins when

a. energy reserves are mobilized
b. the body prepares to escape from the source of stress
c. homeostatic regulation breaks down
d. there are increases in heart and respiratory rates

Objective 9 Discuss the results of abnormal hormone production.

_____ 1. Excessive secretion of growth hormone prior to puberty will cause

a. dwarfism
b. gigantism
c. acromegaly
d. diabetes

_____ 2. The inability of the pancreas to produce insulin results in

a. acromegaly
b. cretinism
c. diabetes mellitus
d. diabetes insipidus

_____ 3. Hyposecretion of glucocorticoids results in

a. Addison's disease
b. Cushing's disease
c. diabetes insipidus
d. goiter

_____ 4. Increased aggressive behavior is associated with increased secretion of

a. progesterone
b. growth hormone
c. miraeralocorticoids
d. testosterone

Objective 10 Discuss the functional relationships between the endocrine system and other body systems.

1. The system which controls the adrenal medulla and secretes ADH and oxytocin is the _____________system.

2. The system which distributes hormones throughout the body is the _____________ system.

3. The system that includes the endocrine cells of the pancreas which secrete insulin and glucagon is the ____________ system.

4. The system which releases renin and erythropoietin and produces calcitrol is the ____________ system.

5. The system primarily effected by endocrine hormones which affect energy production and growth is the ____________ system.

Part II: Chapter Comprehensive Exercises

A. Word Elimination

Circle the term that does not belong in each of the following groupings.

1. hypothalamus prostate pituitary thyroid pineal
2. adrenalin norepinephrine thyroxine keratin melatonin
3. ACTH GH PTH FSH LH
4. estrogen testosterone progesterone inhibin elastin
5. aldosterone cortisol hydrocortisone corticosterone cortisone
6. calcitrol erythropoietin renin hemoglobin angiotensin II
7. alpha cells glucagon beta cells insulin calcitonin
8. testes follicles estrogen corpus luteum progesterone
9. growth hormone melatonin insulin thyroid hormone gonadal hormones
10. exhaustion alarm GAS resistance permissive

B. Matching

Match the terms in Column B with the terms in Column A. Use letters for answers in the spaces provided.

COLUMN A	COLUMN B
___ 1. prostaglandin	A. cyclic - AMP
___ 2. amino acid derivative	B. testosterone
___ 3. second messenger	C. milk let-down
___ 4. pituitary gland	D. red blood cell production
___ 5. androgen	E. fatty acid based compound
___ 6. prolactin	F. produce calcitonin
___ 7. oxytocin	G. milk production
___ 8. C cells	H. epinephrine
___ 9. erythropoietin	I. melatonin
___ 10. pineal gland	J. hypophysis

C. Concept Map I - Endocrine Glands

Using the following terms, fill in the circled, numbered, blank spaces to complete the concept map. Follow the numbers that comply with the organization of the map.

Male/female	**Pituitary**	**Epinephrine**
Pineal	**Hormones**	**Bloodstream**
Testosterone	**Heart**	**Parathyroids**
Peptide hormones		

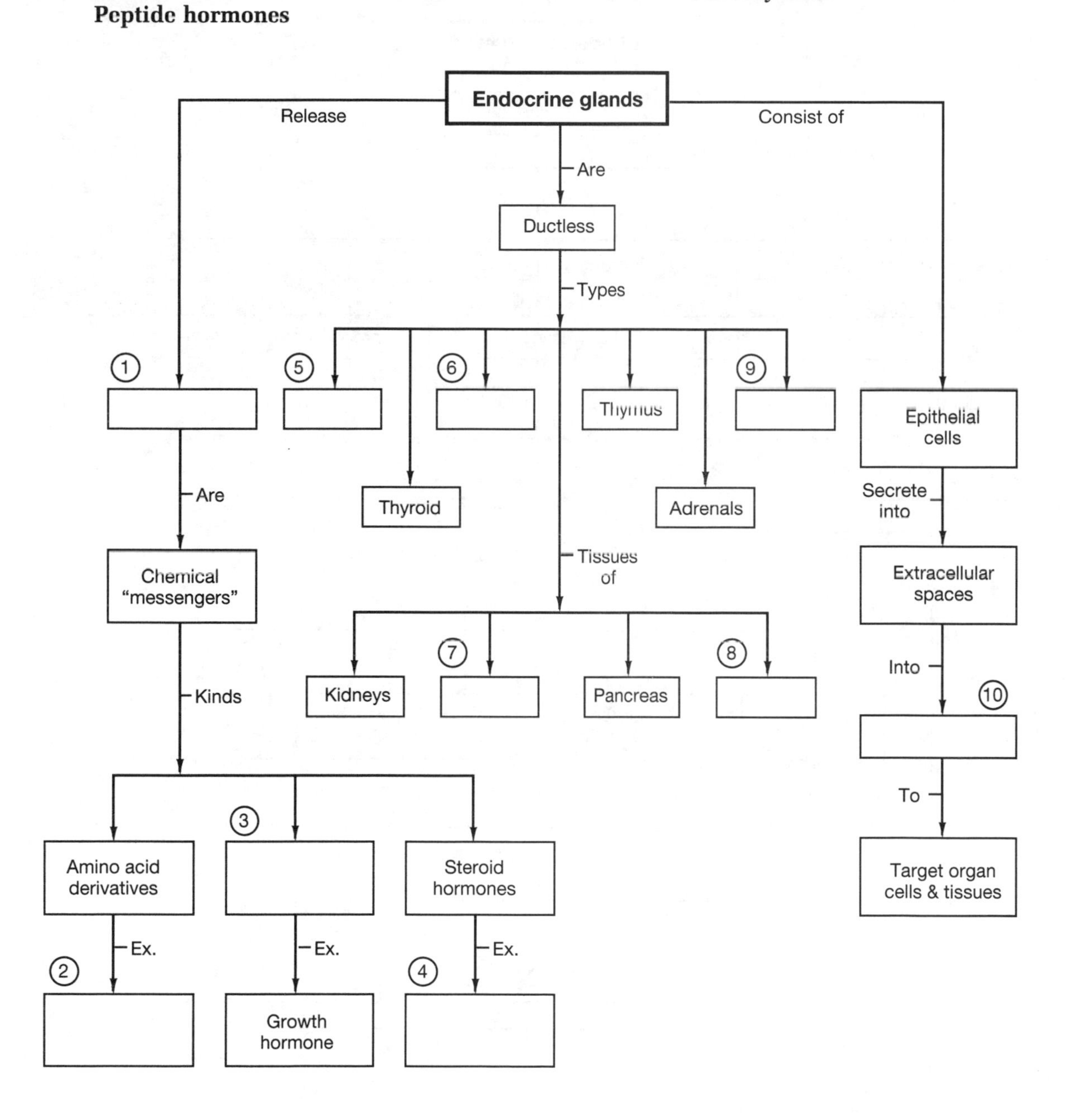

Concept Map II - Endocrine System Functions

Using the following terms, fill in the circle, numbered, blank spaces to complete the concept map. Follow the numbers that comply with the organization of the map.

Homeostasis	**Ion channel opening**	**Contraction**
Target cells	**Hormones**	**Cellular communication**

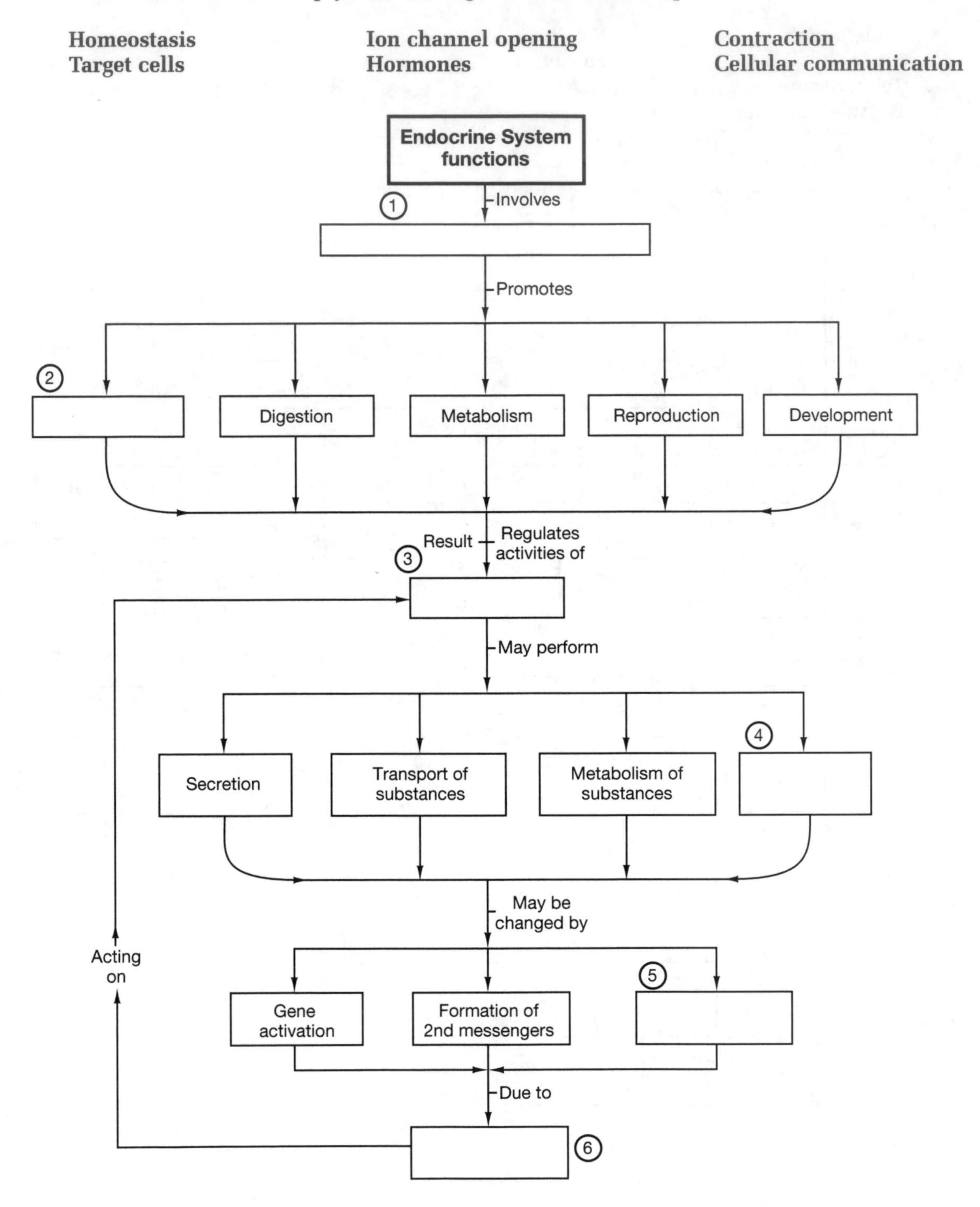

D. Crossword Puzzle

The following crossword puzzle reviews the material in Chapter 11. To complete the puzzle, you must know the answers to the clues given, and must be able to spell the terms correctly.

ACROSS

4. A hormone that prevents the loss of sodium ions.
5. A hormone produced by adipose cells.
7. A hormone that prepares the body for pregnancy.
8. A hormone involved in secondary sex characteristics in males.
10. A hormone involved in secondary sex characteristics in females.
13. A dietary ingredient necessary to make thyroxine.

DOWN

1. A steroid hormone that reduces inflammation.
2. After a hormone leaves a gland, it travels to a(n) __________ organ.
3. The _________ is an endocrine and an exocrine organ.
4. A hormone that causes the loss of sodium ions.
6. A type of gland that secretes hormones into the blood.
9. The hormone that lowers blood glucose levels.
11. The hormone that raises blood glucose levels.
12. A gland involved with the immune system.

E. Short Answer Questions

Briefly answer the following questions in the spaces provided below.

1. What three mechanisms does the hypothalamus use to regulate the activities of the nervous and endocrine systems?

2. What hypothalamic control mechanisms are used to effect endocrine activity in the anterior pituitary?

3. How does the calorigenic effect of thyroid hormones help us to adapt to cold temperatures?

4. How does the kidney hormone erythropoietin cause an increase in blood pressure?

5. How do you explain the relationship between the pineal gland and circadian rhythms?

6. What are the possible results when a cell receives instructions from two different hormones at the same time?

CHAPTER

12

Blood

Overview

Blood is a specialized fluid connective tissue consisting of two basic components; the formed elements and cell fragments, and the fluid plasma in which the elements and fragments are carried. The blood is a greater part of a transportation system that supplies every cell in the body with a continuous supply of vital nutrients and oxygen and provides the means by which the body disposes of metabolic wastes. The life-sustaining blood makes up approximately seven per cent of the body's total weight and serves to maintain the integrity of cells by regulating the pH and electrolyte composition of interstitial cells throughout the body. It also defends against toxins and pathogens and helps to stabilize body temperature.

The exercises in Chapter 12 provide opportunities for you to identify and study the components of blood and understand the functional roles of the blood which provide essential homeostatic services to the trillions of human body cells. The phases of the homeostatic process, along with clinical notes citing clotting abnormalities are also considered.

Review of Chapter Objectives

1. Describe the important components and major functions of blood.
2. Discuss the composition and functions of plasma.
3. Describe the origin and production of the formed elements in blood.
4. Discuss the characteristics and functions of red blood cells.
5. Explain what determines a person's blood type and why blood types are important.
6. Categorize the various white blood cells on the basis of their structure and functions.
7. Describe the mechanisms that control blood loss after an injury.

Part I: Objective Based Questions

Objective 1 Describe the important components and major functions of blood.

_____ 1. The formed elements of the blood consist of

a. antibodies, metalloproteins, and lipoproteins
b. red and white blood cells, and platelets
c. albumins, globulins, and fibrinogen
d. electrolytes, nutrients, and organic wastes

_____ 2. Loose connective tissue and cartilage contain a network of insoluble fibers, whereas plasma, a fluid connective tissue, contains

a. dissolved proteins
b. a network of collagen and elastic fibers
c. elastic fibers only
d. collagen fibers only

_____ 3. Blood transports dissolved gases, bringing oxygen from the lungs to the tissues and carrying

a. carbon dioxide from the lungs to the tissues
b. carbon dioxide from one peripheral cell to another
c. carbon dioxide from the interstitial fluid to the cell
d. carbon dioxide from the tissues to the lungs

_____ 4. The "patrol agents" in the blood that defend the body against toxins and pathogens are

a. hormones and enzymes
b. albumins and globulins
c. white blood cells and antibodies
d. red blood cells and platelets

_____ 5. The combination of plasma and formed elements is called

a. serum
b. interstitial fluid
c. intracellular fluid
d. whole blood

Objective 2 Discuss the composition and functions of plasma.

_____ 1. In addition to water and proteins, the plasma consists of

a. erythrocytes, leukocytes, and platelets
b. electrolytes, nutrients, and organic wastes
c. albumins, globulins, and fibrinogen
d. all of the above

_____ 2. The three primary classes of plasma proteins are

a. antibodies, metalloproteins, lipoproteins
b. serum, fibrin, fibrinogen
c. albumins, globulins, fibrinogen
d. heme, porphyrin, globin

_____ 3. The primary function(s) of plasma is (are)

a. it absorbs and releases heat as needed by the body
b. it transports ions
c. it transports red blood cells
d. all the above

_____ 4. The plasma proteins involved in fighting infections are

a. albumins
b. fibrinogens
c. immunoglobulins
d. lipoproteins

_____ 5. The plasma protein involved in blood clotting is

a. fibrinogen
b. albumin
c. hemoglobin
d. immunoglobulin

Objective 3 Describe the origin and production of the formed elements in blood.

_____ 1. Formed elements in the blood are produced by the process of

a. hemolysis
b. hemopoiesis
c. diapedesis
d. erythrocytosis

_____ 2. The stem cells that produce all the blood cells are called

a. erythroblasts
b. rouleaux
c. hemocytoblasts
d. plasma cells

_____ 3. Platelets are derived from

a. bone marrow
b. megakarocytes
c. stem cells
d. thrombocytes

_____ 4. Red blood cells formation is initiated by

a. a condition of hypoxia causes kidney cells to release EPO.
b. High oxygen detected by kidney cells causes the release of EPO.
c. Low oxygen detected by kidney cells causes the release of EPO.
d. Both a. and c. are correct

5. The term used to describe the formation of red blood cells is ____________________.

6. The term used to describe the formation of white blood cells is ____________________.

7. The term used to describe the formation of platelets is ____________________.

Objective 4 Discuss the characteristics and functions of red blood cells.

_____ 1. The primary function(s) of a mature red blood cell is

a. transport of respiratory gases
b. delivery of enzymes to target tissues
c. defense against toxins and pathogens
d. all the above

_____ 2. Circulating mature red blood cells back

a. mitochondria
b. ribosomes
c. nuclei
d. all the above

_____ 3. RBC production is regulated by the hormone

a. thymosin
b. angiotensin
c. renin
d. erythropoietin

_____ 4. The average lifespan of a red blood cell is

a. 7 days
b. 1 month
c. 120 days
d. 6 months

_____ 5. The function of hemoglobin is to

a. carry oxygen
b. protect the body against infectious agents
c. aid in the process of blood clotting
d. all the above

_____ 6. Aged and damaged erythrocytes are broken down by

a. spleen
b. kidneys
c. liver
d. bone marrow

_____ 7. The hemoglobin molecule is composed of

a. three protein chains and four heme groups
b. two protein chains and two heme groups
c. four protein chains and four heme groups
d. four protein chains and four heme groups

_____ 8. Red blood cells are called

a. leukocytes
b. thrombocytes
c. erythrocytes
d. none of the above

Objective 5 Explain what determines a person's blood type and why blood types are important.

_____ 1. A person's blood type is determined by the

a. shape and size of the red blood cells
b. presence or absence of specific antigens on the cell membrane
c. number of specific antigens in the cell membrane
d. chemical nature of the hemoglobin

_____ 2. A person with type A blood has

a. A agglutinins on their RBC
b. A agglutinins in their plasma
c. B agglutinogens on their RBC
d. B agglutinins in their plasma

_____ 3. Agglutinogens are contained (on, in) the _______________, while the agglutinins are found (on, in) the _______________.

a. plasma; cell membrane of RBCs
b. nucleus of the RBC; mitochondria
c. cell membrane of RBC; plasma
d. mitochondria; nucleus of RBCs

_____ 4. A person with type O blood contains

a. anti-A and anti-B agglutinins
b. anti-O agglutinins
c. anti-A and anti-B agglutinogens
d. type O blood and lack agglutinins

_____ 5. The blood type that does not contain plasma antibodies is

a. A
b. B
c. AB
d. O

_____ 6. A type O person can donate blood to a type A person because

a. a type O person does not have any antigens to be attacked by the type A blood
b. a type O person does not have any antigens to attack the type A blood
c. a type O person does not have any plasma antibodies to be attacked by the type A blood
d. a type O person is a universal donor

_____ 7. When blood types are incompatible the blood will

a. clot
b. clump
c. agglutinate
d. b and c are correct

8. A person with AB type blood has A antigens and B antigens. In Figure 12-1, identify which letters represents the A antigen and which letter represents the B antigen of type AB blood. Place your answers in the spaces below the illustrations.

Figure 12-1 Blood Types and Antigens

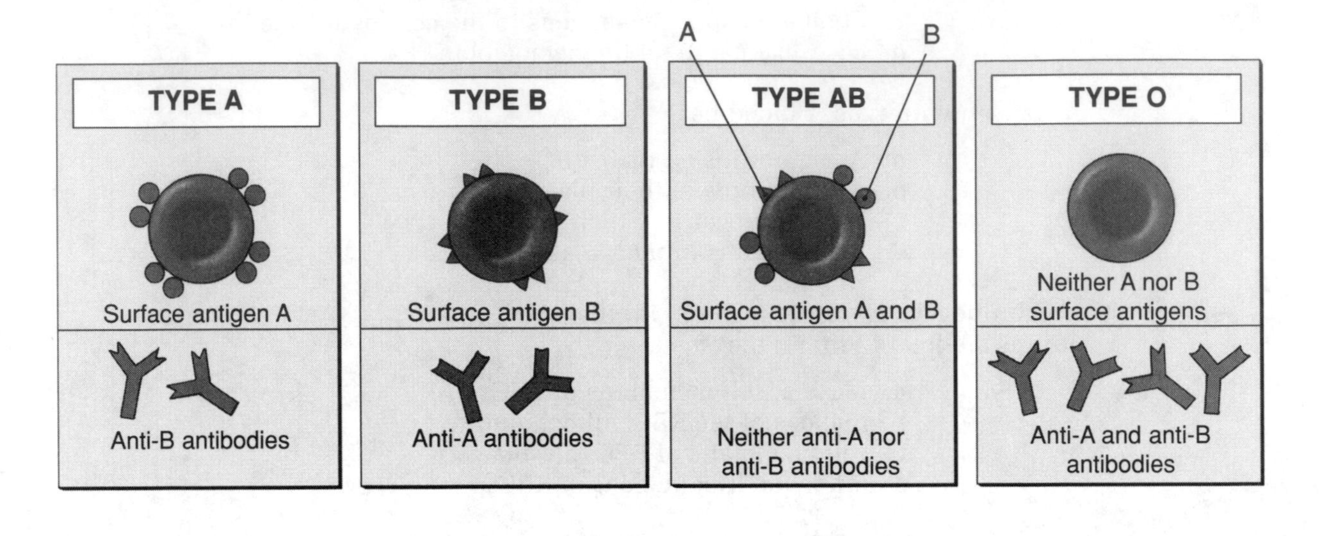

A ____________________ B ____________________

9. In Figure 12-2, identify which letters represent a safe blood donation and which letters represent an unsafe blood donation. In the corresponding blanks, write the words "compatible" or "incompatible."

Figure 12-2 Cross-Reactions between Different Blood Types

Recipient's blood type	Donor's blood type A	B	AB	O
A	A	B	C	D
B	E	F	G	H
AB	I	J	K	L
O	M	N	O	P

A ______________________

B ______________________

C ______________________

D ______________________

E ______________________

F ______________________

G ______________________

H ______________________

I ______________________

J ______________________

K ______________________

L ______________________

M ______________________

N ______________________

O ______________________

P ______________________

Objective 6 Categorize the various white blood cells on the basis of their structure and functions.

_____ 1. The two types of agranular leukocytes found in the blood are

a. neutrophils, eosinophils
b. leukocytes, lymphocytes
c. monocytes, lymphocytes
d. neutrophils, monocytes

_____ 2. Based on their staining characteristics, the types of granular leukocytes found in the blood are

a. lymphocytes, monocytes, erythrocytes
b. neutrophils, monocytes, lymphocytes
c. eosinophils, basophils, lymphocytes
d. neutrophils, eosinophils, basophils

_____ 3. The number of eosinophils increases dramatically during

a. an allergic reaction or a parasitic infection
b. an injury to a tissue or a bacterial infection
c. tissue degeneration or cellular deterioration
d. all the above

_____ 4. The type of leukocyte responsible for the red swollen condition in inflamed tissue is the

a. basopil
b. lymphocyte
c. monocyte
d. neutrophil

_____ 5. The multilobed white blood cell which typically fights bacteria is the

a. basophil
b. lymphocyte
c. monocyte
d. neutrophil

_____ 6. The leukocyte which fuses with another of its kind to create a giant phagocytic cell is the

a. lymphocyte
b. monocyte
c. neutrophil
d. basophil

_____ 7. The most numerous WBC in a normal WBC differential are the

a. neutrophils
b. eosinophils
c. lymphocytes
d. basophil

_____ 8. WBCs that release histamine at the site of an injury are

a. neutrophils
b. eosinophils
c. basophils
d. lymphocytes

_____ 9. The WBC that are important in producing antibodies are the

a. eosinophils
b. lymphocytes
c. neutrophils
d. basophils

_____ 10. The normal number of WBCs in a healthy person is _______________/mm3.

a. 6000 - 9000
b. 100,000 - 150,000
c. 1000 - 2000
d. 25,000 - 50,000

11. Identify the white blood cells in the following illustrations. Place your answers in the spaces provided below each example.

Figure 12-3 White Blood Cells

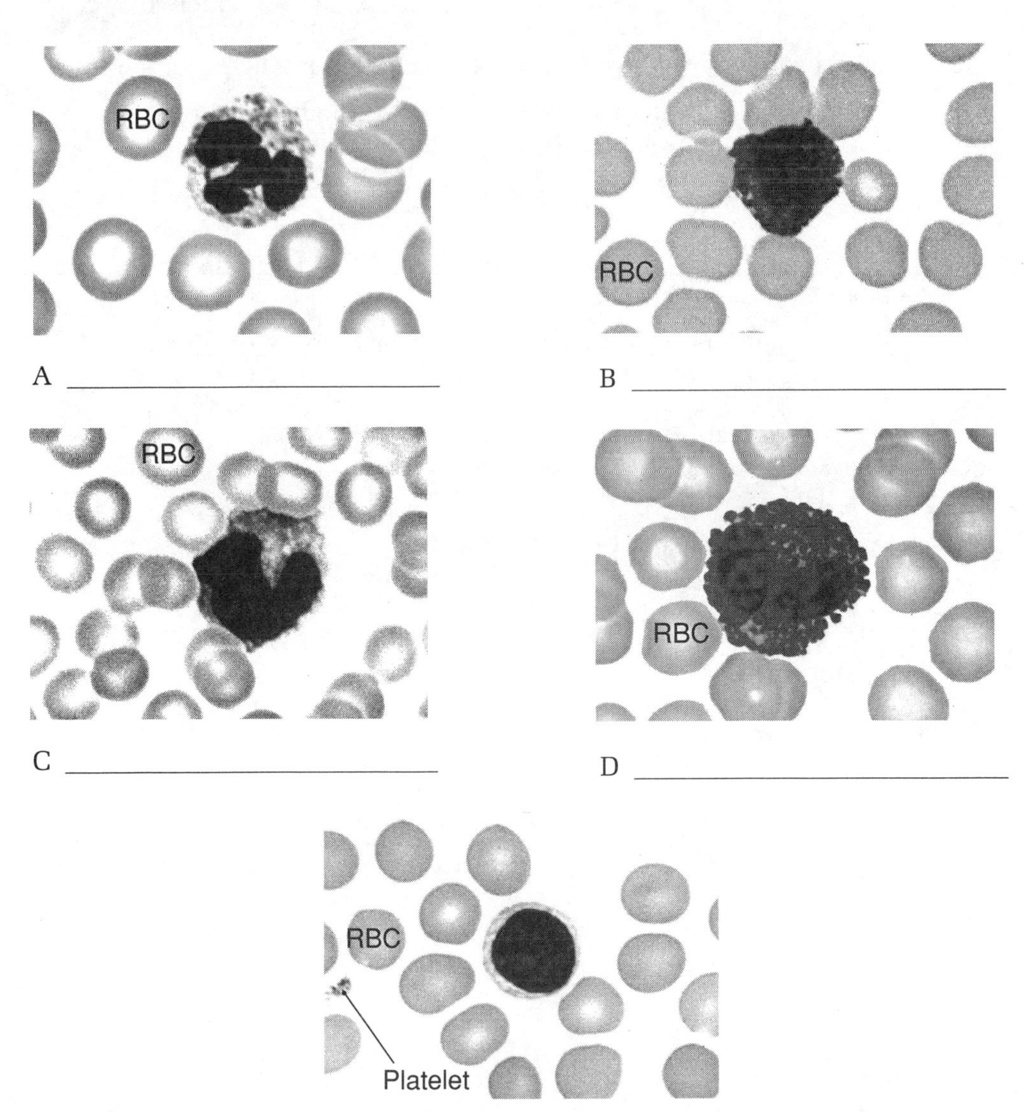

A _______________ B _______________

C _______________ D _______________

E _______________

Objective 7 Describe the mechanisms that control blood loss after an injury.

_____ 1. Basophils are specialized in that they

a. contain microphages that engulf invading bacteria
b. contain histamine that exaggerates the inflammation response at the injury site
c. are enthusiastic phagocytes, often attempting to engulf items as large or larger than themselves
d. produce and secrete antibodies that attack cells or proteins in distant portions of the body

_____ 2. The process of hemostasis includes five phases. The correct order of the phases as they occur after injury is as follows

a. vascular, coagulation, platelet, clot retractions, clot destruction
b. coagulation, vascular, platelet, clot destructions, clot retraction
c. platelet, vascular, coagulation, clot retraction, clot destruction
d. vascular, platelet, coagulation, clot retraction, clot destruction

_____ 3. The extrinsic pathway involved in blood clotting involves the release of

a. platelet factors and platelet thromboplastin
b. Ca^{++} and clotting factors VIII, IX, XI, XIII
c. tissue factors and tissue thromboplastin
d. prothrombin and fibrinogen

_____ 4. The "common pathway"in blood clotting involves the following events in sequential order as follows

a. tissue factors → Ca^{++} → plasminogen → plasmin
b. prothrombin → thrombin → fibrinogen → fibrin
c. platelet factors → Ca^{++} → fibrinogen → fibrin
d. prothrombin and fibrinogen

_____ 5. During the clotting process, *platelets* function in

a. transporting chemicals important for clotting
b. contraction after clot formation
c. initiating the clotting process
d. all the above

_____ 6. The complex sequence of steps leading to the conversion of fibrinigen to fibrin is called

a. coagulation
b. retraction
c. fibrinolysis
d. agglulination

_____ 7. The process of fibrinolysis

a. activates fibrinogen
b. forms emboli
c. dissolves clots
d. forms thrombi

_____ 8. The vitamin needed for the formation of clotting factors is

a. vitamin A
b. vitamin K
c. vitamin E
d. vitamin D

_____ 9. A drifting blood clot is called a (an)

a. thrombus
b. plaque
c. platelet plug
d. embolus

_____ 10. A clotting protein found in the bloodstream and made by the liver is

a. fibrin
b. fibrinogen
c. prothrombin
d. thrombin

Part II: Chapter Comprehensive Exercises

A. Word Elimination

Circle the term that does not belong in each of the following groupings.

1. transportation regulation analysis defense stabilization
2. plasma lymph WBC RBC platelets
3. serum albumins globulins antibodies transport proteins
4. amino acids iron vitamin B 12 folic acid vitamin D
5. Hct Hb urinalysis MCV MCHC
6. Rh factor Type A Type B Type AB Type O
7. neutrophil eosinophil platelets basophil monocyte
8. T cells B cells NK cells RBC plasma cells
9. hemocytoblast stem cell erythroblast reticulocyte platelet
10. prothrombin thrombin transferrin fibrinogen fibrin

B. Matching

Match the terms in Column A with the terms in Column B. Use letters for answers in the spaces provided.

COLUMN A	COLUMN B
___ 1. hemopoiesis	A. agglutinins
___ 2. hypoxia	B. low platelet count
___ 3. antibodies	C. high member of WBCs
___ 4. granulocyte	D. monocyte
___ 5. leukocytosis	E. blood cell formation
___ 6. thrombocytopenia	F. hemocytoblasts
___ 7. agranulocyte	G. agglutinogens
___ 8. stem cells	H. low oxygen concentration
___ 9. surface antigens	I. WBC migration
___ 10. diapedesis	J. neutrophil

C. Concept Map I

Using the following terms, fill in the numbered, blank spaces to complete the concept map. Follow the numbers which comply with the organization of the map.

Solutes	**Leukocytes**	**Plasma**	**Albumins**
Oxygen	**Monocytes**	**Neutrophils**	

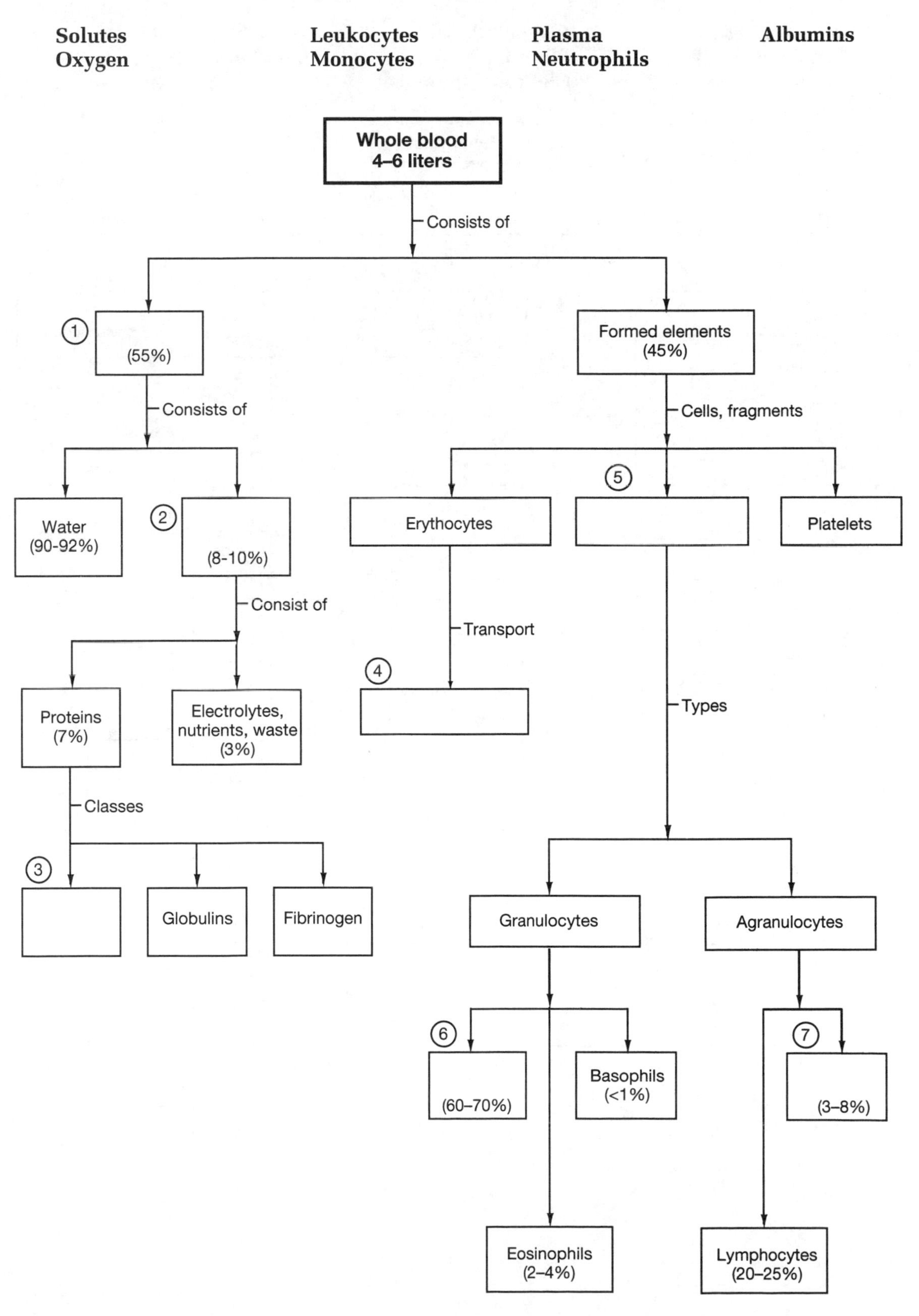

Concept Map II

Using the following terms, fill in the numbered, blank spaces to complete the concept map. Follow the numbers which comply with the organization of the map.

Clot dissolves
Plasminogen
Forms plug
Platelet phase
Plasmin
Clot retraction
Forms blood clot
Vascular spasm

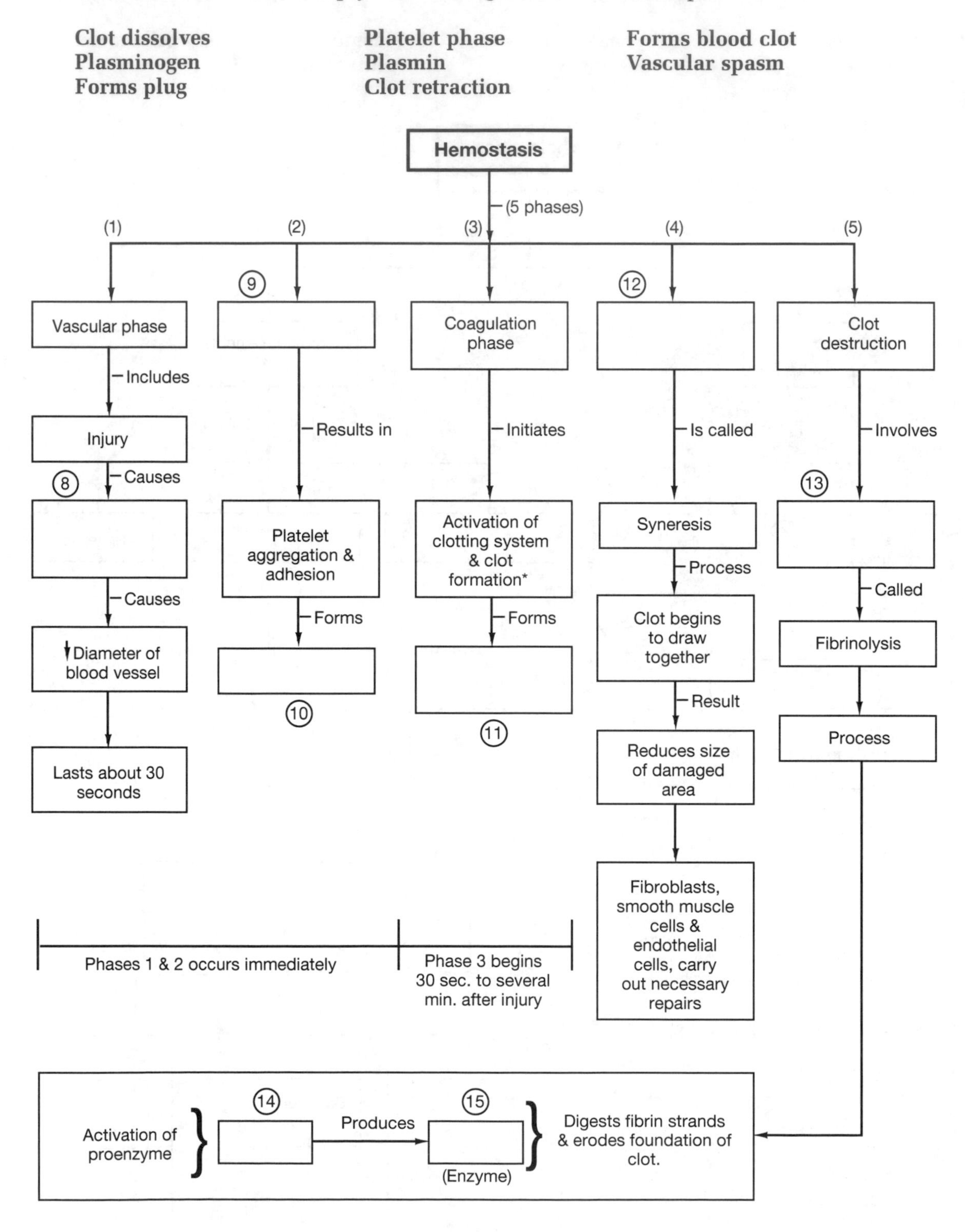

D. Crossword Puzzle

The following crossword reviews the material in Chapter 12. To complete the puzzle, you must know the answers to the clues given, and must be able to spell the terms correctly.

ACROSS

1. The rarest WBC in healthy person.
3. A moving clot.
5. __________ cells contain only the antigens and not the antibodies.
6. Type B whole blood contains antigens and ________.
11. The process of blood clotting.
12. The fluid portion of blood.
15. If type A blood was given to a person with type B blood, the type B blood would begin to ________.
16. A blood clotting protein made in the liver.
17. Term for the formation of red blood cells.
18. Term for the formation of blood.

DOWN

2. Term for the formation of white blood cells.
4. The most common WBC in a healthy person.
7. After about 120 days, red blood cells begin to _______.
8. Type AB blood has A and B __________.
9. Type AB people are universal _________.
10. A normal count of white blood cells is 6000-9000 per cubic __________.
13. Some white blood cells produce these proteins that attack invading organisms.
14. Type O people are considered to be universal __________.

E. Short Answer Questions

Briefly answer the following questions in the spaces provided below.

1. List the primary functions of the blood.

2. What are the major components of the plasma?

3. List and describe the three (3) kinds of granular leukocytes and the two (2) kinds of agranular WBCs.

4. What are the primary functions of platelets?

5. List the events in the clotting response and summarize the results of each occurrence.

6. What is the difference between an embolus and a thrombus?

CHAPTER 13

The Heart

Overview

The heart is an efficient and durable double pump. Every single day it beats 60-80 times per minute, supplying oxygen and other essential nutrients to every cell in the body and removes wastes for elimination from the lungs and the kidneys. The pumping action of this muscular organ pushes blood through a closed network of blood vessels consisting of a pulmonary circuit to the lungs and a systemic circuit serving the other regions of the body. The one way direction of blood flow through the heart is maintained by a system of valves. Special coronary arteries that encircle the heart like a crown deliver nutrient and oxygen supplies to the tissues of the heart. The heart's electrical system is of a group of specialized pacemaker cells located in a sinus node in the right atrium which are responsible for maintaining normal cardiac rhythm. When you have successfully completed the exercises in Chapter 13 you should be able to describe the general features of the heart both internally and externally, identify the major blood vessels, chambers and heart valves, and understand the conducting system of the heart, including the cardiac cycle. The activities which include heart dynamics are designed to help you master the factors that control cardiac output and the effects of autonomic innervation in the heart.

Review of Chapter Objectives

1. Describe the location and general features of the heart.
2. Trace the flow of blood through the heart, identifying the major blood vessels, chambers, and heart valves.
3. Identify the layers of the heart wall.
4. Describe the differences in the action potential and twitch contractions of skeletal muscle fibers and cardiac muscle cells.
5. Describe the components of and functions of the conducting system of the heart.
6. Explain the events of the cardiac cycle and relate the heart sounds to specific events in this cycle.
7. Define stroke volume and cardiac output and describe the factors that influence these values.

Part I: Objective Based Questions

Objective 1 Describe the location and general features of the heart.

_____ 1. The "double pump" function of the heart includes the right side, which serves as the __________ circuit pump, while the left side serves as the __________ pump.

a. systemic; pulmonary
b. pulmonary; hepatic portal
c. hepatic portal; cardiac
d. pulmonary; systemic

_____ 2. The major difference between the left and right ventricles relative to their role in heart function is:

a. the L.V. pumps blood through the short, low-resistance pulmonary circuit
b. the R.V. pumps blood through the low-resistance systemic circulation
c. the L.V. pumps blood through the high-resistance systemic circulation
d. the R.V. pumps blood through the short, high-resistance pulmonary circuit

_____ 3. The average maximum pressure developed in the right ventricle is about:

a. 15-28 mm Hg
b. 50-60 mm Hg
c. 67-78 mm Hg
d. 80-120 mm Hg

_____ 4. Assuming anatomic position, the best way to describe the specific location of the heart in the body is:

a. within the mediastinum of the thorax
b. in the region of the fifth intercostal space
c. just behind the lungs
d. in the center of the chest

_____ 5. The function of the chordae tendinae is to:

a. anchor the semilunar valve flaps and prevent backward flow of blood into the ventricles
b. anchor the AV valve flaps and prevent backflow of blood into the atria
c. anchor the bicuspid valve flaps and prevent backflow of blood into the ventricle
d. anchor the aortic valve flaps and prevent backflow into the ventricles

_____ 6. The expandable extension of the atrium is the

a. ventricle
b. coronary sulcus
c. coronary sinus
d. auricle

_____ 7. The portion of the pericardial membrane that lies on the surface of the heart is the

a. visceral pericardium
b. visceral endocardium
c. parietal pericardium
d. parietal myocardium

_____ 8. The visceral pericardium is the same as the

a. mediastinum
b. epicardium
c. parietal pericardium
d. myocardium

_____ 9. The functions of the pericardium includes which of the following?

a. returning blood to the atria
b. pumping blood into circulation
c. removing excess fluid from the heart chambers
d. anchoring the heart to surrounding structures

10. The chambers in the heart with thin walls that are highly distensible are the _______________.

11. The internal connective tissue network of the heart is called the _______________.

12. Each cardiac muscle fiber contacts several others at specialized sites known as _______________.

13. Identify and label the structures in Figure 13-1. Place your answers in the spaces below the drawing.

Figure 13-1 External View of the Heart

base	**apex**	**left side**
right side	**coronary vessels**	**pulmonary trunk**
aortic arch		

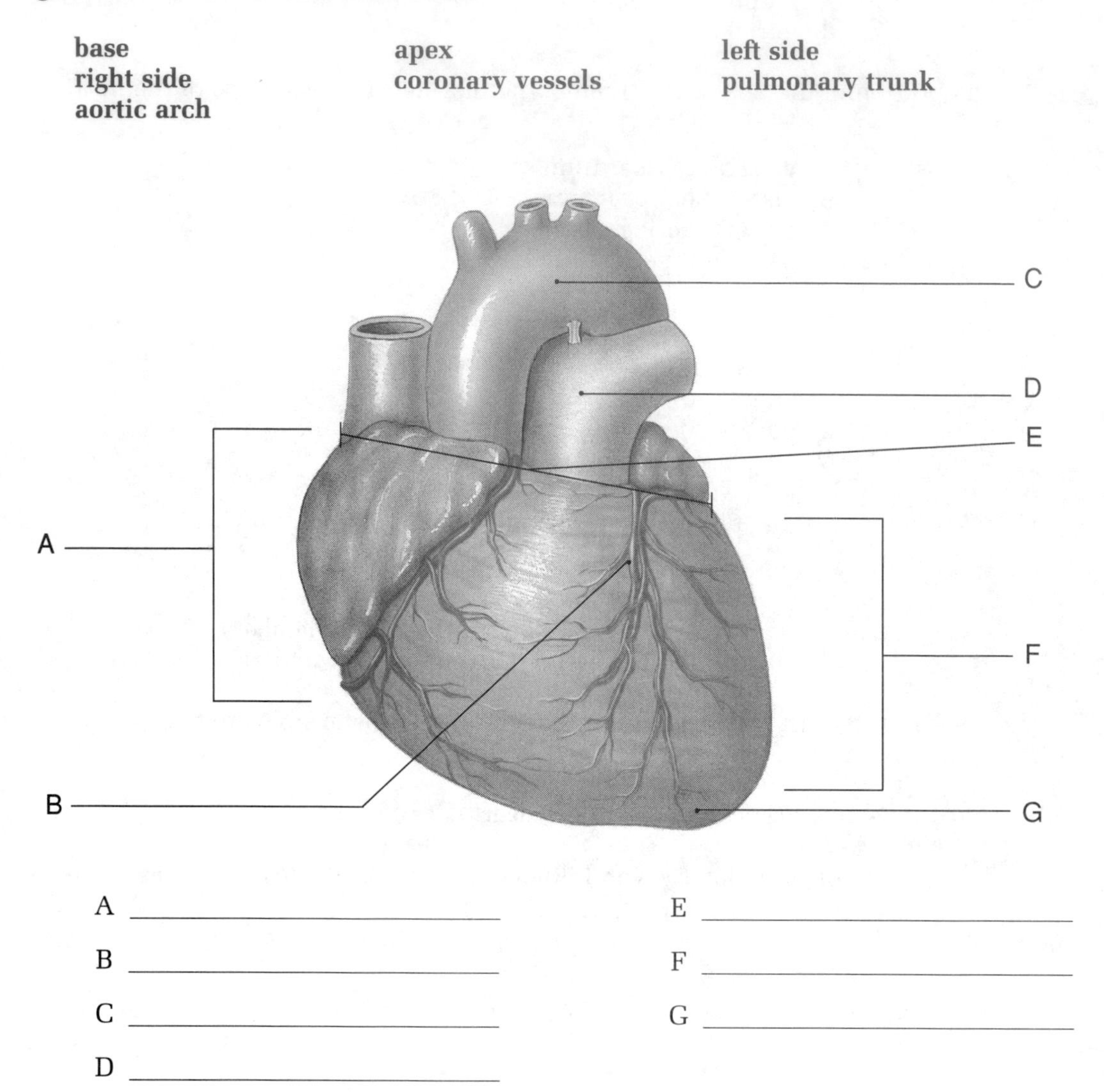

A ______________________

B ______________________

C ______________________

D ______________________

E ______________________

F ______________________

G ______________________

Objective 2 Trace the flow of blood through the heart identifying the major blood vessels, chambers and heart valves.

_____ 1. Blood returning from the systemic circuit first enters the

a. right atrium
b. right ventricle
c. left atrium
d. left ventricle

_____ 2. Blood returning from the lungs enters the

a. right atrium
b. right ventricle
c. left atrium
d. left ventricle

_____ 3. The right ventricle pumps blood to the

a. left ventricle
b. lungs
c. left atrium
d. systemic circuit

_____ 4. The left ventricle pumps blood to the

a. lungs
b. right ventricle
c. right ventricle
d. systemic circuit

_____ 5. The right atrium receives blood from

a. pulmonary veins
b. aorta
c. inferior vena cava
d. pulmonary trunk

_____ 6. The atrioventricular valve located on the right side of the heart is the

a. tricuspid valve
b. mitral valve
c. bicuspid valve
d. aortic semilunar valve

_____ 7. Blood leaving the right ventricle enters the

a. aorta
b. pulmonary artery
c. pulmonary veins
d. inferior vena cava

_____ 8. The pulmonary semilunar valve guards the entrance to the

a. aorta
b. pulmonary veins
c. pulmonary trunk
d. left ventricle

_____ 9. The bicuspid or mitral valve is located

a. in the opening of the aorta
b. in the opening of the pulmonary trunk
c. where the vena cavae join the right atrium
d. between the left atrium and left ventricle

_____ 10. The entrance to the ascending aorta is guarded by

a. an atrioventricular valve
b. the bicuspid valve
c. a semilunar valve
d. the tricuspid valve

_____ 11. The function of an atrium is

a. to collect blood
b. to pump blood to the lungs
c. to pump blood into the systemic circuit
d. to pump blood to the heart muscle
e. all of the above

_____ 12. The following is a list of vessels and structures that are associated with the heart.

1. right atrium
2. left atrium
3. right ventricle
4. left ventricle
5. vena cavae
6. aorta
7. pulmonary trunk
8. pulmonary veins

What is the correct order for the flow of blood entering from the systemic circulation?

a. 1,2,7,8,3,4,6,5
b. 5,1,3,7,8,2,4,6
c. 1,7,3,8,2,4,6,5
d. 5,3,1,7,8,4,2,6

_____ 13. The left and right pulmonary arteries carry blood to the

a. heart
b. intestines
c. lungs
d. brain

_____ 14. The left and right pulmonary veins carry blood to the

a. heart
b. intestines
c. lungs
d. liver

15. Using the following terms, identify the structures of the heart by labeling Figure 13-2. Place the labels in the spaces below the drawing.

Figure 13-2 Anatomy of the Heart - (Ventral View)

interventricular septum
pulmonary trunk
coronary sinus
myocardium
right pulmonary veins
aortic arch
aortic semilunar valves
left ventricle
left atrium
tricuspid valve
left pulmonary veins
right ventricle
right pulmonary arteries
superior vena cava
chordae tendinae
left pulmonary arteries
pulmonary semilunar valves
bicuspid valve
inferior vena cava
right atrium

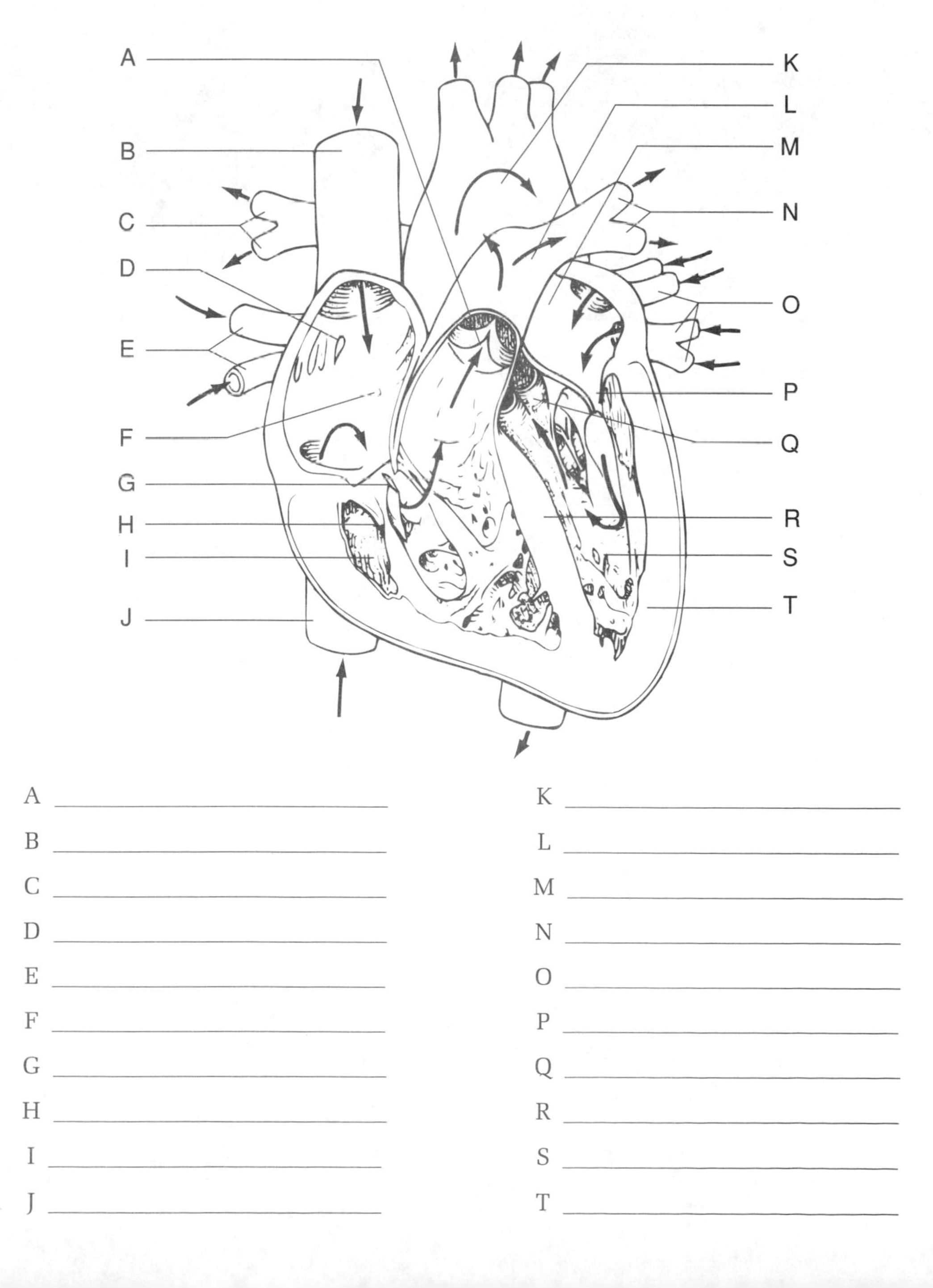

A ______________________________
B ______________________________
C ______________________________
D ______________________________
E ______________________________
F ______________________________
G ______________________________
H ______________________________
I ______________________________
J ______________________________
K ______________________________
L ______________________________
M ______________________________
N ______________________________
O ______________________________
P ______________________________
Q ______________________________
R ______________________________
S ______________________________
T ______________________________

16. In the drawing below identify the valves A thru H and indicate whether the valve is open (O) or closed (C). Place your answers in the spaces below the drawing.

Figure 13-3 Valves of the Heart

pulmonary semilunar valve (O)
tricuspid (right AV) valve (C)
tricuspid valve (O)
pulmonary semilunar (C)

bicuspid AV valve (O)
aortic semilunar valve (O)
bicuspid valve (C)
aortic semilunar valve (C)

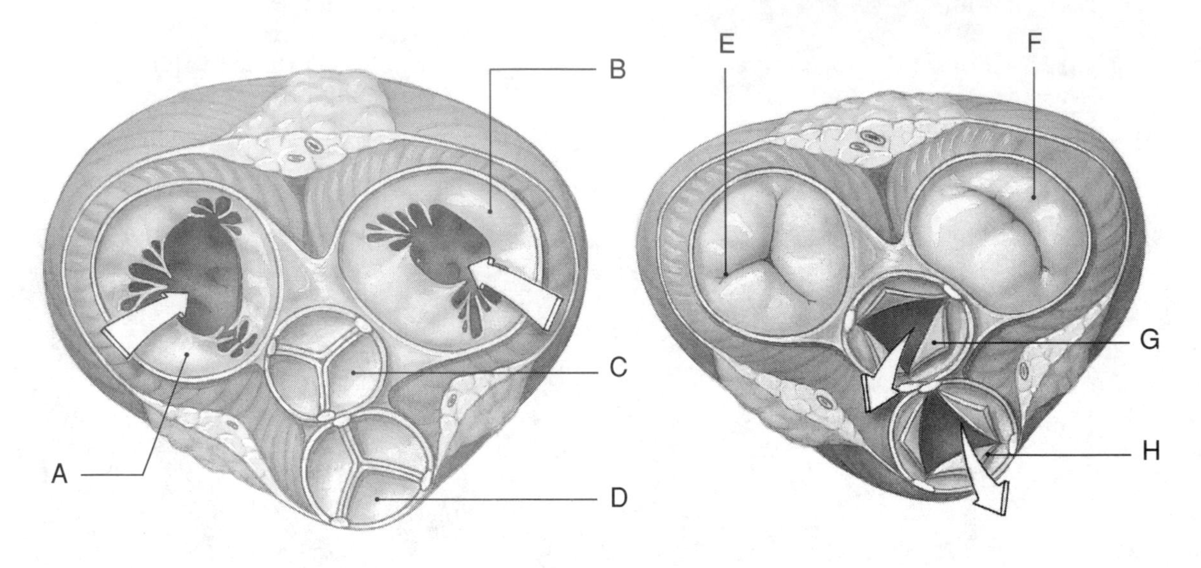

(a) Ventricular relaxation **(b) Ventricular contraction**

A ______________________ E ______________________

B ______________________ F ______________________

C ______________________ G ______________________

D ______________________ H ______________________

Objective 3 Identify the layers of the heart wall.

_____ 1. The outer wall of the heart is called the

a. endocardium
b. myocardium
c. epicardium
d. cardiac layer

_____ 2. The muscular portion of the heart wall is called the

a. myocardium
b. epicardium
c. endocardium
d. intercalation

_____ 3. The cardiac cells are found in the

a. epicardium
b. myocardium
c. endocardium
d. all of the above

_____ 4. The kind of tissue making up the endocardium and epicardium is

a. muscle tissue
b. serous membrane
c. connective tissue
d. nervous tissue

5. Using the selections below, identify and label the areas in Figure 13-4. Place your answers in the spaces provided below the drawing.

Figure 13-4 Layers of the Heart Wall

intercalated disc
myocardium
endocardium
epicardium

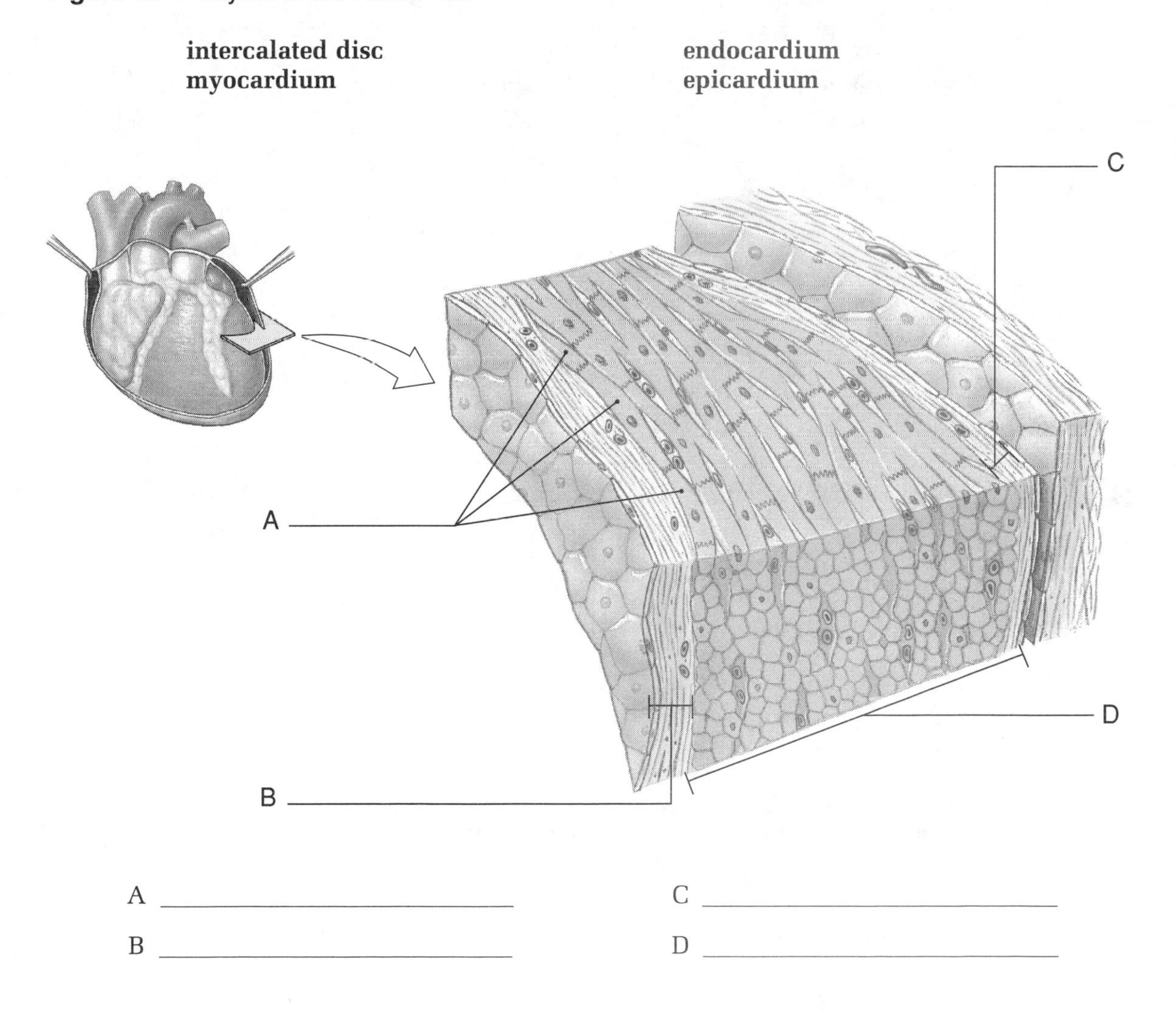

A _______________________ C _______________________

B _______________________ D _______________________

Objective 4 Describe the differences in the action potential and twitch contractions of the skeletal muscle fibers and cardiac muscle cells.

_____ 1. The correct sequential path of a normal action potential in the heart is

a. SA node → AV bundle → AV node → Purkinje fibers
b. AV node → SA node → AV bundle → bundle of His
c. SA node → AV node → bundle of His → bundle branches → Purkinje fibers
d. SA node → AV node → bundle branches → AV bundle → Purkinje fibers

_____ 2. In a cardiac muscle cell an action potential lasts approximately _____________ times longer than the duration of an action potential in a skeletal muscle cell.

a. 2
b. 10
c. 20
d. 30

_____ 3. The maximum rate of contraction in normal cardiac muscle fibers is ___________ per minute.

a. 80
b. 140
c. 200
d. 300+

_____ 4. In cardiac muscle

a. neither summation nor tetany can occur
b. both summation and tetany can occur
c. only summation can occur
d. only tetany can occur

Objective 5 Describe the components and functions of the conducting system of the heart.

_____ 1. The pacemaker cells of the heart are located in the

a. Bundle of His
b. SA node
c. AV node
d. wall of the left ventricle

_____ 2. The following are the components of the conducting system of the heart. (1) Purkinje cells (2) AV bundle (3) AV node (4) SA node (5) bundle branches. The sequence in which an action potential would move through this system is

a. 1,4,3,2,5
b. 4,2,3,5,1
c. 3,2,4,5,1
d. 4,3,2,5,1

_____ 3. Spontaneous depolarization of the nodal tissue in the SA node occurs at __________ action potentials per minute.

a. 20-30
b. 40-50
c. 70-80
d. 100-120

_____ 4. The heart is innervated by both sympathetic and parasympathetic nerves

a. true
b. false

_____ 5. Ventricular contraction occurs when the impulse travels through the

a. AV node
b. SA node
c. AV bundle
d. Purkinje fibers

6. Identify and label the following structures in Figure 13-5. Place the labels next to the letters in the spaces provided below the drawing.

Figure 13-5 The Conducting System of the Heart

AV bundle	**Purkinje fibers**	**SA node**
bundle branches	**AV node**	

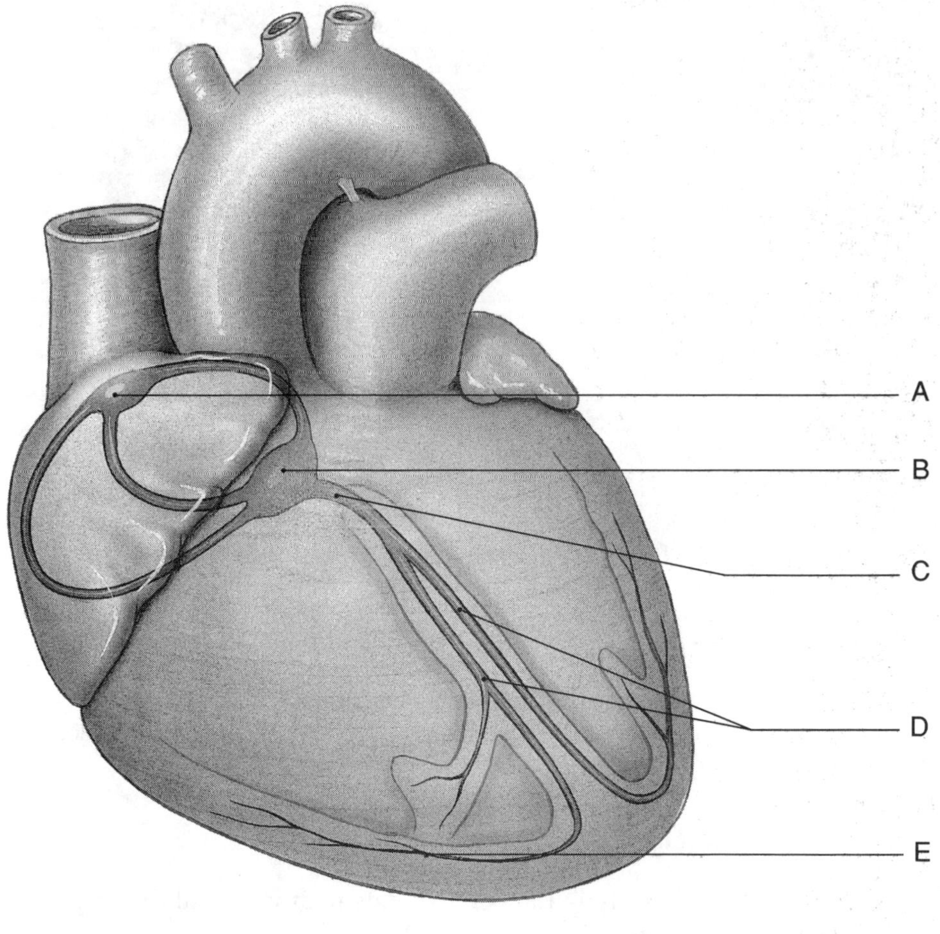

A ______________________ D ______________________

B ______________________ E ______________________

C ______________________

7. In the normal electrocardiogram below identify which part of the ECG recording (1,2,3) correlates with heart A and which parts correlates with part B.

Figure 13-6 An Electrocardiogram

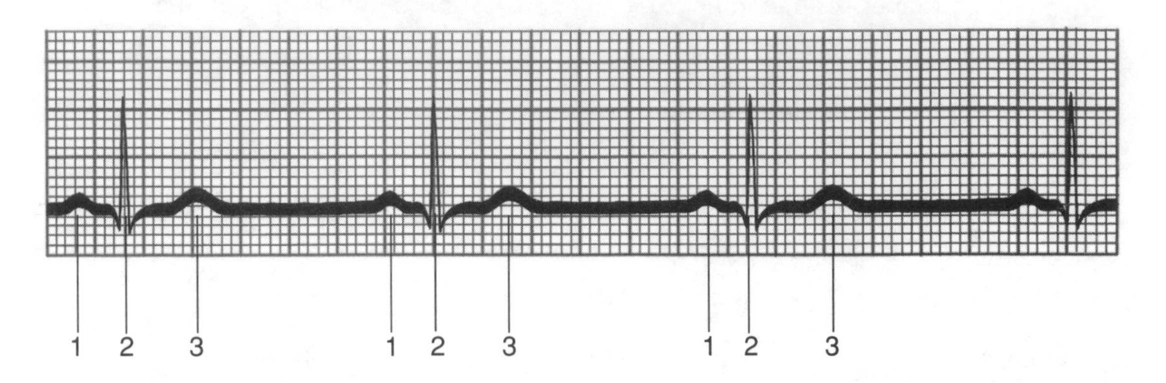

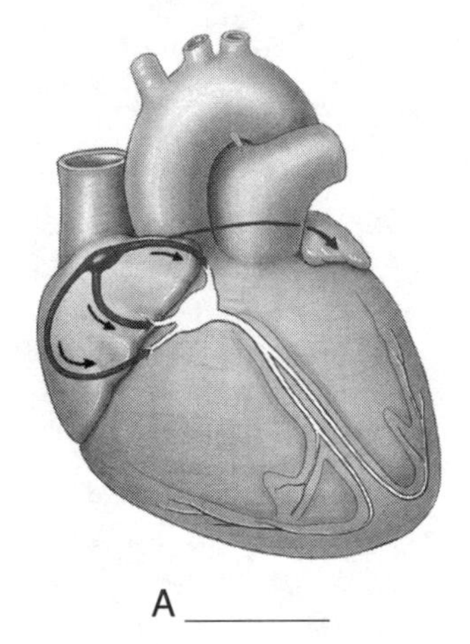

A _______

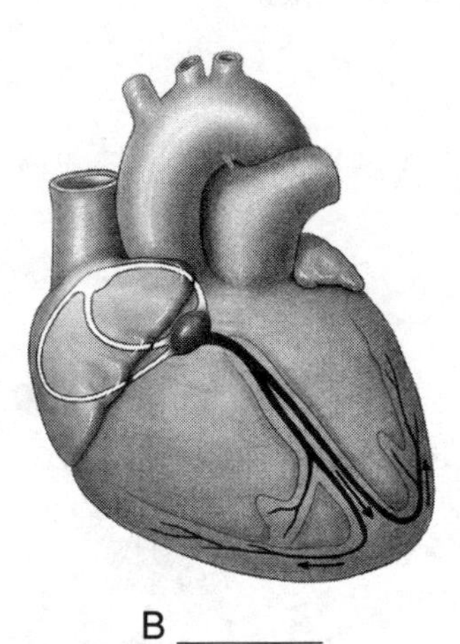

B _______

8. In Figure 13-6, number 1 represents the _______ wave of an ECG recording.

9. In Figure 13-6, number 2 represents the _______ wave of an ECG recording.

10. In Figure 13-6, number 3 represents the _____ wave of an ECG recording.

Objective 6 Explain the events of the cardiac cycle and relate the heart sounds to specific events in this cycle.

_____ 1. The "lubb-dubb" sounds of the heart have practical clinical value because they provide information concerning the

a. strength of ventricular contraction
b. strength of the pulse
c. efficiency of the heart valves
d. relative time the heart spends in systole and diastole

_____ 2. When a chamber of the heart fills with blood and prepares for the start of the next cardiac cycle, the heart is in

a. systole
b. ventricular ejection
c. diastole
d. isovolumetric contraction

_____ 3. At the start of atrial systole, the ventricles are filled to approximately

a. 10 percent of capacity
b. 30 percent of capacity
c. 50 percent of capacity
d. 70 percent of capacity

_____ 4. The first heart sound is heard when the

a. AV valves open
b. AV valves close
c. semilunar valves close
d. blood enters the aorta

_____ 5. Systole and diastole refer to

a. contraction and relaxation of the heart
b. relaxation and contraction of the heart
c. atrial and ventricular contraction
d. ventricular and atrial contraction

Objective 7 Define stroke volume and cardiac output and describe the factors that influence these values.

_____ 1. The amount of blood ejected by the left ventricle per minute is the

a. stroke volume
b. cardiac output
c. end-diastolic volume
d. end-systolic volume

_____ 2. The amount of blood pumped out of each ventricle during a single beat is the

a. stroke volume
b. EDV
c. cardiac output
d. ESV

_____ 3. The cardiac output is equal to the

a. difference between the diastolic volume and the systolic volume
b. product of heart rate and stroke volume
c. difference between the stroke volume at rest and the stroke volume during exercise
d. stroke volume less the systolic volume

_____ 4. Each of the following factors will increase cardiac output except one. Identify the exception.

a. increased venous return
b. increased parasympathetic stimulation
c. increased sympathetic stimulation
d. increased heart rate

_____ 5. According to Starling's law of the heart, the cardiac output is directly related to the

a. size of the ventricle
b. heart rate
c. venous return
d. thickness of the myocardium

Part II: Chapter Comprehensive Exercises

A. Word Elimination

Circle the term that does not belong in each of the following groupings.

1. L. atrium R. atrium L. pulmonary artery L. ventricle R ventricle
2. tricuspid bicuspid mitral R. (AV) valve semilunar
3. chordae tendinae endocardium myocardium epicardium
4. SA node AV bundle AV node anastomoses Purkinje cells
5. EKG systole P wave QRS complex T wave
6. blood volume reflex autonomic innervation tachycardia hormones ECF ions
7. arteries auricles veins capillaries efferent vessels
8. pulmonary artery pulmonary vein superior vena cava aortic arch AV valve
9. single nucleus central nucleus syncytium multinucleated intercalated discs
10. R. coronary artery aortic arch L. coronary artery Anterior cardiace vein Great cardiac vein

B. Matching

Match the terms in Column B with the terms in Column A. Use letters for answers in the spaces provided.

COLUMN A	COLUMN B
I 1. serous membrane	A. decreased heart rate
F 2. myocardium	B. AV valves close; semilunar valves open
G 3. tricuspid valve	C. monitor blood pressure
J 4. aortic valve	D. AV conducting fibers
H 5. cardiac pacemaker	E. semilunar valve closes
D 6. Bundle of His	F. muscular wall of the heart
C 7. baroreceptors	G. right atrioventricular
A 8. increased extracellular	H. SA node
B 9. "Lubb" sound	I. pericardium
E 10. "Dubb" sound	J. semilunar

C. Concept Map I - The Heart

Using the following terms, fill in the circled, numbered, blank spaces to complete the concept map. Follow the numbers that comply with the organization of the map.

Epicardium	**Pacemaker cells**	**Endocardium**
Aortic	**Tricuspid**	**Blood from atria**
Two semilunar	**Two atria**	**Deoxygenated blood**
Oxygenated blood		

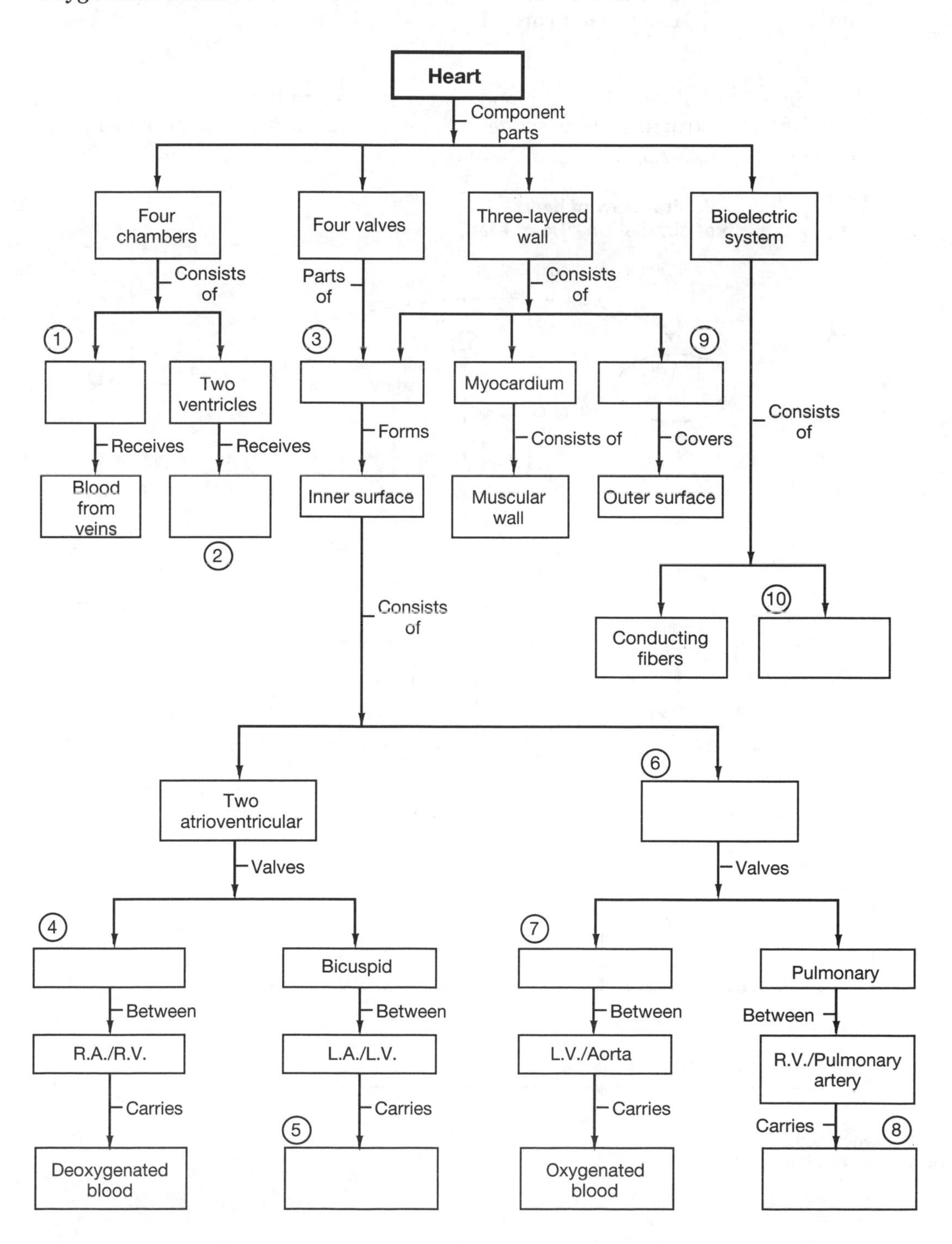

Concept Map II - Path of Blood Flow through the Heart

Using the terms below, fill in the blanks to complete the flow of blood through the heart and peripheral vessels.

aortic semilunar valve	**tricuspid valve**	**aorta**
inferior vena cava	**pulmonary semilunar valve**	**pulmonary arteries**
pulmonary veins	**systemic arteries**	**superior vena cave**
left ventricle	**systemic veins**	**right ventricle**
right atrium	**L. common carotid**	**bicuspid valve**
left atrium		

Use the "heart map" to follow a drop of blood as it goes through the heart and peripheral blood vessels. Identify each structure at the sequential, numbered locations 1 through 16.

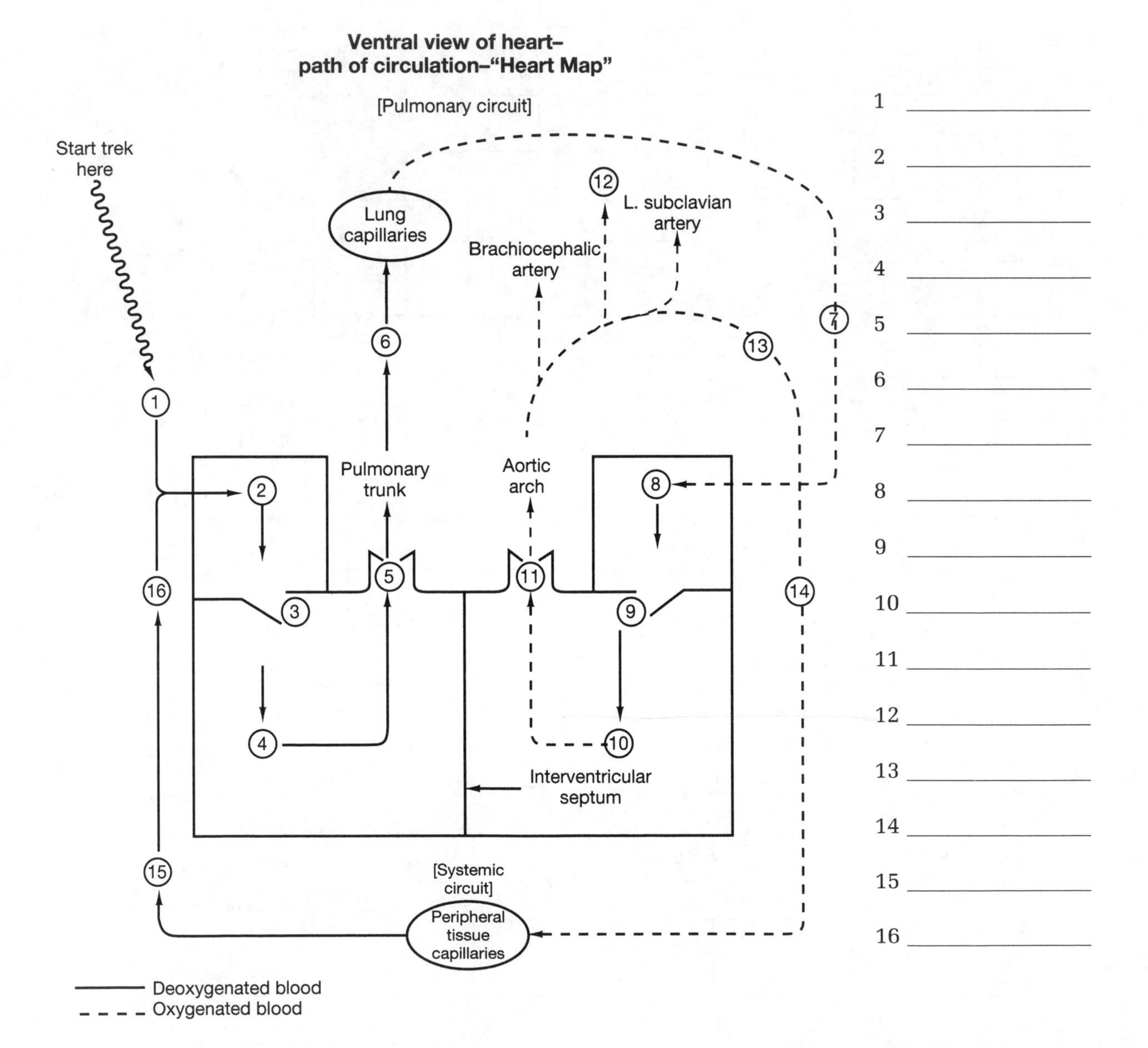

D. Crossword Puzzle

The following crossword puzzle reviews the material in Chapter 13. To complete the puzzle, you must know the answers to the clues given, and must be able to spell the terms correctly.

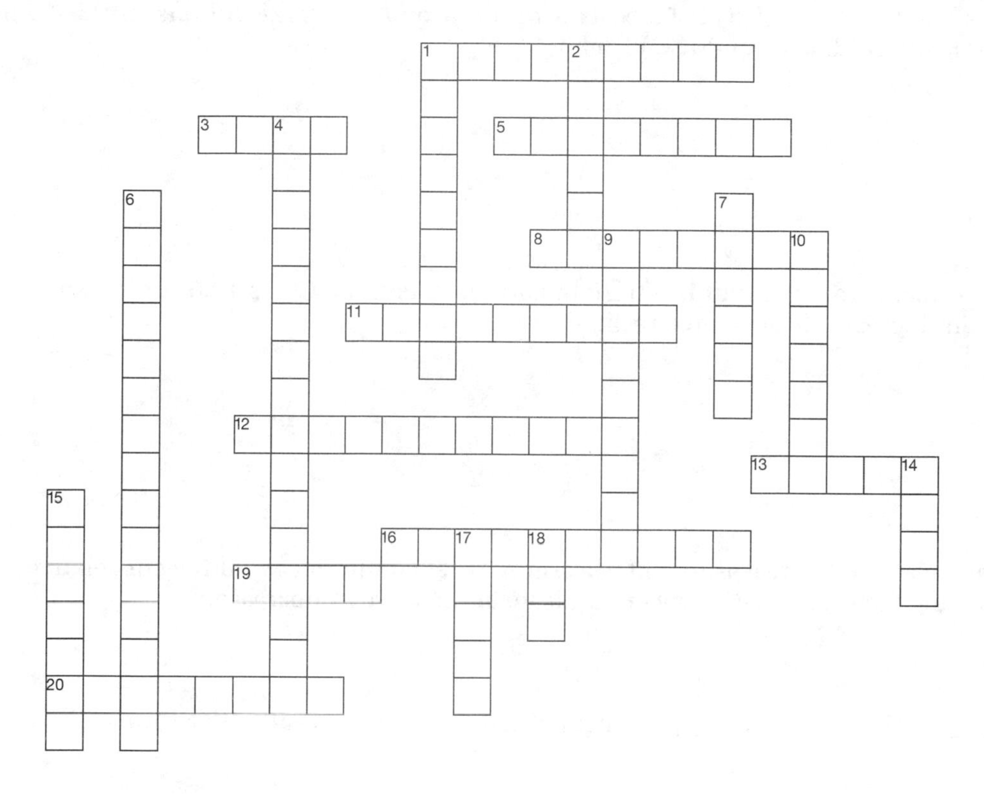

ACROSS

1. The sinoatrial node is the heart's __________.
3. Is the aortic arch closer to the base or to the apex of the heart?
5. When these fibers are activated, the ventricles will contract.
8. Vessels that carry blood away from the heart.
11. The 80 in 120/80 represents the ________ value of blood pressure.
12. Arteries flow into ______ then into venules.
13. Blood entering the inferior vena cava could have come from the ______ vein.
16. The main muscular portion of the heart.
19. Blood leaving the placenta and going to the fetus travels in the umbilical _____.
20. Blood in the _______ artery enters the brachial artery.

DOWN

1. Blood leaving the femoral artery enters the _______ artery.
2. A heart _______ is a condition in which the AV valves do not close properly.
4. The name of the blood pressure apparatus.
6. The jugular vein and the subclavian vein join to form the _________ vein.
7. Another name for the bicuspid valve.
9. The valve located between the right atrium and the right ventricle.
10. Contraction of the heart.
14. The aortic arch arches to the ____________ as it emerges from the heart.
15. Blood in the popliteal vein enters the ______ vein.
17. An opening between the two atria of a fetal heart is called the foramen _________.
18. The hormone that helps lower blood pressure.

E. Short Answer Questions

Briefly answer the following questions in the spaces provided below.

1. If the heart beat rate (HR) is 80 beats per minute and the stroke volume (SV) is 75 ml, what is the cardiac output (CO) (l/min)?

2. If a normal cardiac output is 5.0 l/min and the maximum CO is 10.0 l/min, what is the percent increase in CO above resting?

3. If the heart rate increases to 250 beats/min, what conditions would occur relative to the following factors? (Use arrows to indicate an increase or decrease.)

   _____ CO, _______SV, _____ length of diastole, _____ ventricular filling

4. If the end-systolic volume (ESV) is 60 ml and the end-diastolic volume (EDV) is 140 ml, what is the stroke volume (SV)?

5. If the cardiac output (CO) is 5 l/min and the heart rate (HR) is 100 beats/min, what is the strike volume (SV)?

6. What is the difference between the visceral pericardium and the parietal pericardium?

7. What is the purpose of the chordae tendinae, papillary muscles, trabeculae carneae, and pectinate muscles? Where are they located?

8. What three (3) distinct layers comprise the tissues of the heart wall?

9. What are the seven (7) important functions of fibrous skeleton of the heart?

10. Because of specialized sites known as intercalated discs, cardiac muscle is a functional syncytium. What does this statement mean?

11. Beginning with the SA node, trace the pathway of an action potential through the conducting network of the heart. (Use arrows to indicate direction.)

12. What two common clinical problems result from abnormal pacemaker function? Explain what each term means.

13. What three important factors have a direct effect on the heart rate and the force of contraction?

Blood Vessels and Circulation

Overview

Blood is transported through a system of vessels which includes arteries, veins, and capillaries. The arteries are the high pressure conduits, the veins are the low pressure vessels, and the capillaries allow exchange of gases, nutrients, and wastes between the blood and the cells throughout the body. The general plan of the circulatory system consists of a pulmonary circuit which carries deoxygenated blood from the right side of the heart to the lung capillaries where the blood is oxygenated. The oxygenated blood is transported to the left side of the heart via pulmonary veins and carried into the aorta where it will be circulated throughout the systemic circuit to all parts of the body. The cycle nature of blood flow results in deoxygenated blood returning to the right atrium to continue the cycle of circulation.

Blood vessels in the muscles, the skin, the cerebral circulation, and the hepatic portal circulation are specifically adapted to serve the functions of organs and tissues in these specialized areas of cardiovascular activity.

The major topics of the student activities in this chapter include the anatomy of blood vessels, cardiovascular physiology, regulations and patterns of response. You will be expected to be able to identify the major blood vessels and know the structural and functional interactions among the cardiovascular system and other body systems.

Review of Chapter Objectives

1. Distinguish among the types of blood vessels on the basis of their structure and function.
2. Explain the mechanisms that regulate blood flow through arteries, capillaries, and veins.
3. Discuss the mechanisms and various pressures involved in the movement of fluids between capillaries and interstitial spaces.
4. Describe the factors that influence blood pressure and how blood pressure is regulated.
5. Describe how central and local control mechanisms interact to regulate blood flow and pressure in tissues.
6. Explain how the activities of the cardiac, vasomotor, and respiratory centers are coordinated to control the blood flow through the tissues.
7. Explain how the circulatory system responds to the demands of exercise and hemorrhaging.

8. Identify the major arteries and veins and the areas they serve.
9. Describe the age-related changes that occur in the cardiovascular system.
10. Discuss the structural and functional interactions among the cardiovascular system and other body systems.

Part I: Objective Based Questions

Objective 1 Distinguish among the types of blood vessels on the basis of their structure and function.

_____ 1. The layer of vascular tissue that consists of an endothelial lining and an underlying layer of connective tissue dominated by elastic fibers is the

a. tunica interna
b. tunica media
c. tunica externa
d. tunica adventitia

_____ 2. Smooth muscle fibers in arteries and veins are found in the

a. endothelial lining
b. tunica externa
c. tunica interna
d. tunica media

_____ 3. One of the major characteristics of arteries supplying peripheral tissues is that they are

a. elastic
b. muscular
c. rigid
d. all of the above

_____ 4. The blood vessels that play the most important role in the regulation of blood flow to a tissue and blood pressure are the

a. arteries
b. veins
c. arterioles
d. capillaries

_____ 5. The blood vessels which have valves are the

a. arteries
b. arterioles
c. capillaries
d. veins

_____ 6. The vessels that carry blood to the heart are the

a. veins
b. arteries
c. capillaries
d. arterioles

7. In Figure 14-1 identify the following blood vessels on the basis of their structural features. Place your answers in the spaces provided below the drawings.

Figure 14-1 Types of Blood Vessels

artery **vein** **capillaries**
arterioles **venules**

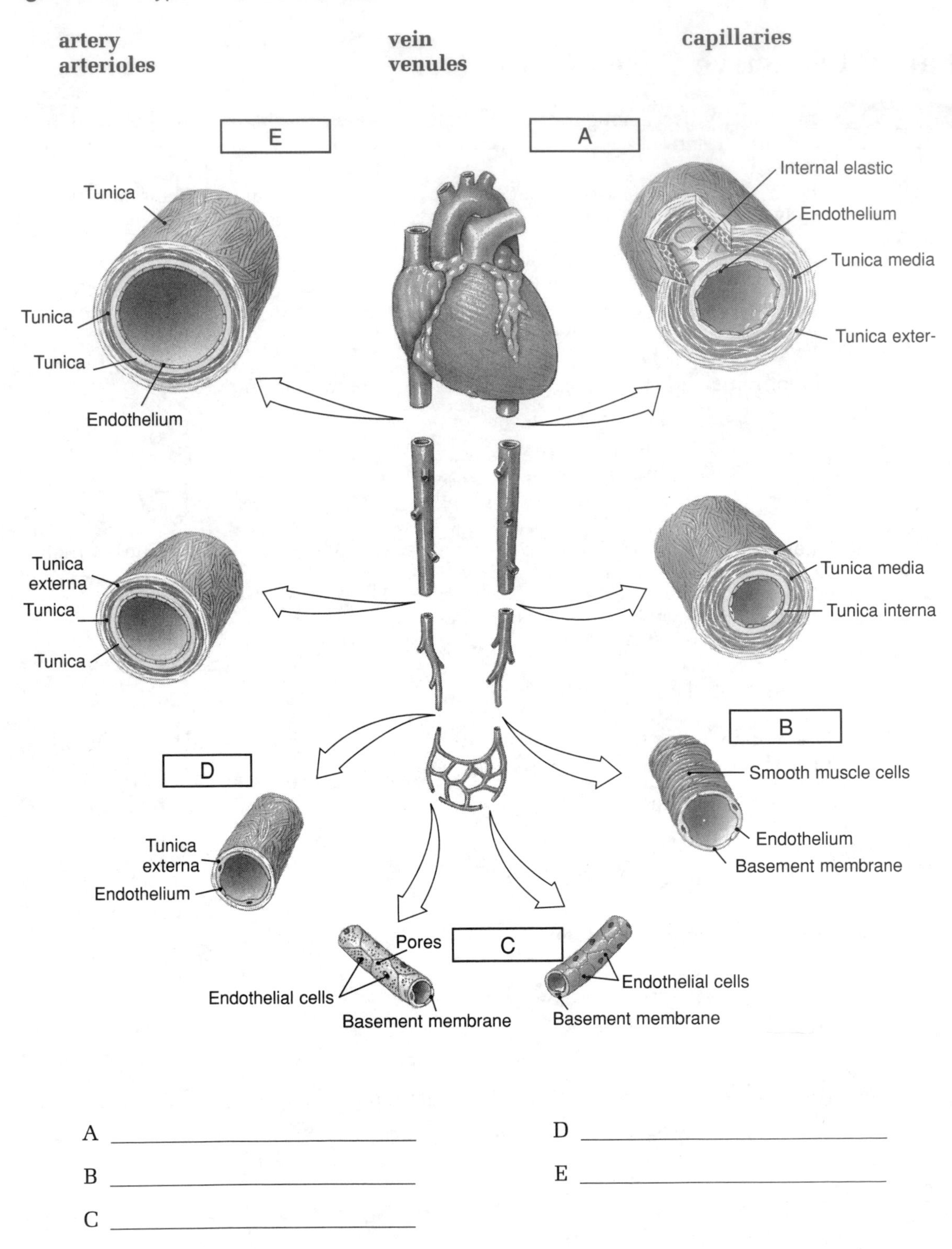

A ______________________ D ______________________

B ______________________ E ______________________

C ______________________

Objective 2 Explain the mechanisms that regulate blood flow through arteries, capillaries, and veins.

___ 1. Blood flow through a capillary is regulated by the

a. endothelium
b. action of the heart
c. precapillary sphincter
d. smooth muscle tissue

___ 2. Blood moves back to the heart through veins

a. because the pressure in the veins is lower than in the arteries
b. with the aid of contractions of skeletal muscles
c. with the aid of thoracic cavity pressure changes
d. all of the above

___ 3. Blood flow through the circulatory system is affected by

a. pressure differences
b. the viscosity of the blood
c. the length and diameter of the blood vessels
d. all of the above

___ 4. Of the following blood vessels, the greatest resistance to blood flow occurs in the

a. veins
b. capillaries
c. arterioles
d. venules

___ 5. The most important determinant of vascular resistance is

a. a combination of neural and hormonal mechanisms
b. differences in the length of the blood vessels
c. friction between the blood and the vessel walls
d. the diameter of the arterioles

Objective 3 Discuss the mechansims and various pressures involved in the movement of fluids between capillaries and interstitial spaces.

___ 1. Blood osmotic pressure is most affected by changes in the concentration of plasma

a. sodium ions
b. proteins
c. metabolic wastes
d. excessive glucose

___ 2. Blood hydrostatic pressure forces fluid from the capillary into the interstitial space.

a. true
b. false

___ 3. The blood hydrostatic pressure and the blood osmotic pressure are equal in magnitude but opposite in direction.

a. true
b. false

_____ 4. Blood osmotic pressure moves fluid from the interstitial space into the capillary.

a. true
b. false

5. The force that pushes water molecules out of solution is ___________.

6. A force that pulls water into a solution is ___________.

Objective 4 Describe the factors that influence blood pressure and how blood pressure is regulated.

_____ 1. Venous pressure is produced by

a. the skeletal muscle pump
b. increasing sympathetic activity to the veins
c. increasing respiratory movements
d. all of the above

_____ 2. From the following selections, choose the one that correctly identifies all the factors which would increase blood pressure. (Note: CO=Cardiac Output; SV=Stroke Volume; VR=Venous Return; PR=Peripheral Resistance; BV=Blood Volume)

a. increased CO; increased SV; decreased VR; decreased PR; increased BV
b. increased CO; increased SV; increased VR; increased PR; increased BV
c. increased CO; increased SV; decreased VR; increased PR; decreased BV
d. increased CO; decreased SV; increased VR; decreased PR; increased BV

_____ 3. As blood travels from the aorta towards the capillaries the

a. pressure increases
b. resistance increases
c. viscosity increases
d. all of the above

_____ 4. Blood pressure is determined by measuring the

a. rate of the pulse
b. pressure in the left ventricle
c. degree of turbulence in a closed blood vessel
d. force exerted by blood in a vessel against air in a closed cuff

_____ 5. Blood pressure increases with increased

a. cardiac output
b. peripheral resistance
c. blood volume
d. all of the above

_____ 6. The difference between the systolic and diastolic pressure is called the

a. mean arterial pressure
b. pulse pressure
c. blood pressure
d. circulatory pressure

Objective 5 Describe how control and local control mechanisms interact to regulate blood flow and pressure in tissues.

_____ 1. The two major factors affecting blood flow rates are

a. diameter and length of blood vessels
b. pressure and resistance
c. neural and hormonal control mechanisms
d. turbulence and viscosity

_____ 2. The formula F = P/R means: (note: F=Flow; P=Pressure; R=Resistance)

a. increasing P, decreasing R, increasing F
b. decreasing P, increasing R, increasing F
c. increasing P, increasing R, decreasing F
d. decreasing P, decreasing R, increasing F

_____ 3. Atrial natruiretic peptide (ANP) reduces blood volume and pressure by

a. blocking release of ADH
b. stimulating peripheral vasodilation
c. increased water loss by kidneys
d. all of the above

_____ 4. Of the following selections, the one which does not cause an increase in blood pressure is

a. increased levels of aldosterone
b. increased levels of angiotensin II
c. increased blood volume
d. increased levels of ANF

_____ 5. Of the following selections, the one which will not result in increased blood flow to tissues is

a. increased blood volume
b. increased vessel diameter
c. increased blood pressure
d. decreased peripheral resistance

Objective 6 Explain how the activities of the cardiac, vasomotor, and respiratory centers are coordinated to control the blood flow through the tissues.

_____ 1. The central regulation of cardiac output primarily involves the activities of the

a. somatic nervous system
b. autonomic nervous system
c. central nervous system
d. all of the above

_____ 2. An increase in cardiac output normally occurs during

a. widespread sympathetic stimulation
b. widespread parasympathetic stimulation
c. the process of vasomotion
d. stimulation of the vasomotor center

_____ 3. Stimulation of the vasomotor center in the medulla causes _________ , and inhibition of the vasomotor center causes ___________.

a. vasodilation; vasoconstruction
b. increased diameter of arterioles; decreased diameter of arterioles
c. hyperemia; ischemia
d. vasoconstruction; vasodilation

_____ 4. Hormonal regulation by vasopressin, epinephrine, angiotensin II, and norepinephrine results in

a. increasing peripheral vasodilation
b. decreasing peripheral vasoconstruction
c. increasing peripheral vasoconstruction
d. all of the above

_____ 5. Cardiovascular function are regulated by

a. neural factors
b. venous return
c. endocrine factors
d. all of the above

_____ 6. Baroreceptors that function in the regulation of blood pressure are located in the

a. carotid sinus
b. left ventricle
c. brain stem
d. common iliac artery

Objective 7 Explain how the circulatory system responds to the demands of exercise and hemorrhaging.

_____ 1. The three primary interrelated changes that occur as exercise begins are

a. decreased vasodilation; increased venous return; increased cardiac output
b. increased vasodilation; decreased venous return; increased cardiac output
c. increased vasodilation; increased venous return; increased cardiac output
d. decreased vasodilation; decreased venous return; decreased cardiac output

_____ 2. The only area of the body where the blood supply is unaffected exercising at maximum levels is the

a. hepatic portal circulation
b. pulmonary circulation
c. brain
d. peripheral circulation

_____ 3. In response to hemorrhage, there is

a. decreased vasomotor tone
b. increased parasympathetic stimulation of the heart
c. mobilization of the venous reserve
d. all of the above

_____ 4. Symptoms of shock include

a. decreased urine formation
b. rapid, weak, pulse
c. acidosis
d. all of the above

_____ 5. Homeostatic mechanisms can compensate for circulatory shock during the

a. ischemic stage
b. progressive stage
c. compensated stage
d. reversible stage

Objective 8 Identify the major arteries and veins and the areas they serve.

1. The ring-shaped anastomosis which forms the cerebral arterial circle is known as the ____________.

2. The four large blood vessels, two from each lung, that empty into the left atrium, completing the pulmonary circuit are the ____________.

3. The blood vessels that provide blood to capillary networks that surround the alveoli in the lungs are the ____________.

4. Blood from the brain returns to the heart by way of the ____________.

5. After passing the first rib, the subclavian artery becomes the ____________ artery.

6. In the upper arm, the axillary artery becomes the ____________ artery.

7. The brachial artery branches to form the __________ and __________ arteries.

8. The two vertebral arteries fuse to form the large __________ artery.

9. The __________ divides the aorta into a superior thoracic aorta and an inferior abdominal aorta.

10. The inferior abdominal aorta branches to form the ____________ arteries.

11. The external iliac artery branches to form the ________ and __________ artery.

12. The blood vessel that receives blood from the head, neck, chest, shoulders, and arms is the ________.

13. The dural sinuses collect blood from the ____________.

14. The radial and ulnar veins fuse to form the __________ vein.

15. The vein that is formed from the fusion of the subclavian with the internal and external jugulars is the ____________ vein.

16. The fusion of the brachiocephalic veins forms the ____________.

17. When the popliteal vein reaches the femur, it becomes the ____________ vein before penetrating the abdominal wall.

18. The two common iliac veins form the ____________.

19. Nutrients from the digestive tract enter the ____________ vein.

20. The large blood vessel that collects most of the venous blood from organs below the diaphragm is the ____________.

21. In the figure below identify and label the major *arteries*. Place your answers in the spaces on page 263.

Figure 14-2 Overview of the Arterial System

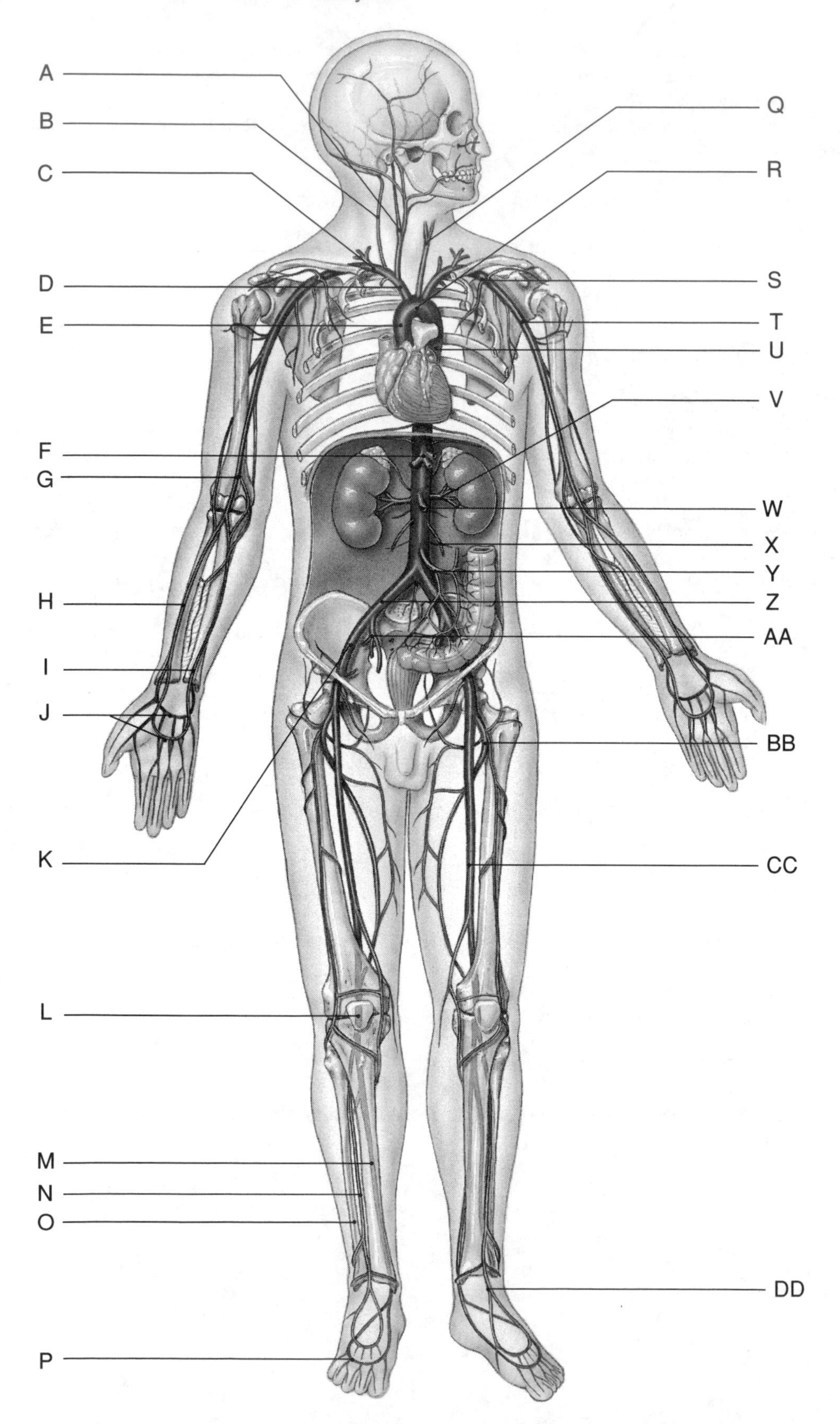

A ______________________________

B ______________________________

C ______________________________

D ______________________________

E ______________________________

F ______________________________

G ______________________________

H ______________________________

I ______________________________

J ______________________________

K ______________________________

L ______________________________

M ______________________________

N ______________________________

O ______________________________

P ______________________________

Q ______________________________

R ______________________________

S ______________________________

T ______________________________

U ______________________________

V ______________________________

W ______________________________

X ______________________________

Y ______________________________

Z ______________________________

AA ______________________________

BB ______________________________

CC ______________________________

DD ______________________________

22. In the figure below, identify and label the major *veins*. Place your answers in the spaces on page 265.

Figure 14-3 Overview of the Venous System

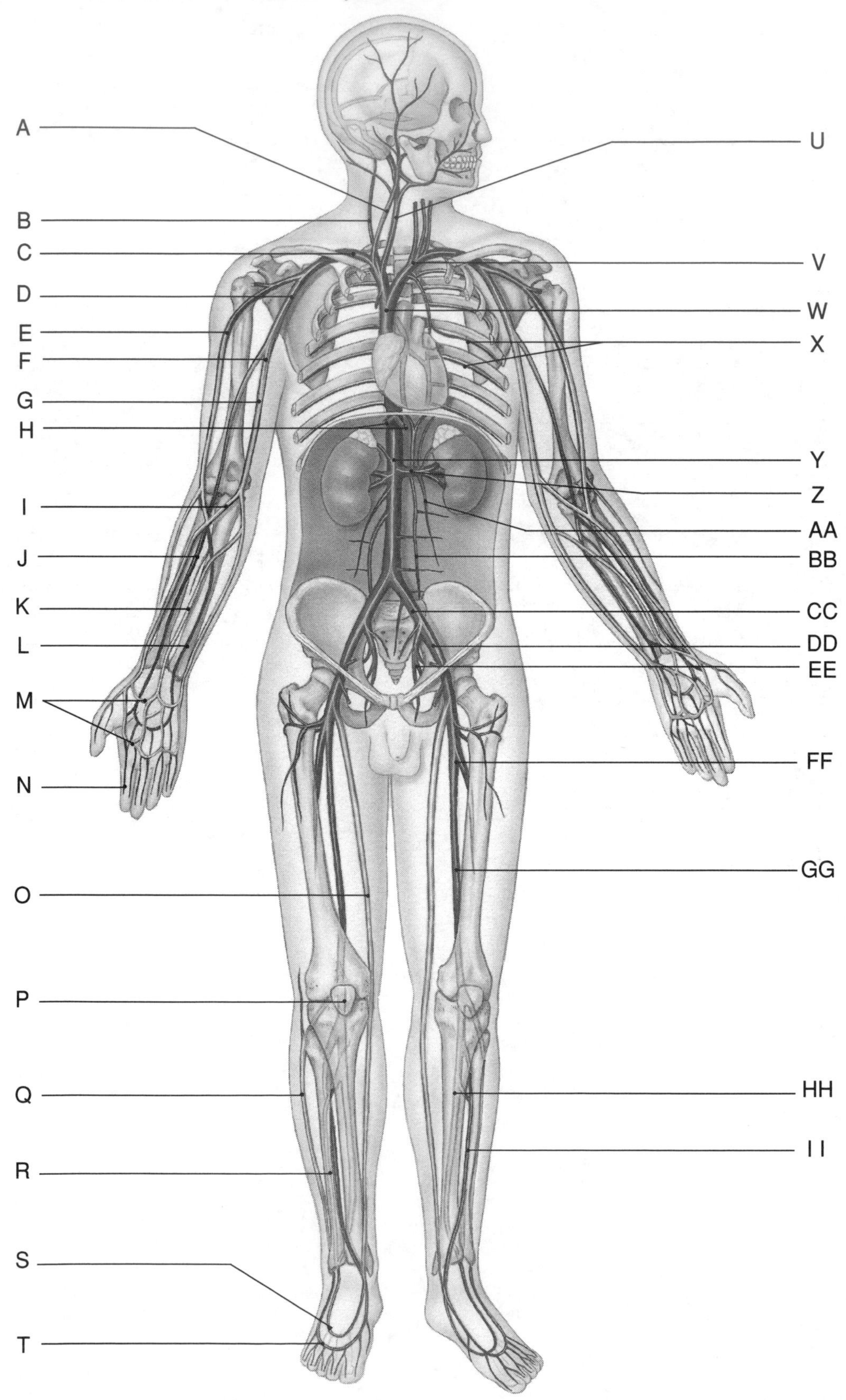

A ______________________

B ______________________

C ______________________

D ______________________

E ______________________

F ______________________

G ______________________

H ______________________

I ______________________

J ______________________

K ______________________

L ______________________

M ______________________

N ______________________

O ______________________

P ______________________

Q ______________________

R ______________________

S ______________________

T ______________________

U ______________________

V ______________________

W ______________________

X ______________________

Y ______________________

Z ______________________

AA ______________________

BB ______________________

CC ______________________

DD ______________________

EE ______________________

FF ______________________

GG ______________________

HH ______________________

II ______________________

23. In the figure below, identify and label the *arteries* which supply blood to the brain. Place your answers in the spaces provided below the drawing. Make your selections from the following choices.

Figure 14-4 Arterial Supply to the Brain

Anterior communicating
Internal carotid
Posterior cerebral
Basilar
Circle of Willis
Anterior cerebral
Posterior communicating
Middle cerebral
Vertebral
Anterior cerebral

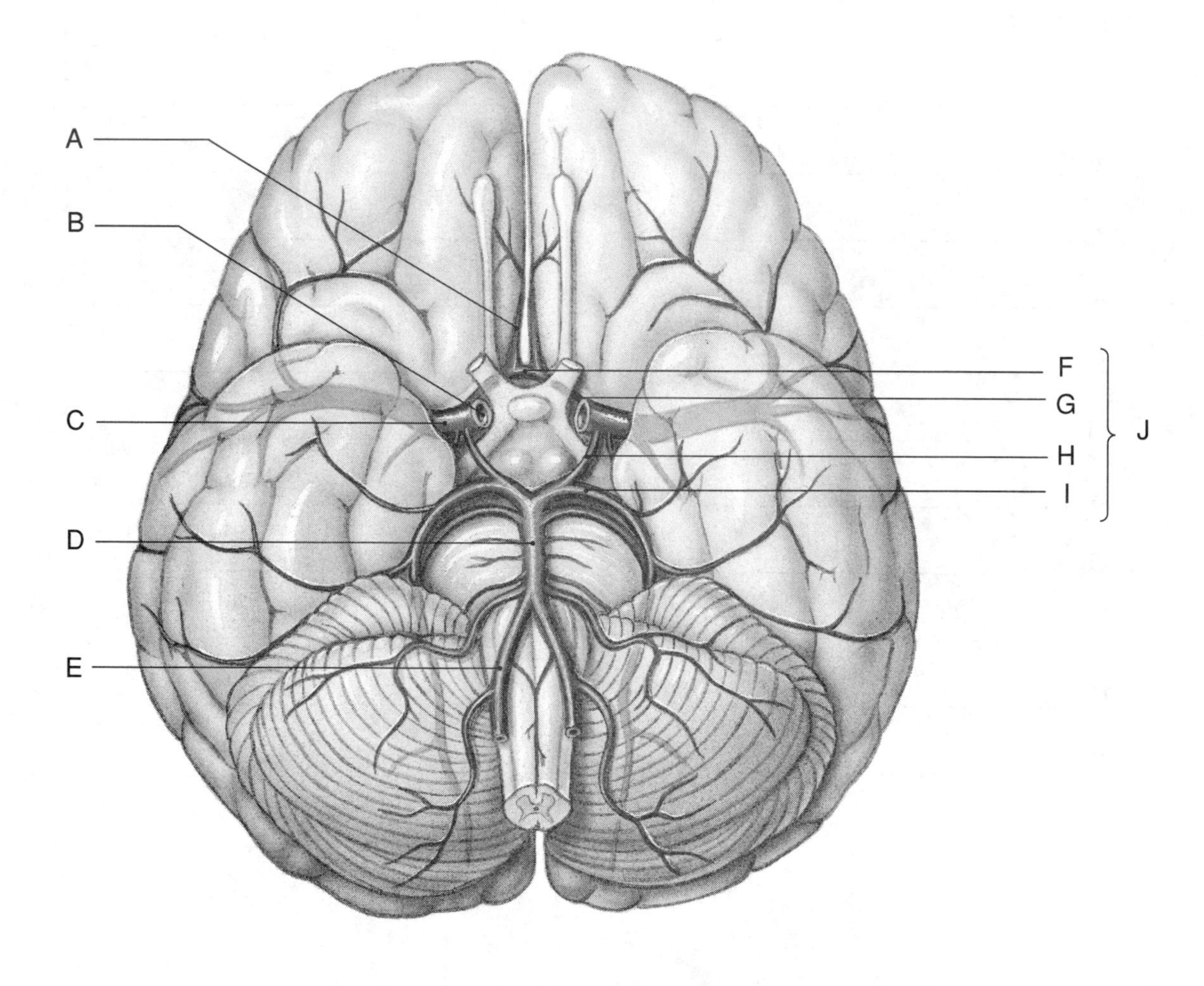

A ______________________

B ______________________

C ______________________

D ______________________

E ______________________

F ______________________

G ______________________

H ______________________

I ______________________

J ______________________

24. In the figure below identify and label the blood vessels and structures of the hepatic portal system. Make your selections from the following list of blood vessels and structures of the digestive system. Place your answers in the spaces below the drawing.

Figure 14-5 The Hepatic Portal System

Aorta	**Liver**	**Spleen**
Pancreas	**Stomach**	**Esophagus**
Colic veins	**Inferior vena cava**	**Left colic vein**
Hepatic portal vein	**Ascending colon**	**Descending colon**
Splenic vein	**Hepatic veins**	**Cystic vein**
Superior mesenteric vein	**Left gastric vein**	**Gastroepiploic veins**
Inferior mesenteric vein	**Sigmoid branches**	**Small intestine**
Superior rectal vein		

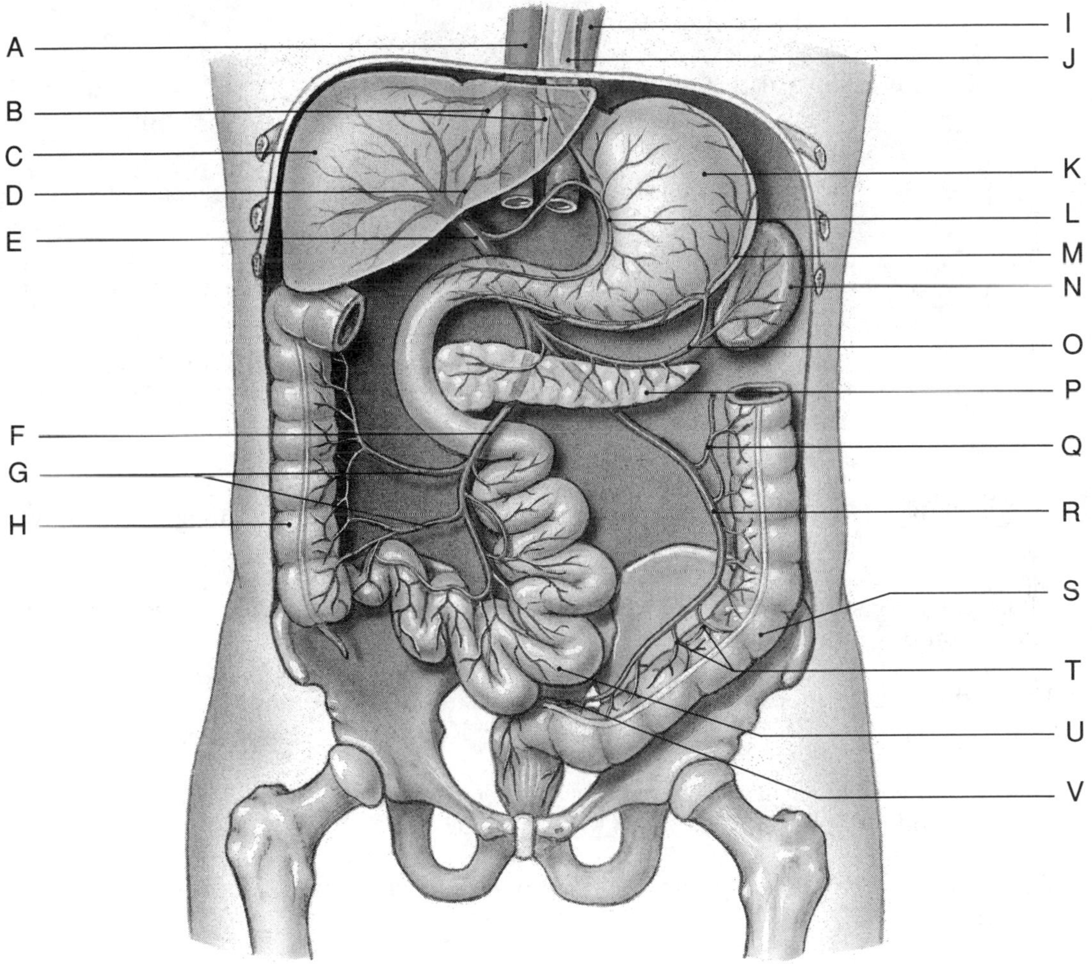
A
B
C
D
E
F
G
H
I
J
K
L
M
N
O
P
Q
R
S
T
U
V

A ______________ I ______________ P ______________

B ______________ J ______________ Q ______________

C ______________ K ______________ R ______________

D ______________ L ______________ S ______________

E ______________ M ______________ T ______________

F ______________ N ______________ U ______________

G ______________ O ______________ V ______________

H ______________

Objective 9 Describe the age-related changes that occur in the cardiovascular system.

_____ 1. Elderly individuals usually have

a. elevated hematocrits
b. stiff inelastic arteries
c. decreased blood pressure
d. all of the above

_____ 2. Elderly individuals are more prone to suffer from __________ than younger individuals.

a. hypertension
b. venous thrombosis
c. arteriosclerosis
d. all of the above

_____ 3. The primary cause of varicose veins is

a. improper diet
b. aging
c. swelling due to edema in the veins
d. inefficient venous valves

_____ 4. The primary effect of a decrease in the hematocrit of elderly individuals is

a. thrombus formation in the blood vessels
b. a lowering of the oxygen-carrying capacity of the blood
c. a reduction in the maximum cardiac output
d. damage to ventricular cardiac muscle fibers

Objective 10 Discuss the structural and functional interactions among the cardiovascular system and the other body systems.

1. Of all the body systems, the most extensive communication occurs between the cardiovascular and __________ systems.

2. The system which modifies heart rate and regulates blood pressure is the __________ system.

3. The system which releases renin to elevate blood pressure and erythropoietin to accelerate RBC production is the __________ system.

4. The system which may slow development of arteriosclerosis with age is the __________ system.

5. The system which stores calcium needed for normal cardiac muscle contraction is the __________ system.

Part II: Chapter Comprehensive Exercises

A. Word Elimination

Circle the term that does not belong in each of the following groupings.

1. arteries valves arterioles venules veins
2. tunica interna tunica media tunica externa tunica adventitia tunica lumen
3. autoregulation ANS compression endocrine factors hormones
4. baroreceptors increased acidity rise in CO_2 decreased plasma O_2 decreased pH
5. NE ACh ADH EPO ANP
6. hypotension acidosis "clammy" skin increased pH disorientation
7. aortic arch pulmonary vein L. common carotid brachiocephalic L. subclavian
8. carotid A. axillary A. brachial A. radial A. ulnar A.
9. median cubital cephalic radial phrenic median antebrachial
10. decreased CO increased hematocrit decreased elasticity increased scar tissue athersclerosis

B. Matching

Match the terms in Column B with the terms in Column A. Use letters for answers in the spaces provided.

COLUMN A	COLUMN B
___ 1. vasa vasorum	A. changes in blood pressure
___ 2. metarteriole	B. preferred channel
___ 3. systole	C. vasomotor center
___ 4. diastole	D. weakened vascular wall
___ 5. baroreceptors	E. vasodilation
___ 6. mean arterial pressure	F. brain circulation
___ 7. atrial natriuretic peptide	G. peak blood pressure
___ 8. circle of Willis	H. "vessels of vessels"
___ 9. medulla	I. single value for blood pressure
___ 10. aneurysm	J. minimum blood pressure

C. Concept Map I: The Cardiovascular System

Using the following terms, fill in the numbered, blank spaces to complete the concept map. Follow the numbers that comply with the organization of the map, moving in a "clock-wise" direction from top to bottom and back to top.

Systemic circuit
Pulmonary veins
Veins and venules
Arteries and arterioles
Pulmonary arteries

The cardiovascular system

(CO_2) Lung capillaries (O_2)

4

1

Pulmonary circuit

Pulmonary trunk

Heart

Vena cavae

Aortic arch

5

Aorta

3

2

(CO_2) Systemic capillaries (O_2)

—— Oxygenated blood

- - - - Deoxygenated blood

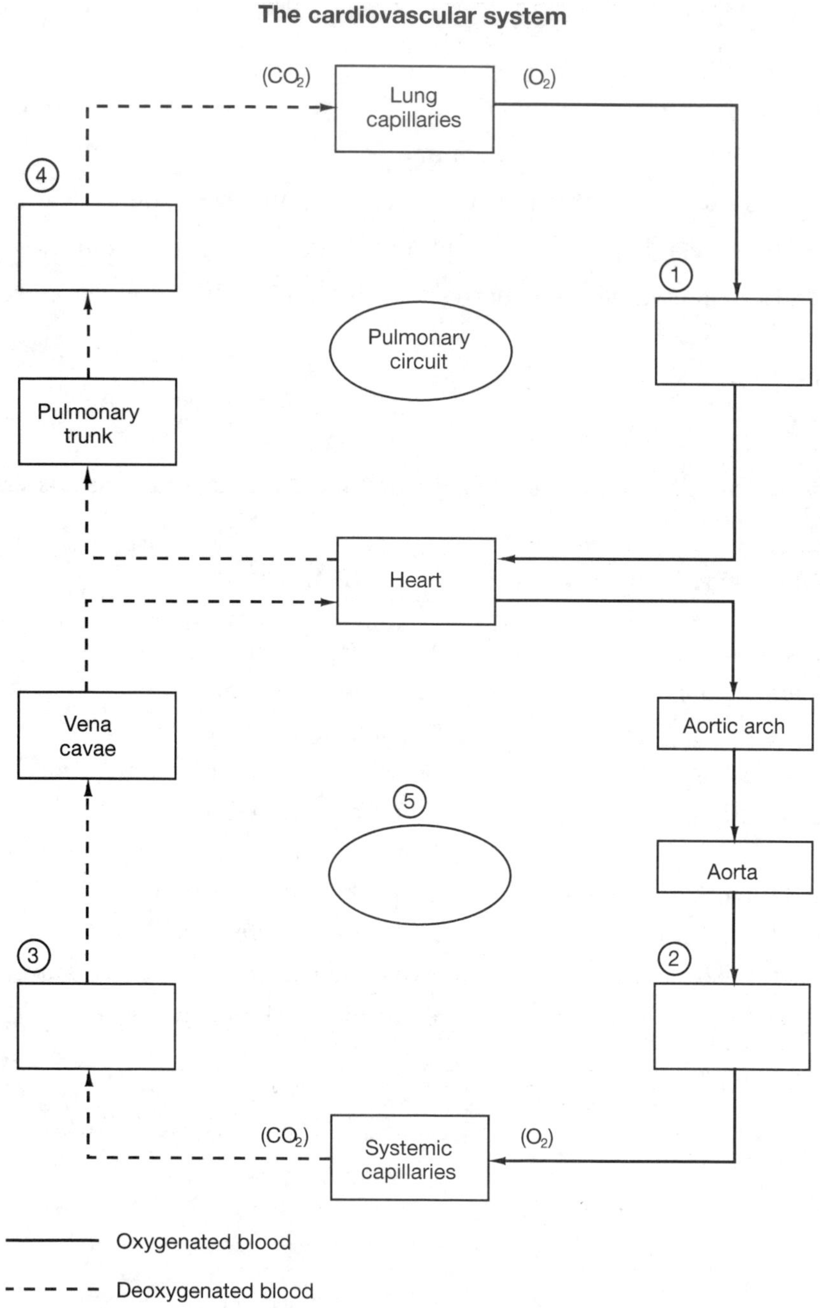

Concept Map II: Major Branches of the Aorta

Using the following terms, fill in the circled, numbered, blank spaces to complete the concept map. Follow the numbers that comply with the organization of the map.

L. Subclavian artery
Ascending aorta
Superior mesenteric artery
R. Gonadal artery
L. common iliac artery
Celiac trunk
Thoracic aorta
Brachiocephalic artery

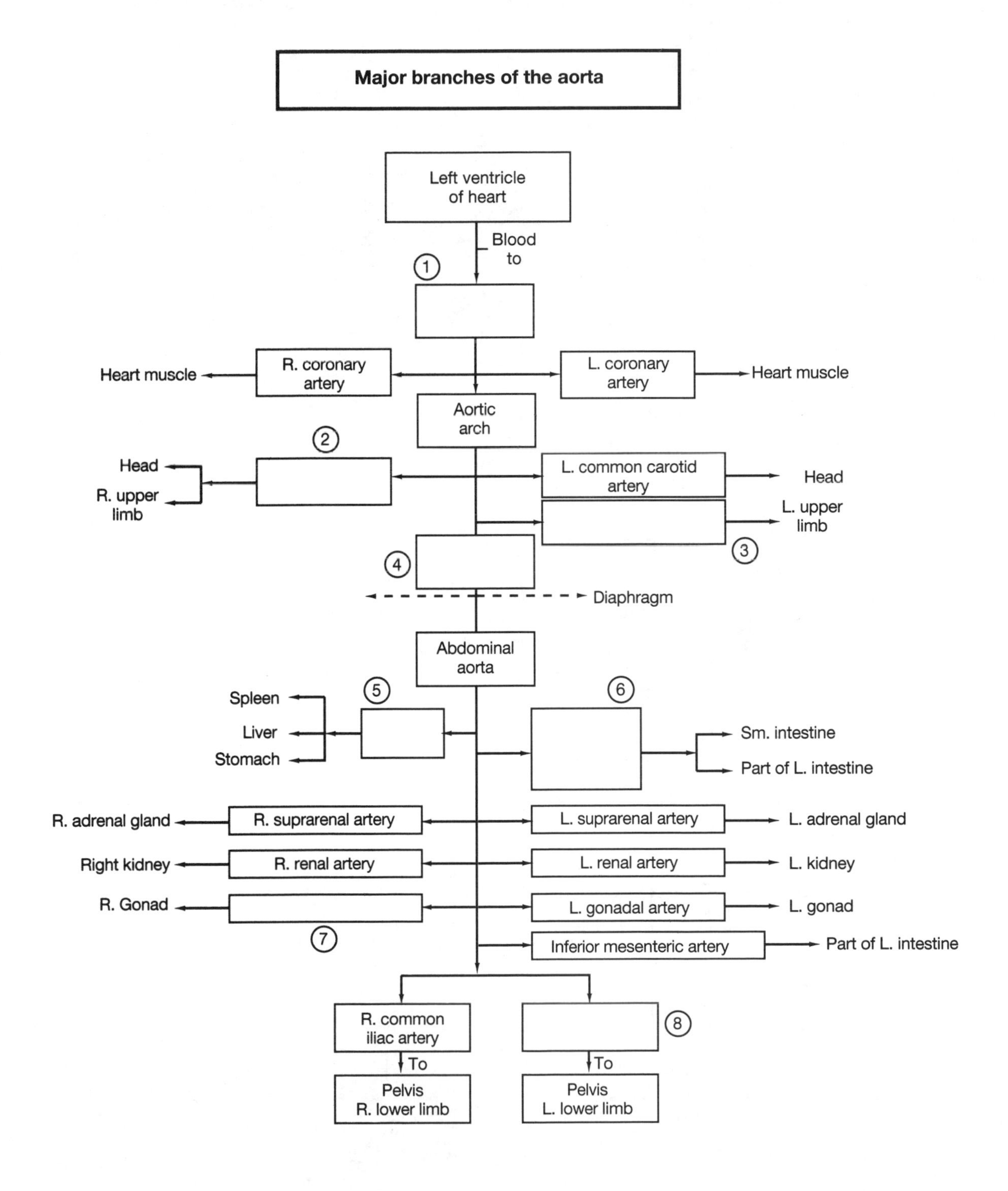

Concept Map III: Major Veins

Using the following terms, fill in the circled, numbered, blank spaces to complete the concept map. Follow the numbers that comply with the organization of the map. (Note the direction of the arrows on this map.)

Azygous vein	**L. hepatic veins**	**L. common iliac vein**
superior vena cava	**L. renal vein**	**R. suprarenal vein**

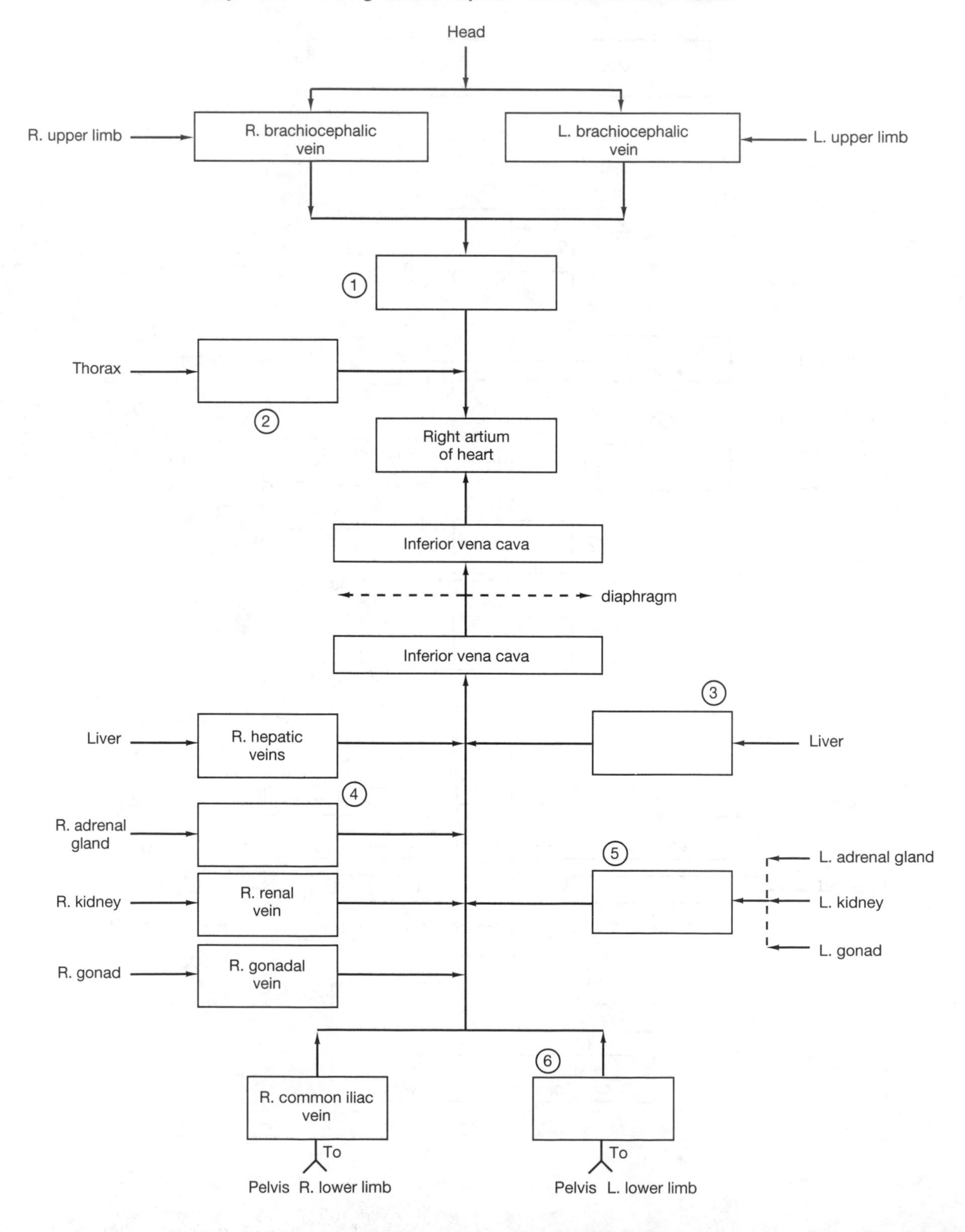

D. Crossword Puzzle

This crossword puzzle reviews the material in Chapter 14. To complete the puzzle you must know the answers to the clues given, and you must be able to spell the terms correctly.

ACROSS

2. Small air pockets surrounded by capillary networks.
4. Stimulates water reabsorption from urine.
6. Carry blood away from heart.
7. Alternating contraction and relaxation.
8. Minimum blood pressure.
9. Peak blood pressure.
10. Reflexes which respond to O_2 levels in the blood.
12. Movement of water across a semipermeable membrane.
13. Hormone produced by specialized cardiac cells.

DOWN

1. Aortic and carotid sinuses.
3. Permit exchanges between blood and interstitial fluid.
5. Only blood vessels which contain valves.
6. Stimulates water conservation at kidneys (abbreviation).
7. Carry blood to the heart.
11. Stimulates red blood cell production.

E. Short Answer Questions

Briefly answer the following questions in the spaces provided below.

1. List the types of blood vessels in the cardiovascular tree and briefly describe their anatomical associations (use arrows to show this relationship).

2. (a) What are sinusoids?

   (b) Where are they found?

   (c) Why are they important functionally?

3. Relative to gaseous exchange, what is the primary difference between the pulmonary circuit and the systemic circuit?

4. (a) What are the three (3) primary sources of peripheral resistance?

(b) Which one can be adjusted by the nervous or endocrine system to regulate blood flow?

5. Symbolically summarize the relationship among blood pressure (BP), peripheral resistance (PR), and blood flow (F). State what the formula means.

6. Explain what is meant by: BP = 120 mm Hg/80 mm Hg

7. What is the mean arterial pressure (MAP) if the systolic pressure is 110 mm Hg and the diastolic pressure is 80 mm Hg?

8. What are the three (3) primary factors that influence blood pressure and blood flow?

9. What three (3) major baroreceptor populations enable the cardiovascular system to respond to alterations in blood pressure?

10. What hormones are responsible for long-term and short-term regulation of cardiovascular performance?

11. How does arteriosclerosis affect blood vessels?

CHAPTER

15

The Lymphatic System and Immunity

Overview

All the organ systems in the human body structurally and functionally interact in an effort to keep us alive and healthy. In this battle for survival, the lymphatic system plays a major role. The major enemies of the body's system of defense include an assortment of viruses, bacteria, fungi, and parasites. These potential disease causing pathogens are responsible for many human diseases. The lymphatic system protects against disease causing organisms primarily by the activities of lymphocytes. It is also responsible for return of fluid and solutes from peripheral tissues to the blood, and distribution of hormones, nutrients, and waste products from their tissues of origin to venous circulation.

Structurally, the lymphatic system consists of (1) a fluid called *lymph*; (2) a network of *lymphatic vessels*; (3) specialized cells called *lymphocytes*; and (4) *lymphoid organs* and *tissues* distributed throughout the body. Chapter 15 provides exercises which focus on topics that include the organization of the lymphatic system, the body's defense mechanisms, patterns of immune response, and the interactions between the lymphatic system and the other systems of the body.

Review of Chapter Objectives

1. Identify the major components of the lymphatic system and explain their functions.
2. Discuss the importance of lymphocytes and describe where they are found in the body.
3. List the body's nonspecific defenses and explain how each functions.
4. Define specific resistance and identify the forms and properties of immunity.
5. Distinguish between cell-mediated immunity and antibody-mediated (humoral) immunity.
6. Discuss the different types of T cells and the role played by each in the immune response.
7. Describe the structure of antibody molecules and explain how they function.
8. Describe the primary and secondary responses to antigen exposure.
9. Relate allergic reactions and autoimmune disorders to immune mechanisms.
10. Describe the changes in the immune system that occur with aging.
11. Discuss the structural and functional interactions among the lymphatic system and other body systems.

Part I: Objective Based Questions

Objective 1 Identify the major components of the lymphatic system and explain their functions.

_____ 1. The major components of the lymphatic system include

a. lymph nodes, lymph, lymphocytes
b. spleen, thymus, tonsils
c. thoracic duct, R. lymphatic duct, lymph nodes
d. lymphatic vessels, lymph, lymphatic organs

_____ 2. Lymphatic *organs* found in the lymphatic system include

a. thoracic duct, R. lymphatic duct, lymph nodes
b. lymphatic vessels, tonsils, lymph nodes
c. spleen, thymus, lymph nodes
d. all of the above

_____ 3. The *primary* function of the lymphatic system is

a. transporting of nutrients and oxygen to tissues
b. removal of carbon dioxide and waste products from tissues
c. regulation of temperature, fluid, electrolytes, and pH balance
d. production, maintenance, and distribution of lymphocytes

_____ 4. The lymphatic system

a. helps maintain normal blood volume
b. fights infection
c. eliminates variations in the composition of interstitial fluid
d. all of the above

_____ 5. Anatomically, lymph vessels resemble

a. arterioles
b. veins
c. elastic arteries
d. the vena cava

6. Most of the lymph returns to the venous circulation by way of the _____________.

7. The structures in the lymphatic system which act as a "way station" for cancer cells are the _____________.

8. Identify and label the following structures of the lymphatic system. Place your answers in the spaces provided below the drawing.

Figure 15-1 The Lymphatic System

inguinal lymph nodes	**R. lymphatic duct**	**axillary lymph nodes**
lumbar lymph nodes	**thymus**	**spleen**
thoracic duct	**L. lymphatic duct**	

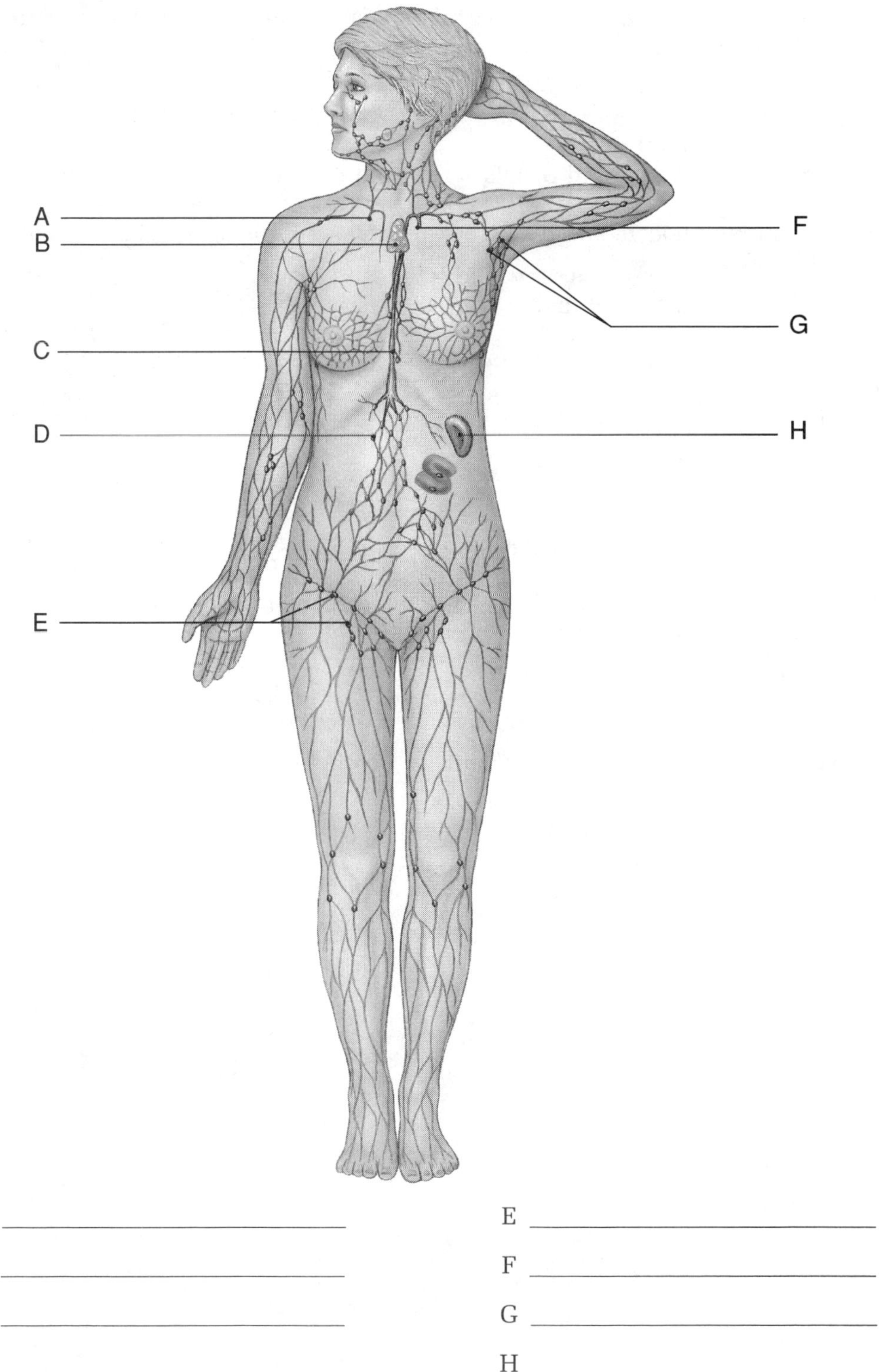

A ______________________________ E ______________________________

B ______________________________ F ______________________________

C ______________________________ G ______________________________

D ______________________________ H ______________________________

Objective 2 Discuss the importance of lymphocytes and describe where they are found in the body.

_____ 1. The white pulp of the spleen is composed primarily of

a. lymphocytes
b. neutrophils
c. red blood cells
d. platelets

_____ 2. Lymphocytes that assist in the regulation and coordination of the immune response are

a. plasma cells
b. helper T and suppressor T cells
c. B cells
d. NK and B cells

_____ 3. Normal lymphocyte population are maintained through lymphpoiesis in the

a. bone marrow and lymphatic tissues
b. lymph in the lymphatic tissues
c. blood and the lymph
d. spleen and liver

_____ 4. The type(s) of lymphocytes that produce antibodies are

a. NK cells
b. T cells
c. B cells
d. all of the above

_____ 5. B cells, NK cells, and T cells can leave the circulatory system to enter the body tissues to "fight" invaders.

a. true
b. false

_____ 6. The red pulp of the spleen contains large numbers of

a. red blood cells
b. macrophages
c. neutrophils
d. antibodies

Objective 3 List the body's nonspecific defenses and explain how each functions.

_____ 1. Of the following selections, the one that includes only nonspecific defenses is

a. T and B cell activation, complement, inflammation, phagocytosis
b. hair, skin, mucous membranes, antibodies
c. hair, skin, complement, inflammation, phagocytosis
d. antigens, antibodies, complement, macrophages

_____ 2. NK (natural killer) cells sensitive to the presence of abnormal cell membranes are primarily involved with

a. defenses against specific threats
b. complex and time-consuming defense mechanisms
c. phagocytic activity for defense
d. immunological surveillance

_____ 3. The protein that interferes with the replication of viruses is

a. complement proteins
b. heparin
c. interferon
d. pyrogens

_____ 4. Of the following selections, the one which *is not* a physical barrier is

a. hair
b. complement
c. epithelium
d. secretions

_____ 5. Of the selections, the one which *is not* a part of the non-specific immunity strategy is

a. B cells
b. fever
c. interferon
d. macrophages

6. Fill in the blanks (A through G) with the type of nonspecific defense strategy. Place your answers in the spaces provided below the drawing.

Figure 15-2 Nonspecific Defenses

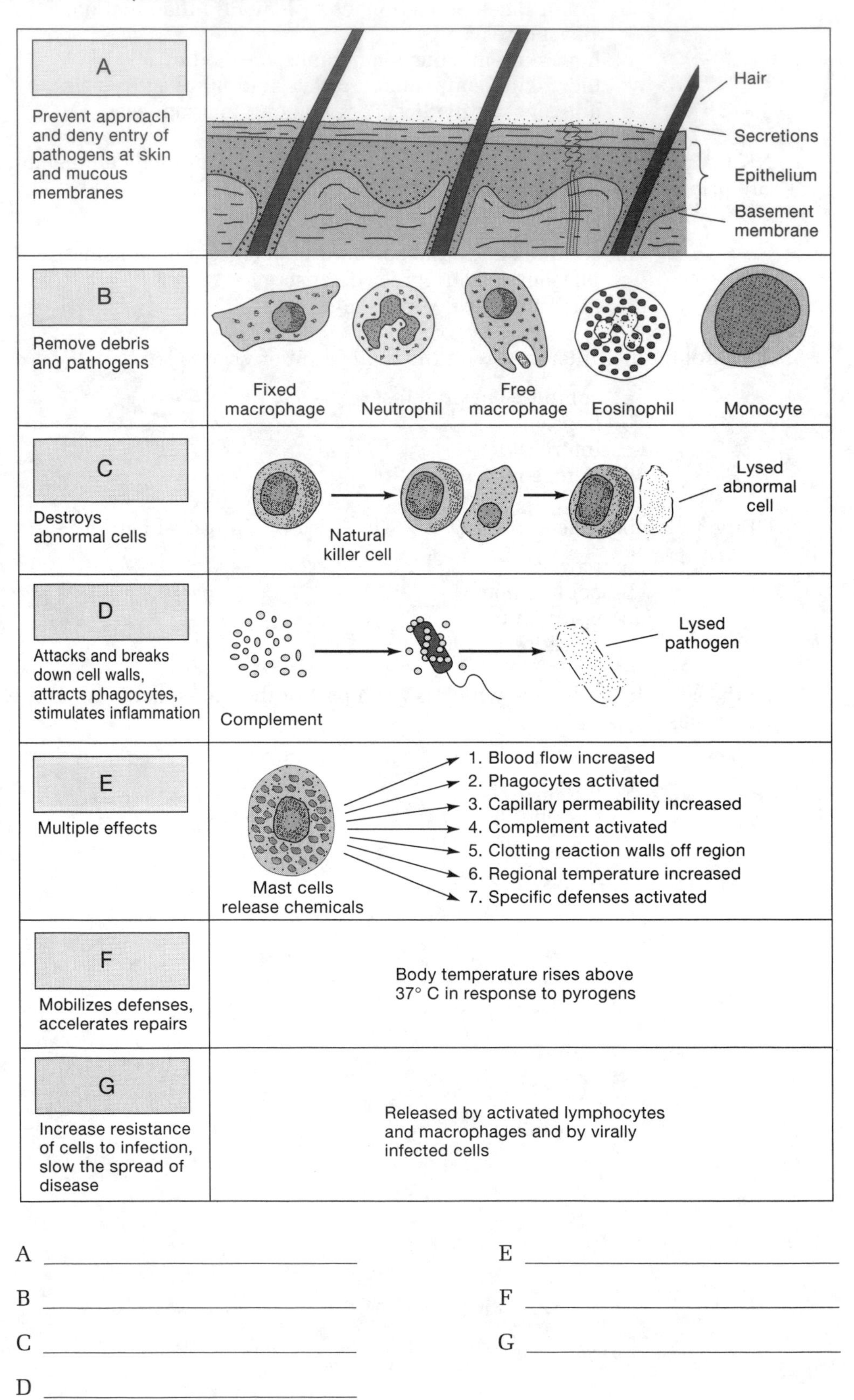

A ______________________________ E ______________________________

B ______________________________ F ______________________________

C ______________________________ G ______________________________

D ______________________________

Objective 4 Define specific resistance and identify the forms and properties of immunity.

_____ 1. The four general characteristics of specific defenses include

a. specificity, versatility, memory, and tolerance
b. innate, active, acquired, passive
c. accessibility, recognition, compatibility, immunity
d. all of the above

_____ 2. The two major ways the body "carries out" the immune response are

a. phagocytosis and inflammation
b. immunological surveillance and fever
c. direct attack by T cells and circulating antibodies
d. physical barriers and complement

_____ 3. A specific defense mechanism is always activated by

a. inflammation
b. an antibody
c. fever
d. an antigen

_____ 4. The first line of cellular defense against pathogens is

a. T cells
b. B cells
c. phagocytes
d. NK cells

_____ 5. Immunity resulting from natural exposure to an antigen in the environment is called

a. passive natural
b. active natural
c. autoimmunity
d. innate

_____ 6. Immunity that results from antibodies that pass the placenta from mother to fetus is called

a. active artificial
b. autoimmunity
c. passive natural
d. active natural

_____ 7. In active *artificial immunity* the

a. immune system attacks normal body cells
b. body is deliberately exposed to an antigen
c. body receives antibodies produced by another person
d. body receives antibodies produced by another animal

Objective 5 Distinguish between cell-mediated immunity and antibody-mediated (humoral) immunity.

_____ 1. When an antigen appears, the immune response begins with the

a. presence of immunoglobulins in body fluids
b. release of endogenous pyrogens
c. activation of the complement system
d. activation of specific T cells and B cells

_____ 2. When the immune "recognition" system malfunctions, activated B cells begin to

a. manufacture antibodies against other cells and tissues
b. activate cytotoxic T killer cells
c. secrete lymphotoxins to destroy foreign antigens
d. recall memory T cells to initiate the proper response

_____ 3. T cells are involved with _______ and ______ attack pathogens

a. cell-mediated responses; directly
b. cell-mediated responses; indirectly
c. humoral responses; directly
d. humoral responses; indirectly

_____ 4. B cells are involved with ________ and create a chemical attack on ________.

a. cell-mediated responses; antigens
b. cell-mediated responses; pathogens
c. humoral responses; antibodies
d. humoral responses; antigens

_____ 5. The cells responsible for the production of circulating antibodies are

a. NK cells
b. plasma cells
c. helper T cells
d. cytotoxic T cells

Objective 6 Discuss the different types of T cells and the role played by each in the immune response.

_____ 1. T-cell activation leads to the formation of cytotoxic T cells and memory T cells that provide

a. humoral immunity
b. cellular immunity
c. phagocytosis and immunological surveillance
d. stimulation of inflammation and fever

_____ 2. Stem cells that will form T cells are modified in the

a. bone marrow
b. spleen
c. liver
d. thymus

_____ 3. Suppressor T cells act to

a. suppress antigens
b. limit the degree of memory in memory T cells
c. limit antigen proliferation
d. depress the responses of other T cells and B cells

_____ 4. Cells that help to regulate the immune response are

a. B cells
b. helper T cells
c. plasma cells
d. cytotoxic T cells

Objective 7 Describe the structure of antibody molecules and explain how they function.

_____ 1. An active antibody is shaped like a(n)

a. T
b. Y
c. A
d. B

_____ 2. The most important antibody action(s) in the body is/are

a. alterations in the cell membrane to increase phagocytosis
b. to attract macrophages and neutrophils to the infected areas
c. activation of the complement systems
d. cell lysis and digestion of the cell membrane

_____ 3. The specificity of an antibody is determined by the

a. fixed segment
b. size of the antibodies
c. antigenic determinants
d. variable region

_____ 4. The binding of an antigen to an antibody can result in

a. neutralization of the antigen
b. agglutination or precipitation
c. complement activation
d. all of the above

Objective 8 Describe the primary and secondary responses to antigen exposure.

_____ 1. The antigenic determinant site is the certain portion of the antigens exposed surface where

a. the foreign "body" attacks
b. phagocytosis occurs
c. the antibody attacks
d. the immune surveillance system is activated

_____ 2. In order for an antigenic molecule to be a complete antigen it must

a. be a large molecule
b. be immunogenic and reactive
c. contain a hapten and a small organic molecule
d. be subject to antibody activity

_____ 3. Memory B cells will differentiate into plasma cells if they are exposed to the same antigen a second time.

a. true
b. false

_____ 4. The first exposure to an antigen produces the highest number of circulatory antibodies.

a. true
b. false

_____ 5. The primary response produces more antibodies than the secondary response.

a. true
b. false

6. Small organic molecules that are not antigens by themselves are called _____________.

7. The ability to demonstrate an immune response upon exposure to an antigen is called _____________.

Objective 9 Relate allergic reactions and autoimmune disorders to immune mechanisms.

_____ 1. Autoantibodies

a. are produced by activated T-cells
b. are produced during an allergic reaction
c. function against the body's normal antigens
d. are produced during immunodeficiency diseases

_____ 2. Inappropriate or excessive immune responses to antigens are

a. autoimmune diseases
b. allergies
c. immunodeficiency diseases
d. the result of stress

_____ 3. When an immune response mistakenly targets normal body cells and tissues, the result is

a. immune system failure
b. the development of an allergy
c. depression of the inflammatory response
d. an autoimmune disorder

_____ 4. When the immune system fails to develop normally or the immune response is blocked in some way, the condition is called an

a. allergy
b. inflammation
c. immunodeficiency disease
d. autoimmune disorder

Objective 10 Describe the changes in the immune system that occur with aging.

_____ 1. With advancing age the immune system becomes

a. increasingly susceptible to viral infection
b. increasingly susceptible to bacterial infection
c. less effective at combating disease
d. all of the above

_____ 2. With age B cells become less active due to a reduced number of

a. antibodies
b. cytotoxic T cells
c. helper T cells
d. plasma cells

_____ 3. B cells, helper T cells, and cytotoxic T cells all have reduced activity with age.

a. true
b. false

_____ 4. The aging process is accompanied by more cytotoxic T cells responding to infections.

a. true
b. false

Objective 11 Discuss the structural and functional interactions among the lymphatic system and other body systems.

1. The lymphatic system provides IgA for secretion by epithelial glands in the ______________ system.

2. The lymphatic system provides specific defenses against infection and immune surveillance against cancer for all the body systems.

   a. true
   b. false

3. The tonsils protect against infection at entrance to the ______________ tract.

4. Protection of superficial lymph nodes and the lymphatic vessels in the abdominopelvic cavity is provided by the ______________ system.

5. Lymphocytes and other cells involved in the immune response are produced and stored in the ______________ system.

Part II: Chapter Comprehensive Exercises

A. Word Elimination

Circle the term that does not belong in each of the following groupings.

1. thymus lymph nodes spleen pineal gland tonsils
2. B cells plasma cells antigens T cells NK cells
3. pharyngeal salivary adenoid palatine lingual
4. capsule medulla sinus cortex nephron
5. antibodies phagocytosis complement fever interferons
6. mucus acid blood urine glandular secretions
7. complement innate active acquired passive
8. specificity versatility memory tolerance compatibility
9. IgG IgM IgA IgE IgB
10. MCF interleukins interferons enzymes CSFs

B. Matching

Match the terms in Column B with the terms in Column A. Use letters for answers in the spaces provided.

COLUMN A	COLUMN B
___ 1. macrophages	A. active and passive
___ 2. microphages	B. humoral immunity
___ 3. mast cells	C. coating of antibodies
___ 4. acquired immunity	D. transfer of antibodies
___ 5. specific immunity	E. neutrophils and eosinophils
___ 6. B cells	F. cellular immunity
___ 7. passive immunity	G. innate or acquired
___ 8. cytotoxic T cells	H. monocytes
___ 9. opsonization	I. decline in immune surveillance
___ 10. cancer	J. nonspecific immune response

C. Concept Map I

Using the following terms, fill in the numbered, blank spaces to complete the concept map. Follow the numbers that comply with the organization of the map.

Active immunization	**Passive immunization**	**Acquired**
Nonspecific-immunity	**Inflammation**	**Specific immunity**
Active	**Phagocytic cells**	**Innate**
Transfer of antibodies via placenta		

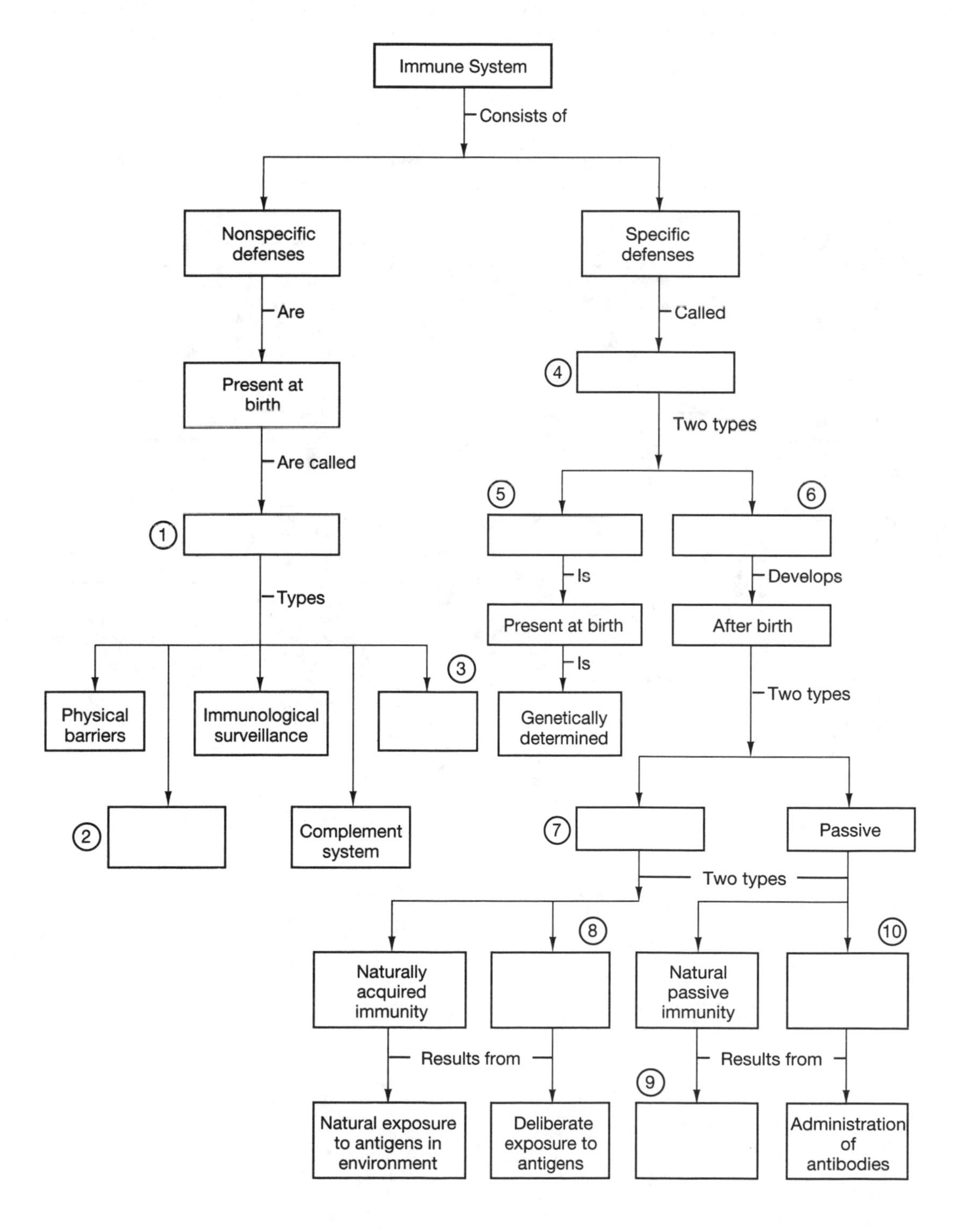

Concept Map II

Using the terms below, fill in the blanks to complete a map through the body's department of defense.

killer T cells	**antibodies**	**natural killer cells**
viruses	**helper T cells**	**suppressor T cells**
B cells	**macrophages**	**memory T and B cells**

You will proceed from statement 1 (question 11) to statement 9 (question 19). At each site you will complete the event that is taking place by identifying the type of cells or proteins participating in the specific immune response. Place your answers in the spaces provided.

The body's department of defense

1 The race is on; (11) ______ try to replicate before the immune system can gear up. Two have already taken over cells in the body.

2 (12) ______ quickly recognize the viruses as a foreign threat. They begin destroying viruses by engulfing them.

3 Stimulated by the release of interleukins from macrophages, Helper T cells, and interferons, (13) ______ join the attack on virally-infected cells. They also fight cancer cells.

4 (14) ______ the battle managers of the immune system), emit signals to B cells and Killer T cells to join the attack.

5 (15) ______ (produced in bones) mature into plasma cells, which in turn produce antibodies.

6 (16) ______ are Y-shaped proteins designed specifically to recognize a particular viral or bacterial invader. Antibodies bind to the virus and neutralize it.

7 (17) ______ wage chemical warfare on virally- infected cells by firing lethal proteins at them.

8 As the body begins to conquer the viruses, (18) ______ help the immune system gear down. Otherwise, it might attack the body.

9 As the viruses are being defeated, the body creates (19) ______ and ______ that circulate permanently in the bloodstream, ensuring that next time, that particular virus will be swiftly conquered.

B B P P Y Y Y Y Y Y T T T T T T T T T N N V V V V V V V V V T B

STOP STOP STOP

D. Crossword Puzzle

The following crossword puzzle reviews the material in Chapter 15. To complete the puzzle, you must know the answers to the clues given, and must be able to spell the terms correctly.

ACROSS

4. Antibodies target and "attack" specific ___________.
6. The type of white blood cell that can differentiate into T or B cells.
7. Inborn immunity is called _________ immunity.
8. A compound that promotes a fever.

DOWN

1. Vaccination is a type of artificially _________ immunity.
2. These cells produce antibodies.
3. Injecting an antigen from a weakened or dead pathogen into the body is a process called ___________.
4. The pharyngeal tonsils are also called the ___________.
5. The _________ gland converts lymphocytes to T cells.
6. After tissue fluid drains into the lymphatic vessels, it is called _____________.

E. Short Answer Questions

Briefly answer the following questions in the spaces provided below.

1. What are the three (3) primary organizational components of the lymphatic system?

2. What three (3) major functions are performed by the lymphatic system?

3. What are the three (3) different classes of lymphocytes found in the blood and where does each class originate?

4. What three (3) kinds of T cells comprise 80 percent of the circulating lymphocyte population? What is each type basically responsible for?

5. What is the primary function of stimulated B cells and what are they ultimately responsible for?

6. What three (3) lymphatic organs are important in the lymphatic system?

7. What are the primary differences between the "recognition" mechanisms of NK (natural killer) cells and T and B cells?

8. What four (4) primary effects result from complement activation?

9. What are the four (4) general characteristics of specific defenses?

10. What is the primary difference between active and passive immunity?

CHAPTER

16

The Respiratory System

Overview

The circulatory and respiratory systems interact to obtain and deliver oxygen to body cells and to remove carbon dioxide from the body. The respiratory system consists of passageways which are divided into *upper* and *lower* tracts. The upper passageways include the (1) *nose*; (2) *nasal cavity*; (3) *paranasal sinuses*; and (4) the *pharynx*.

The lower part includes the (1) *larynx*; (2) *trachea*; (3) *bronchi*; (4) *bronchioles*; and (5) *alveoli*. The terminal bronchioles and alveoli are involved with the gaseous exchange of oxygen and carbon dioxide, whereas all the other structures of the respiratory tracts serve as passageways for incoming and outgoing air and protection of the lungs. The respiratory system also plays an important role in regulating the pH of the body fluids.

The questions and activities in this chapter reinforce the concepts of the structural organization and the primary functions of the respiratory system, control of respiration, and respiratory interactions with other systems.

Review of Chapter Objectives

1. Describe the primary functions of the respiratory system.
2. Explain how the delicate respiratory exchange surfaces are protected from pathogens.
3. Relate respiratory functions to the structural specializations of the tissues and organs of the system.
4. Describe the physical principles governing the movement of air into the lungs and the diffusion of gases into and out of the blood.
5. Describe the actions of muscles responsible for respiratory movements.
6. Describe how oxygen and carbon dioxide are transported in the blood.
7. Describe the major factors that influence the rate of respiration.
8. Identify the reflexes that regulate respiration.
9. Describe the changes that occur in the respiratory system at birth and with aging.
10. Discuss the interrelationships among the respiratory system and other systems.

Part I: Objective Based Questions

Objective 1 Describe the primary functions of the respiratory system.

_____ 1. The primary function(s) of the respiratory system is (are) to

a. move air to and from the exchange surfaces
b. provide an area for gas exchange between air and circulating blood
c. protect respiratory surfaces from dehydration and environmental variations
d. all of the above

_____ 2. The *breathing* process is called

a. cellular respiration
b. pulmonary ventilation
c. respiration
d. systemic ventilation

_____ 3. The two primary gases involved in the respiratory process are

a. oxygen and carbon dioxide
b. carbon dioxide and nitrogen
c. oxygen and nitrogen
d. all of the above

_____ 4. The movement of gases into the blood from the lungs is a process of

a. osmosis
b. metabolism
c. net diffusion
d. cellular respiration

_____ 5. External respiration involves the diffusion of gases between the cells and the blood.

a. true
b. false

_____ 6. Internal respiration involves the diffusion of gases between the lungs and the blood.

a. true
b. false

Objective 2 Explain how the delicate respiratory exchange surfaces are protected from pathogens, debris and other hazards.

_____ 1. The "patrol force" of the alveolar epithelium involved with phagocytosis is composed primarily of alveolar

a. NK cells
b. cytotoxic cells
c. macrophages
d. plasma cells

_____ 2. Pulmonary surfactant is a phospholipid secretion produced by alveolar cells to

a. increase the surface area of alveoli
b. reduce the cohesive force of H_2O molecules and lower surface tension
c. increase the cohesive force of air molecules and raise surface tension
d. reduce the attractive force of O_2 molecules and increase surface tension

_____ 3. Large airborne particles are filtered by the

a. external olfactory meatuses
b. soft palate
c. nasal sinuses
d. nasal hairs in the vestibule of the nose

_____ 4. The function of the nasal conchae is to create turbulence in the air so as to trap small particulates in mucus.

a. true
b. false

_____ 5. Air entering the body is filtered, warmed, and humidified by the

a. upper respiratory tract
b. lower respiratory tract
c. lungs
d. alveoli

6. Identify and label the following structures in Figure 16-1. Place your answers in the spaces provided below the drawing.

Figure 16-1 The Nose, Nasal Cavity, and Pharynx

trachea	**nasopharynx**	**mandible**
oropharynx	**tongue**	**epiglottis**
cricoid cartilage	**internal nares**	**hyoid bone**
pharyngeal tonsil	**thyroid cartilage**	**soft palate**
auditory tube	**oral cavity**	**palatine tonsil**
external nares	**vocal cord**	**hard palate**
nasal vestibule	**laryngopharynx**	**nasal conchae**
glottis	**esophagus**	**frontal sinus**

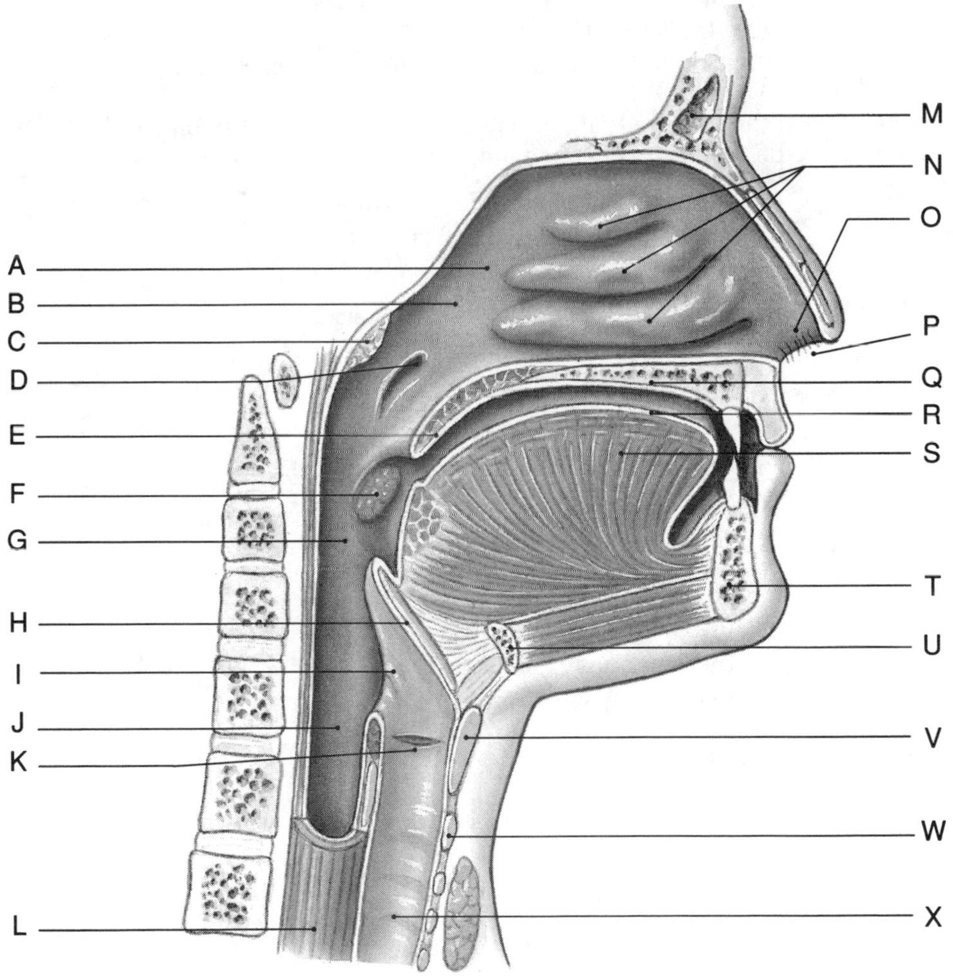

A ________	I ________	Q ________
B ________	J ________	R ________
C ________	K ________	S ________
D ________	L ________	T ________
E ________	M ________	U ________
F ________	N ________	V ________
G ________	O ________	W ________
H ________	P ________	X ________

Objective 3 Relate respiratory functions to the structural specializations of the tissues and organs of the system.

_____ 1. The respiratory system consists of structures that

a. provide defense from pathogenic invasion
b. permit vocalization and production of other sounds
c. regulate blood volume and pressure
d. all of the above

_____ 2. The difference between the true and false vocal cords is that the false vocal cords

a. are highly elastic
b. are involved with the production of sound
c. play no part in sound production
d. articulate with the corniculate cartilages

_____ 3. Structures in the trachea that prevent its collapse or over-expansion as pressure changes in the respiratory system are

a. O-shaped ringed trachael cartilages
b. C-shaped trachael cartilages
c. irregular circular bones
d. S-shaped trachael bones

_____ 4. The hard palate separates the

a. nasal cavity and the oral cavity
b. nasal cavity from the larynx
c. left and right sides of the nasal cavity
d. external nares from the internal nares

_____ 5. The glottis is

a. a flap of elastic cartilage
b. the opening to the pharynx
c. the opening to the larynx
d. part of the hard palate

_____ 6. The cartilage that makes up most of the anterior and lateral surface of the larynx is the

a. cricoid cartilage
b. thyroid cartilage
c. cuneiform cartilage
d. arytenoid cartilage

_____ 7. The trachea can alter its diameter when stimulated by the autonomic nervous system.

a. true
b. false

_____ 8. The following is a list of some of the structures of the respiratory tree. (1) secondary bronchi (2) bronchioles (3) alveolar ducts (4) primary bronchi (5) respiratory bronchioles (6) alveoli (7) terminal bronchioles.

The order in which air passes through the structures is

a. 4, 1, 2, 7, 5, 3, 6
b. 4, 1, 2, 5, 7, 3, 6
c. 1, 4, 2, 5, 7, 3, 6
d. 1, 4, 2, 7, 5, 3, 6

_____ 9. The respiratory membrane consists primarily of

a. pseudostratified ciliated columnar epithelium
b. moist cuboidal epithelium
c. simple squamous epithelium
d. ciliated squamous epithelium

10. The opening to the nostrils are the __________.

11. The portion of the nasal cavity contained within the flexible tissues of the external nose is the ________.

12. The portion of the pharynx that receives both air and food is the ______________.

13. The common passageway shared by the respiratory and digestive systems is the ____________.

14. The openings to the eustachian tube are located in the ____________.

15. The vocal cords are located in the ______________.

16. The airway between the larynx and the primary bronchi is the ______________.

17. Secondary bronchi supply air to the ______________.

18. Structures formed by the branching of the trachea within the mediastinum are ______________.

19. The actual sites of gas exchange within the lungs are the ____________.

20. Identify and label the following structures in Figure 16-2. Place your answers in the spaces provided below the drawing.

Figure 16-2 Components of the Respiratory System

left lung
sphenoidal sinus
left bronchus
tracheal cartilage
tongue
cricoid cartilage
vocal cords
diaphragm
epiglottis
frontal sinus
thyroid cartilage
larynx
nasal conchae
pharynx
right lung
esophagus
hyoid

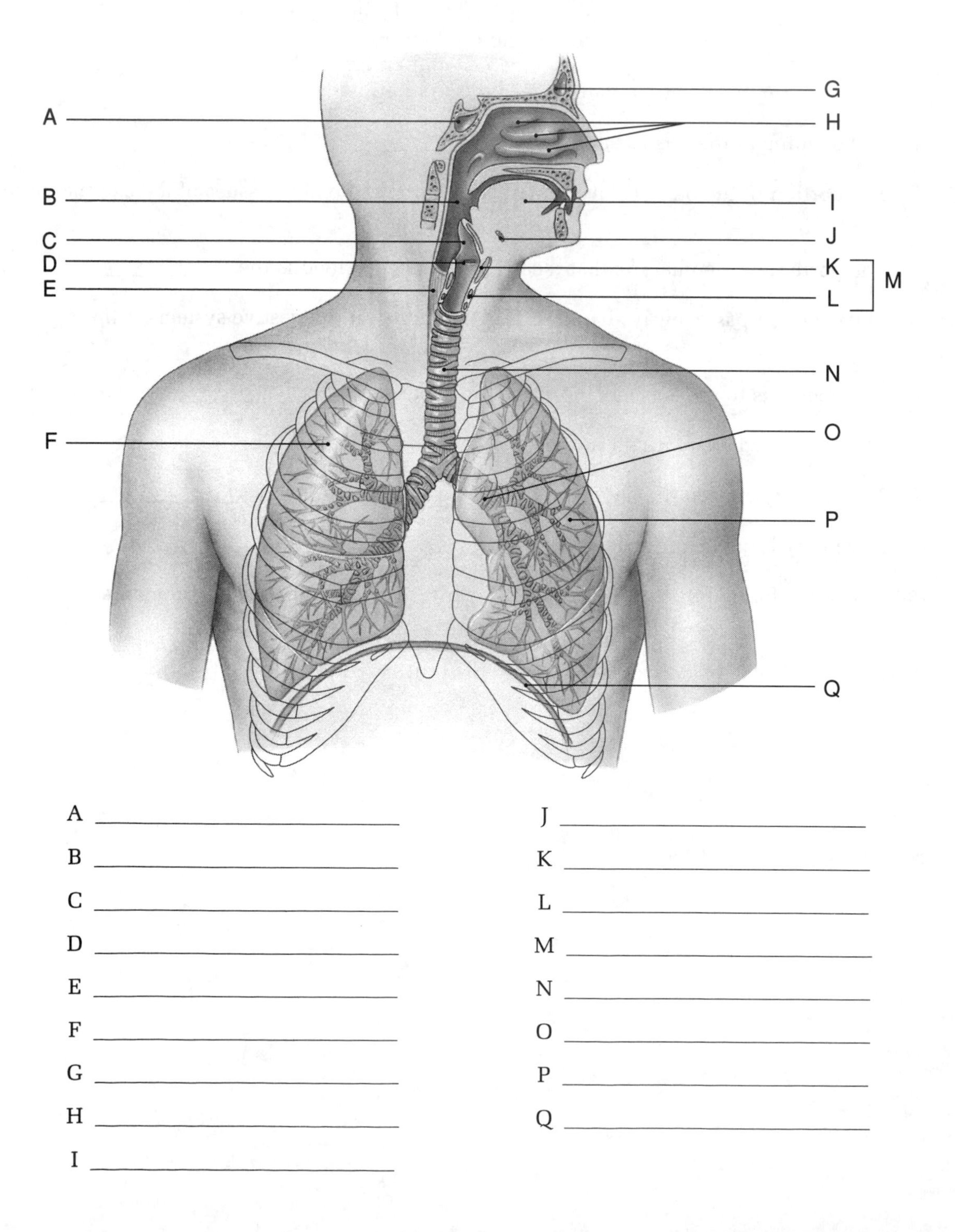

A ______________________

B ______________________

C ______________________

D ______________________

E ______________________

F ______________________

G ______________________

H ______________________

I ______________________

J ______________________

K ______________________

L ______________________

M ______________________

N ______________________

O ______________________

P ______________________

Q ______________________

Objective 4 Describe the physical principles governing the movement of air into the lungs and the diffusion of gases into and out of the blood.

_____ 1. A necessary feature for normal gas exchange in the alveoli is:

a. for the alveoli to remain dry
b. an increase in pressure in pulmonary circulation
c. for fluid to move into the pulmonary capillaries
d. an increase in lung volume

_____ 2. The movement of air into and out of the lungs is primarily dependent on:

a. pressure differences between the air in the atmosphere and air in the lungs
b. pressure differences between the air in the atmosphere and the anatomic dead space
c. pressure differences between the air in the atmosphere and individual cells
d. a, b, and c, are correct

_____ 3. During inspiration there will be an increase in the volume of the thoracic cavity and a(n):

a. decreasing lung volume, increasing intrapulmonary pressure
b. decreasing lung volume, decreasing intrapulmonary pressure
c. increasing lung volume, decreasing intrapulmonary pressure
d. increasing lung volume, increasing intrapulmonary pressure

_____ 4. During expiration there is a(n):

a. decrease in intrapulmonary pressure
b. increase in intrapulmonary pressure
c. increase in atmospheric pressure
d. increase in the volume of the lungs

_____ 5. If there is a Po_2 of 104 mm Hg and Pco_2 of 40 mm Hg in the alveoli, and a Po_2 of 40 mm Hg and a Pco_2 of 45 mm Hg within the pulmonary blood, there will be a net diffusion of:

a. CO_2 into the blood from alveoli; O_2 from the blood into alveoli
b. O_2 and CO_2 into the blood from the alveoli
c. O_2 and CO_2 from the blood into the alveoli
d. O_2 into the blood from alveoli; CO_2 from the blood into the alveoli

_____ 6. Each molecule of hemoglobin has the capacity to carry __________ molecules of oxygen (O_2).

a. 6
b. 8
c. 4
d. 2

_____ 7. What percentage of total oxygen (O_2) is carried within red blood cells chemically bound to hemoglobin?

a. 5%
b. 68%
c. 98%
d. 100%

_____ 8. Factors that cause a decrease in hemoglobin saturation at a given Po_2 are:

a. increasing Po_2, decreasing CO_2, increasing temperature
b. increasing diphosphoglycerate (DPG), increasing temperature, decreasing pH
c. decreasing DPG, increasing pH, increasing CO_2
d. decreasing temperature, decreasing CO_2, decreasing Po_2

Objective 5 Describe the actions of muscles responsible for respiratory movements.

_____ 1. During exhalation the *diaphragm* moves

a. upward and the ribs move downward
b. downward and the ribs move upward
c. downward and the ribs move downward
d. upward and the ribs move upward

_____ 2. When the diaphragm and external intercostal muscles contract the

a. volume of the thorax increases
b. volume of the thorax decreases
c. volume of the lungs decreases
d. lungs collapse

_____ 3. The amount of air moved into or out of the lungs in a single *passive* respiratory cycle is the

a. expiratory reserve volume
b. residual
c. tidal volume
d. vital capacity

_____ 4. The amount of air exhaled with one *forceful* breath is the

a. tidal volume
b. vital capacity
c. residual volume
d. expiratory reserve volume

Objective 6 Describe how oxygen and carbon dioxide are transported in the blood.

_____ 1. Most of the oxygen transported by the blood is

a. dissolved in plasma
b. bound to hemoglobin
c. carried by white blood cells
d. ionic form in the plasma

_____ 2. Most of the carbon dioxide in the blood is transported as

a. carbonic acid
b. bicarbonate ions
c. carbaminohemoglobin
d. a solute dissolved in the plasma

_____ 3. A 10% increase in the level of carbon dioxide in the blood will

a. decrease the rate of breathing
b. decrease pulmonary ventilation
c. increase the rate of breathing
d. decrease the rate of alveolar ventilation

_____ 4. The percentage of CO_2 which binds to hemoglobin in the RBCs is

a. 7%
b. 23%
c. 70%
d. 93%

_____ 5. The percentage of CO_2 converted to bicarbonate ions in RBCs is

a. 7%
b. 23%
c. 70%
d. 98%

_____ 6. An increase in CO_2 will _________ the concentration of H^+ in the RBCs, which will _________ the pH of the blood.

a. increase, decrease
b. increase, increase
c. decrease, decrease
d. decrease, increase

7. In the following diagram, fill in the correct percentages to complete Figure 16-3 correctly. Use the following percentages to complete the exercise and record your answers below.

Figure 16-3 Carbon Dioxide Transport in the Blood

7% 7% 93% 23%

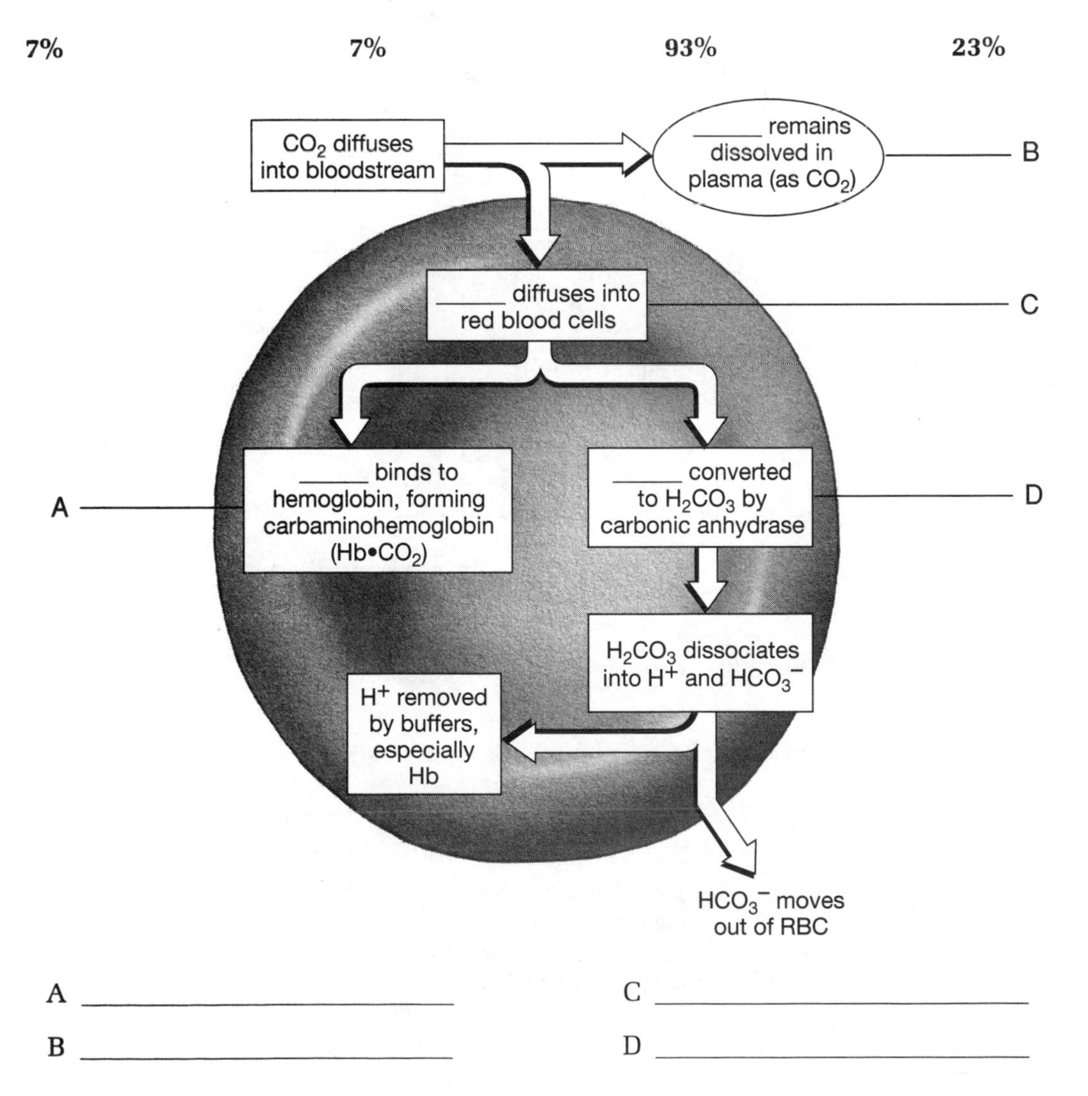

A _______________ C _______________

B _______________ D _______________

Objective 7 Describe the major factors that influence the rate of respiration.

_____ 1. The respiratory centers that regulate breathing are located in the

a. brain stem and cerebrum
b. cerebrum and cerebellum
c. lungs
d. pons and medulla oblongata

_____ 2. The normal rate and depth of breathing is established by the

a. inspiratory center
b. expiratory center
c. vasomotor center
d. dorsal respiratory group

_____ 3. The receptors located in the carotid arteries are

a. chemoreceptors
b. pressure receptors
c. stretch receptors
d. both a and b

_____ 4. The output from baroreceptors affects the respiratory centers causing the respiratory rate to

a. decrease without affecting blood pressure
b. increase with an increase in blood pressure
c. decrease with a decrease in blood pressure
d. increase with a decrease in blood pressure

_____ 5. Under normal conditions the greatest effect on the respiratory centers is initiated by

a. decreases in Po_2
b. increases and decreases in Po_2 and Pco_2
c. increases and decreases in Pco_2
d. increases in Po_2

_____ 6. An elevated body temperature will

a. accelerate respiration
b. decrease respiration
c. increase depth of respiration
d. not affect the respiratory rate

7. The volume of air breathed in during an inhalation is controlled by stretch receptors in the ____________.

8. The gas that establishes the rate of breathing is ____________.

Objective 8 Identify the reflexes that regulate respiration.

_____ 1. Reflexes important in regulating the forced ventilations that accompany strenuous exercise are known as the

a. Hering - Brewer reflexes
b. protective reflexes
c. plantar reflexes
d. chemoreceptor reflexes

_____ 2. Reflexes that are a response to changes in the volume of the lungs or changes in arterial blood pressure are

a. chemoreceptor reflexes
b. baroreceptors reflexes
c. stretch receptor reflexes
d. mechanoreceptor reflexes

_____ 3. Reflexes that are a response to changes in the Po_2 and Pco_2 of the blood and cerebrospinal fluid are ____________ reflexes

a. chemoreceptor
b. stretch receptor
c. mechanoreceptor
d. baroreceptors

_____ 4. The *deflation* reflex stimulates the inspiratory center.

a. true
b. false

_____ 5. The *inflation* reflex prevents overexpansion of the lungs.

a. true
b. false

Objective 9 Describe the changes that occur in the respiratory system at birth and with aging.

_____ 1. At birth, blood is pulled into the pulmonary circulation by the

a. same drop in pressure that pulls air into the lungs
b. pulmonary surfactant
c. pulmonary arterial resistance
d. powerful contractions of the diaphragm

_____ 2. The destruction of alveoli due to age or by respiratory irritants is called

a. asthma
b. emphysema
c. pneumonia
d. pulmonary embolism

_____ 3. The loss of elasticity in the lungs due to aging decreases the

a. residual volume
b. tidal volume
c. vital capacity
d. all of the above

_____ 4. The most probable cause of restricted movements of the chest cavity in the elderly is

a. arthritic changes in the rib joints
b. excessive smoking
c. increased incidence of asthma
d. the presence of pulmonary emboli

Objective 10 Discuss the interrelationships among the respiratory system and other systems.

_____ 1. All the systems of the body are affected by the respiratory system in that the respiratory system provides

a. antigens to trigger specific defenses
b. bicarbonate ions to buffer pH activity in the body
c. increased thoracic pressure for promoting defecation
d. oxygen and eliminate carbon dioxide

2. The _________ system eliminates organic wastes generated by cells of the respiratory system.

3. The _________ system releases epinephrine and norepinephrine which stimulates respiratory activity and dilates respiratory passageways.

4. The _________ system controls the pace and depth of respiration.

5. Protection for the lungs is provided by the _____________ system.

Part II: Chapter Comprehensive Exercises

A. Word Elimination

Circle the term that does not belong in each of the following groupings.

1. pharynx larynx alveoli trachea bronchi
2. frontal mandibular sphenoid ethmoid maxillary
3. glottis epiglottis thyroid cricoid arytenoid
4. apex superior lobe cardiac notch inferior lobe middle lobe
5. primary bronchi respiratory bronchioles alveolar ducts alveolar sacs alveoli
6. scalenes pectoralis minor internal intercostals serratus anterior external intercostals
7. ERV vital capacity IRV residual volume Pco2
8. N_2 O_2 NH_4 H_2O CO_2
9. chemoreceptor inflation deflation Hering-Brewer mechanoreceptor
10. respiratory surfactant nervous endocrine lymphatic

B. Matching

Match the terms in Column B with the terms in Column A. Use letters for answers in the spaces provided.

COLUMN A	COLUMN B
___ 1. apneustic center	A. quiet breathing
___ 2. pneumotaxic center	B. increased arterial Pco_2
___ 3. hyperventilation	C. delivers air to lungs
___ 4. hypoventilation	D. carbon dioxide
___ 5. alveolar macrophages	E. breathing
___ 6. pulmonary ventilation	F. inhibitory effect on respiration
___ 7. eupnea	G. lungs
___ 8. upper respiratory tract	H. dust cells
___ 9. lower respiratory tract	I. stimulatory effect on respiration
___ 10. bicarbonate ions	J. decreased arterial Pco_2

C. Concept Map - The Respiratory System

Using the following terms, fill in the circled, numbered, blank spaces to complete the concept map. Follow the numbers that comply with the organization of the map.

Lungs
Alveoli
Ciliated mucous membrane
Paranasal sinuses
Speech production
Blood
Upper respiratory tract
O_2 from alveoli into blood
Pharynx & larynx

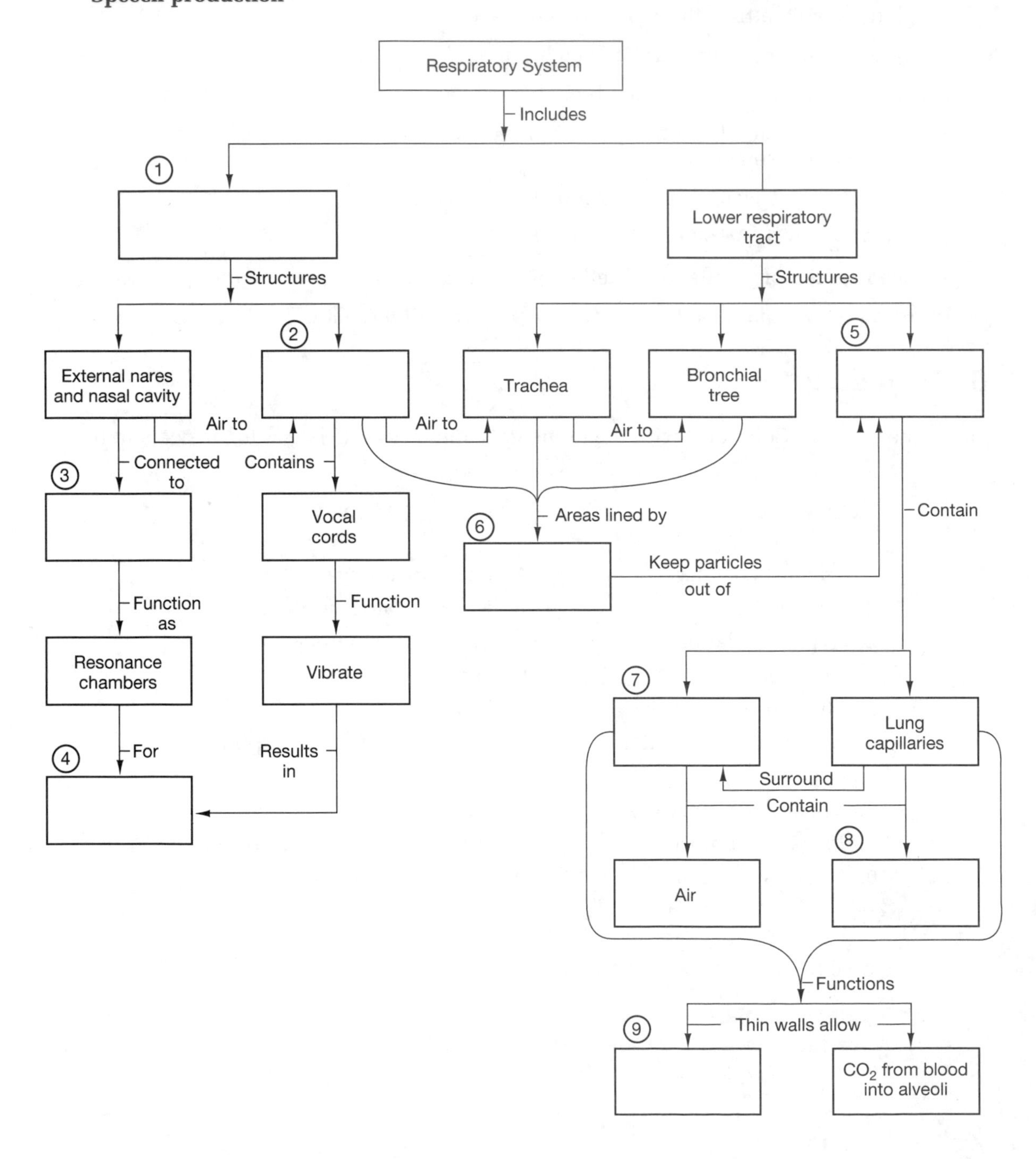

D. Crossword Puzzle

The following crossword puzzle reviews the material in Chapter 16. To complete the puzzle, you must know the answers to the clues given, and must be able to spell the terms correctly.

ACROSS

2. The amount of air we can forcibly exhale.
5. The opening into the trachea.
6. These cells produce mucus to protect the lining of the respiratory tubes.
8. The tracheal tubes that enter each lung.
10. Chemoreceptors are located in the ______ arteries.
11. The amount of air we exhale passively is called __________ volume.
12. Most of the carbon dioxide in the blood is converted to __________.

DOWN

1. Low oxygen concentration in the peripheral tissues.
3. Gas exchange occurs here.
4. About 20.8% of the atmospheric air consists of __________.
7. When the diaphragm muscle is moving upward, we are __________.
9. This volume of air in the lungs is equivalent to about 1200 ml.

E. Short Answer Questions

Briefly answer the following questions in the spaces provided below.

1. What are the primary functions of the respiratory system?

2. How is the inhaled air warmed and humidified in the nasal cavities, and why is this type of environmental condition necessary?

3. What is surfactant and what is its function in the respiratory membrane?

4. What is the difference among external respiration, internal respiration, and cellular respiration?

5. What is vital capacity and how is it measured?

6. What is alveolar ventilation and how is it measured?

7. What are the three (3) methods of CO_2 transport in the blood?

8. What three (3) kinds of reflexes are involved in the regulation of respiration?

9. What is the relationship between a rise in arterial P_{CO_2} and hyperventilation?

CHAPTER

17

The Digestive System

Overview

The average person probably knows more about indigestion than he/she knows about digestion. Perhaps you are one of many who comprise a population know as the "Rolaids Generation." The nutrients in food are not ready for use by cells; that is, the food must be converted (digested) into chemical forms that cells can utilize for their metabolism functions. The digestive organs are responsible for ingestion of food, digestion, and elimination of undigested remains from the body. Accessory digestive organs, such as the teeth, tongue, salivary glands, liver, gallbladder, and pancreas, assist in the digestive process as food moves through the digestive tract which includes the (1) mouth; (2) pharynx; (3) esophagus; (4) stomach; and (5) the small and large intestines.

When the "processing" of the food is completed by the digestive tract, the nutrients are absorbed into the bloodstream and transported to the cells and tissues of the body to be used for energy, tissue growth and repair, and the cell's metabolic activities.

The exercises in this chapter are organized to guide your study of the structure and function of the digestive tract and accessory organs. The review questions will introduce you to the endocrine and nervous systems' influences on the digestive process, and the series of steps necessary to promote absorption and eventual use of nutrients by cells for metabolic activities.

Review of Chapter Objectives

1. Identify the organs of the digestive tract and the accessory organs of digestion.
2. List the functions of the digestive system.
3. Describe the histology of each segment of the digestive tract in relation to the function.
4. Explain how ingested materials are propelled through the digestive tract.
5. Describe how food is processed in the mouth and describe the key events of the swallowing process.
6. Describe the anatomy of the stomach, its histological features, and its roles in digestion and absorption.

7. Explain the functions of the intestinal secretions and discuss the relative significance of digestion in the small intestine.
8. Describe the structure and functions of the pancreas, liver, and gall bladder and explain how their activities are regulated.
9. Describe the structures of the large intestine, its movements, and its absorptive processes.
10. Describe the digestion and absorption of carbohydrates, lipids, and proteins.
11. Describe the changes in the digestive system that occur with aging.
12. Discuss the interactions among the digestive system and other systems.

Part I: Objective Based Questions

Objective 1 Identify the organs of the digestive tract and the accessory organs of digestion.

_____ 1. Which one of the following organs is not a part of the digestive system?

a. liver
b. gall bladder
c. spleen
d. pancreas

_____ 2. Of the following selections, the one which contains only accessory structures is

a. salivary glands, pancreas, liver, gall bladder
b. pharynx, esophagus, small and large intestine
c. oral cavity, stomach, pancreas, and liver
d. tongue, teeth, stomach, small and large intestine

_____ 3. The digestive tube between the pharynx and the stomach is the

a. trachea
b. larynx
c. pylorus
d. esophagus

_____ 4. Food particles pass through the accessory organs to complete the digestive process.

a. true
b. false

5. Identify and label the structures of the digestive tract in Figure 17-1. Place your answers in the spaces provided below the drawing after selecting from the following choices.

Figure 17-1 Components of the Digestive System

gall bladder	**small intestine**	**pharynx**
large intestine	**esophagus**	**stomach**
liver	**pancreas**	**oral cavity/tongue**
salivary glands		

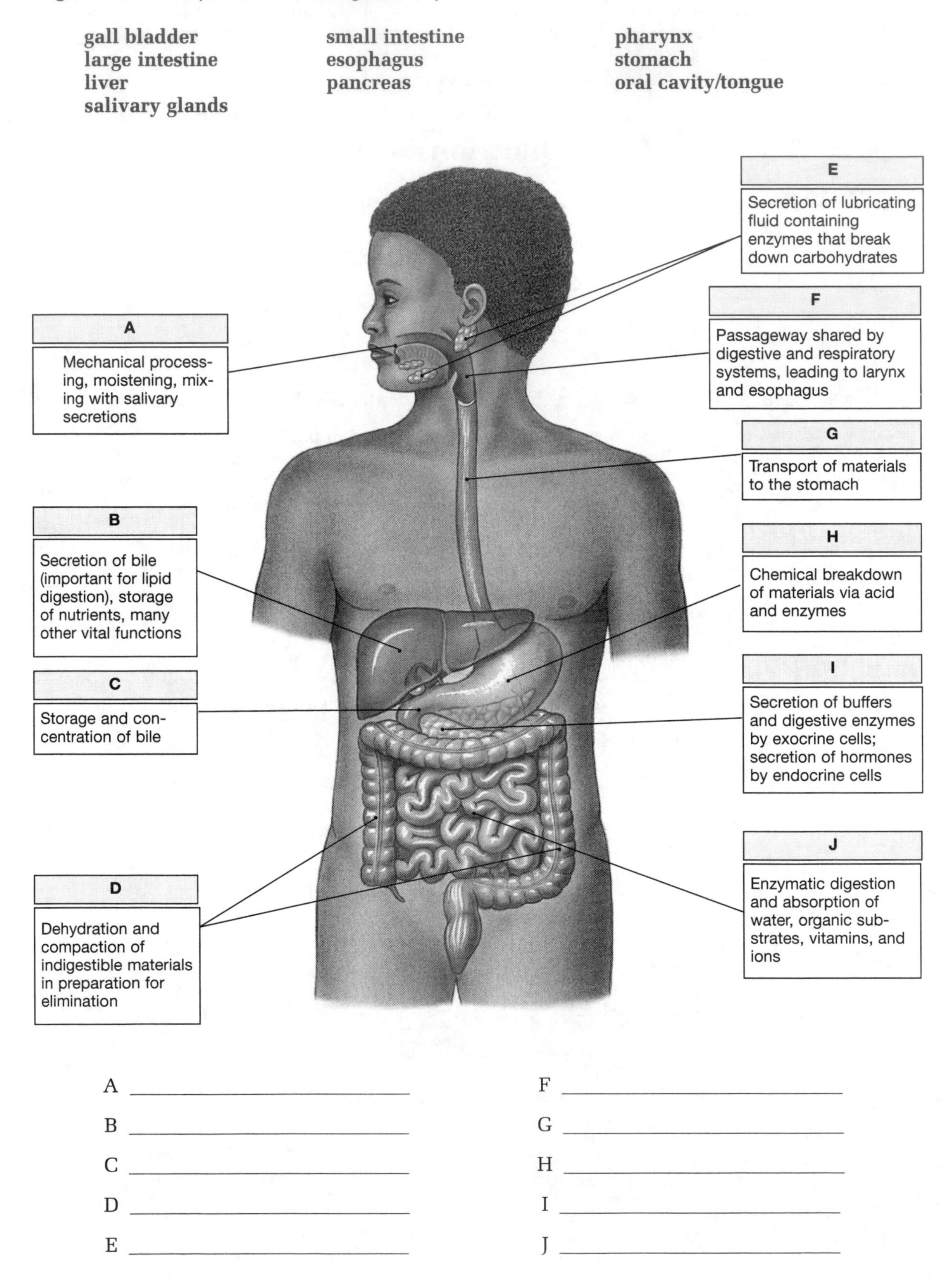

A __________________ F __________________

B __________________ G __________________

C __________________ H __________________

D __________________ I __________________

E __________________ J __________________

Objective 2 List the functions of the digestive system.

_____ 1. The physical manipulation of solid foods by the tongue, teeth, and swirling and mixing motions of the digestive tract is called

a. secretion
b. absorption
c. compaction
d. mechanical processing

_____ 2. Digestion refers to the chemical breakdown of food into small organic fragments that can be absorbed by the digestive epithelium.

a. true
b. false

3. The release of water, acids, enzymes, and buffers by the digestive tract and accessory organs is _______________.

4. The movement of small organic molecules, electrolytes, vitamins and water across the digestive epithelium and into the interstitium is _______________.

5. The elimination of waste products from the body is _______________.

Objective 3 Describe the histology of each segment of the digestive tract in relation to the function.

_____ 1. Sympathetic stimulation of the muscularis externa promotes

a. muscular inhibition and relaxation
b. increased muscular tone and activity
c. muscular contraction and increased excitation
d. increased digestive and gastric motility

_____ 2. The mucous-producing, unicellular glands found in the mucosal epithelium of the stomach and small and large intestine are

a. enteroendocrine cells
b. parietal cells
c. chief cells
d. goblet cells

_____ 3. The primary tissue of the tunica submucosa is

a. lamina propria
b. alveolar connective tissue
c. muscularis externa
d. a circular muscular layer

_____ 4. The layer of loose connective tissue that contains blood vessels, glands and lymph nodes is the

a. circular muscular layer
b. submucosa
c. lamina propria
d. mucosal epithelium

_____ 5. The layer of the intestinal wall which contracts and changes the shape of the intestinal lumen to move food through its length is the

a. muscularis
b. mucosa
c. submucosa
d. serosa

_____ 6. Double sheets of peritoneal membrane that hold some of the visceral organs in their proper position are the

a. adventitia
b. fibrosa
c. serosa
d. mesenteries

_____ 7. Most of the digestive tract is lined by

a. cuboidal epithelium
b. stratified squamous epithelium
c. simple columnar epithelium
d. simple squamous epithelium

_____ 8. A modification of the mucosa of the small intestine that allows for expansion are the

a. plicae
b. mucus glands
c. presence of striations
d. adventitia

9. Identify and label Figure 17-2 from the following selections. Place your answers in the spaces provided below the drawing.

Figure 17-2 Histological Feature of the G.I. Tract Wall

plica	**lumen**
submucosa	**mucosa**
mesenteric artery and vein	**circular muscular layer**
visceral peritoneum (serosa)	**mucosal gland**
mesentery	**muscularis externa**
serosa	

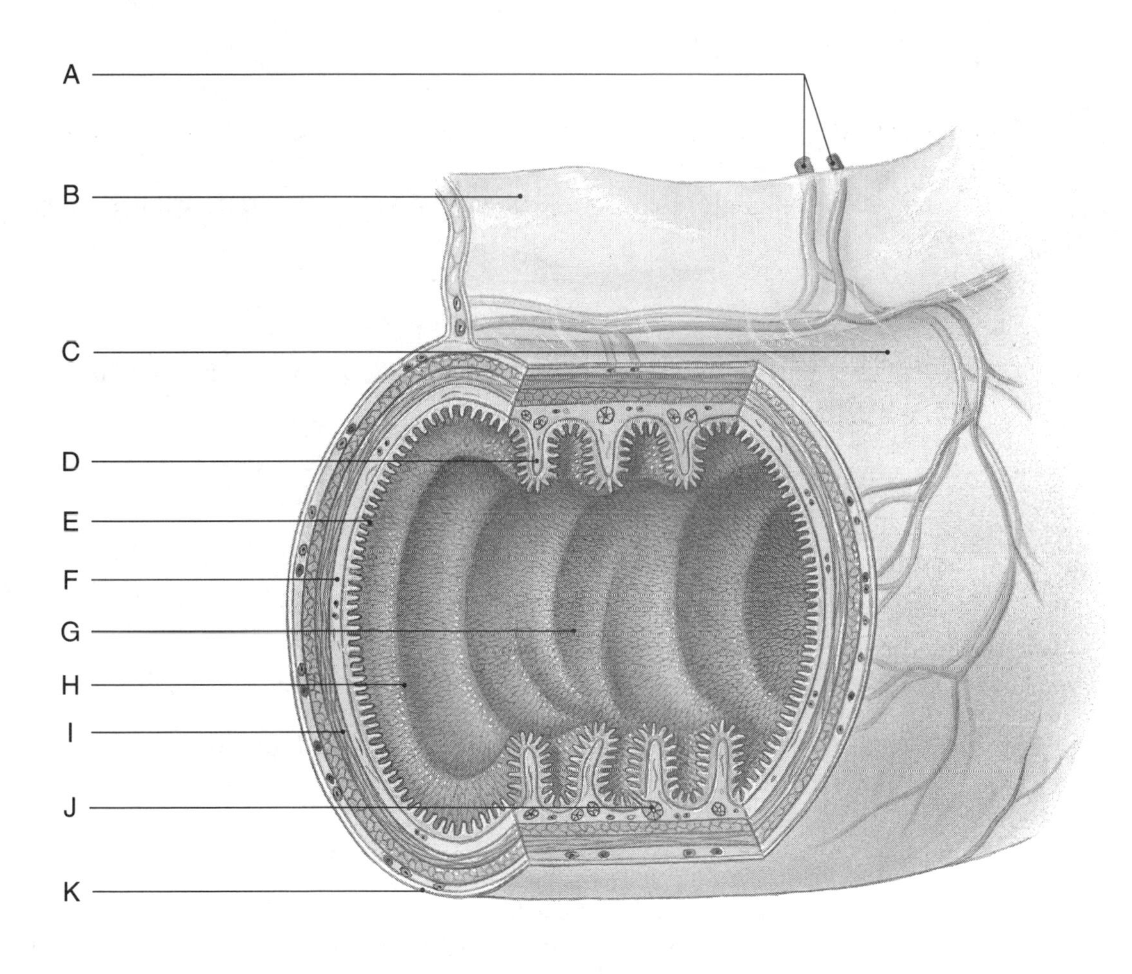

A ______________________ G ______________________

B ______________________ H ______________________

C ______________________ I ______________________

D ______________________ J ______________________

E ______________________ K ______________________

F ______________________

Objective 4 Explain how ingested materials are propelled through the digestive tract.

_____ 1. Waves of muscular contractions that propel the contents of the digestive tract from one point to another are called

a. segmentations
b. peristalsis
c. mastications
d. compactions

_____ 2. Regional movements that occur in the small intestine and function to churn and fragment the digestive materials are called

a. segmentation
b. peristalsis
c. pendulums
d. mastications

_____ 3. The circular and longitudinal layers are most responsible for peristalsis along the esophagus.

a. true
b. false

_____ 4. Once a bolus of food has entered the laryngopharynx, swallowing continues involuntarily due to the

a. swallowing reflex
b. size of the bolus
c. peristalic activity
d. all of the above

_____ 5. Peristalsis in the esophagus is controlled by the endocrine system.

a. true
b. false

_____ 6. Swirling mixing and churning motions of the digestive tract provide

a. action of enzymes, acids, and buffers
b. chemical breakdown of food
c. mechanical processing after ingestion
d. peristalic activity

_____ 7. Strong contractions of the ascending and transverse colon moving the contents of the colon toward the sigmoid colon are called

a. defecation
b. pendular movements
c. mass peristalsis
d. segmentation

Objective 5 Describe how food is processed in the mouth and describe the key events of the swallowing process.

_____ 1. The functions of the oral cavity include

a. partial digestion of proteins and carbohydrates
b. mechanical processing of food
c. lubrication
d. both b and c

_____ 2. Functions of the tongue include

a. mechanical processing of food
b. manipulation of food
c. sensory analysis of food
d. all of the above

_____ 3. During swallowing the

a. soft palate elevates
b. larynx elevates
c. epiglottis closes
d. all of the above

_____ 4. Secretions from the salivary glands

a. are mostly digestive enzymes
b. provide analysis before swallowing
c. help to control bacterial population in the mouth
d. all of the above

_____ 5. Food is initially ground and torn into smaller pieces by the teeth. Typically, adults have _____ incisors, _____ cuspids, _____ bicuspids, and _____ molars.

a. 4, 2, 8, 6
b. 8, 4, 8, 6
c. 8, 4, 8, 12
d. 8, 8, 4, 12

_____ 6. Salivary amylase is an enzyme produced and released by the salivary glands that partially digest

a. carbohydrates
b. fats
c. proteins
d. all of the above

_____ 7. The tongue is controlled by the _____ nerve.

a. abducens
b. facial
c. glossopharyngeal
d. vagus

_____ 8. When the _____ cells of the tongue are activated they send signals to the brain via the _____ nerve to interpret the flavor of food.

a. papillary, trigeminal
b. gustatory, facial
c. tastebuds, facial
d. none of the above

_____ 9. During the swallowing process, the _____ closes off the _____.

a. epiglottis, esophagus
b. glottis, trachea
c. epiglottis, trachea
d. glottis, esophagus

10. Blade-shaped teeth that function in cutting or chopping are _____________.

11. Pointed teeth that are adapted for tearing and shredding are _____________.

12. Teeth with flattened crowns and prominent ridges that are adapted for grinding are the ______________.

13. The gland that empties into the upper regions of the oral cavity is the ______________.

14. The structure at the posterior midregion of the soft palate is the ______________.

15. Identify and label the teeth in Figure 17-3. Place your answers in the spaces provided below the drawing.

Figure 17-3 The Teeth

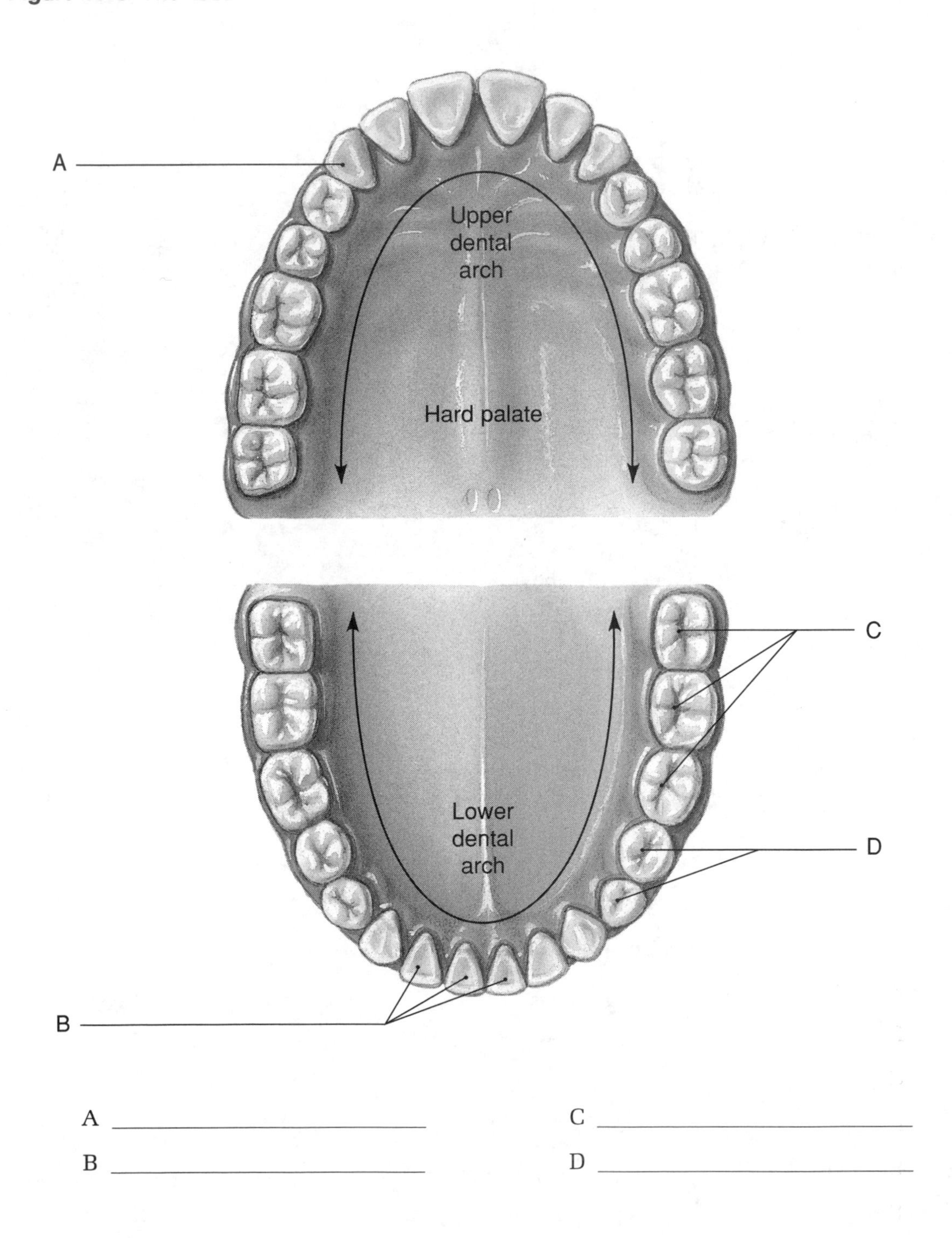

A ______________________

B ______________________

C ______________________

D ______________________

16. Identify and label the parts and areas of a typical tooth. Place your answers in the spaces provided below the drawing after selecting from the following choices.

Figure 17-4 Structure of a Typical Tooth

enamel	**cementum**	**crown**
dentin	**neck**	**root canal**
blood vessels and nerves	**root**	**alveolar bone**
pulp cavity	**gingival sulcus**	**peridontal ligament**

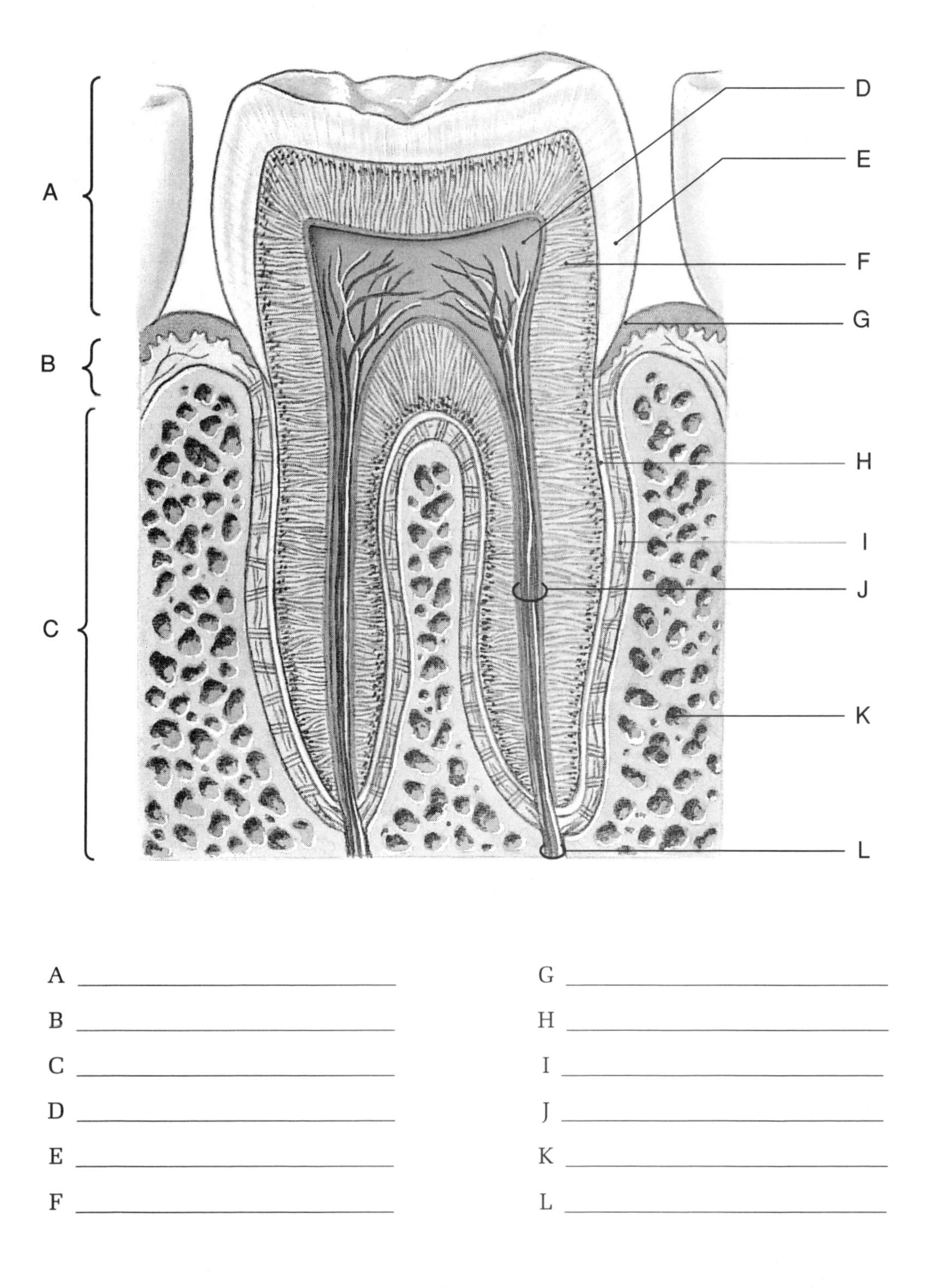

A ______________________ G ______________________

B ______________________ H ______________________

C ______________________ I ______________________

D ______________________ J ______________________

E ______________________ K ______________________

F ______________________ L ______________________

Objective 6 Describe the anatomy of the stomach, its histological features, and its roles in digestion and absorption.

_____ 1. Of the following selections, the one which is not a function of the stomach is

a. storage of ingested food
b. denaturation of protein
c. digestion of carbohydrates and fats
d. mechanical breakdown of food

_____ 2. The greater omentum is

a. the major portion of the stomach
b. attached to the stomach at the lesser curvature
c. important in the digestion of fats
d. a fatty sheet that hangs like an apron over the abdominal viscera

_____ 3. Gastric pits are

a. ridges in the body of the stomach
b. pockets in the lining of the stomach that contains secretory cells
c. involved in absorption of liquids from the stomach
d. areas in the stomach where proteins are denatured

4. Parietal cells secrete _______________.

5. Chief cells secrete _______________.

6. The portion of the stomach that connects to the esophagus is the _______________.

7. The bulge of the greater curvature of the stomach superior to the esophageal junction is the _______________.

8. The curved, tubular portion of the J-shaped stomach is the _______________.

9. The prominent ridges in the lining of the stomach are called _______________.

10. The enzyme pepsin is involved in the digestion of _______________.

11. Identify and label the structures of the stomach in Figure 17-5. Select your answers from the following selections. Place your answers in the spaces provided below the drawing.

Figure 17-5 The Stomach

cardia	**pylorus**	**fundus**
esophagus	**rugae**	**diaphragm**
lesser curvature	**greater curvature**	**lesser omentum**
greater omentum	**body**	

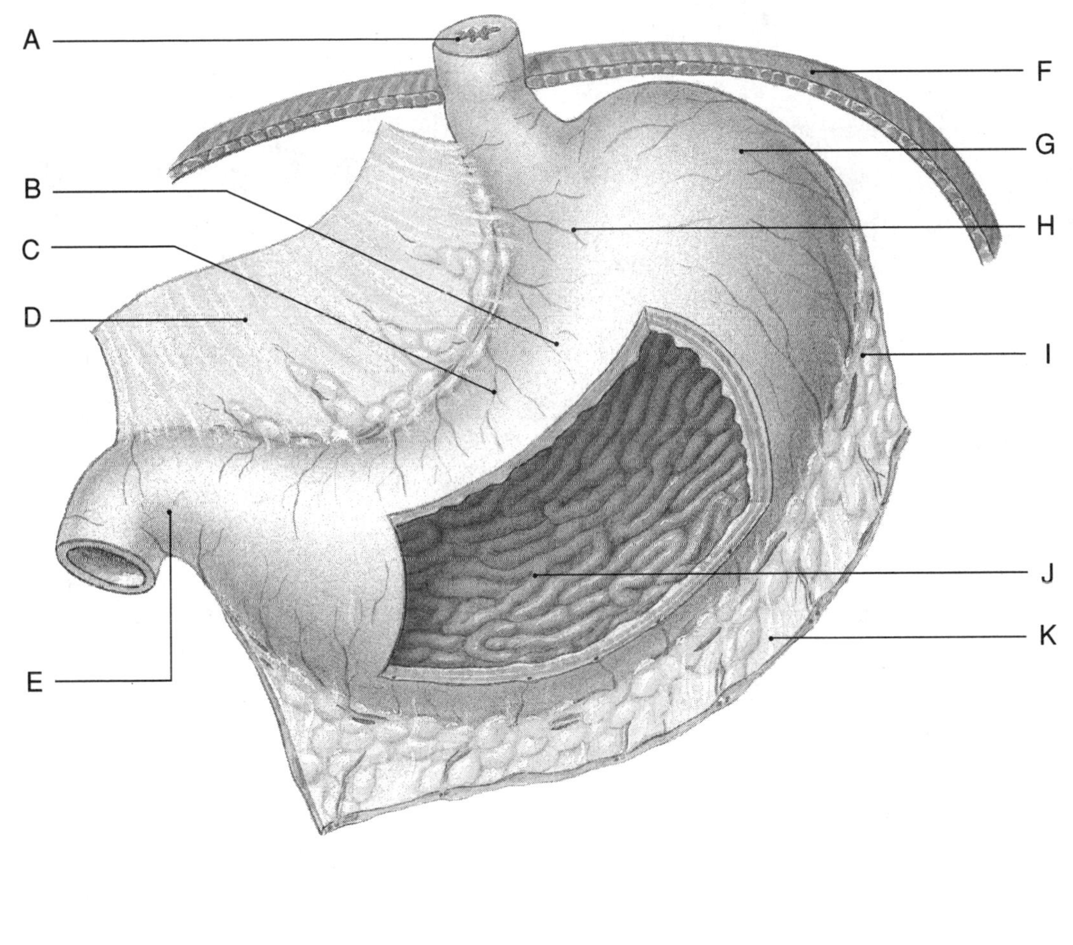

A ______________________ G ______________________

B ______________________ H ______________________

C ______________________ I ______________________

D ______________________ J ______________________

E ______________________ K ______________________

F ______________________

Objective 7 Explain the functions of the intestinal secretions and discuss the relative significance of digestion in the small intestine.

_____ 1. The three divisions of the small intestine are

a. cephalic, gastric, intestinal
b. duodenum, jejunum, ileum
c. fundus, body, pylorus
d. buccal, pharyngeal, esophageal

_____ 2. The majority of digestion in the small intestine occurs in the

a. cecum
b. duodenum
c. ileum
d. jejunum

_____ 3. The majority of the absorption of the nutrients into the bloodstream occurs in the

a. ileum
b. jejunum
c. duodenum
d. cecum

_____ 4. In order for the nutrients to leave the small intestine and enter the bloodstream, the nutrients must be absorbed through the

a. gastric pits
b. mucosal glands
c. villi
d. lacteals

_____ 5. The hormone which causes the release of insulin into the bloodstream when glucose is present in the small intestine is

a. GIP
b. gastrin
c. CCK
d. secretin

_____ 6. The hormones cholecystokinin and secretin are released by the

a. gall bladder
b. pancreas
c. small intestine
d. liver

_____ 7. The intestinal hormone that stimulates the pancreas to release a watery secretion that is high in bicarbonate ions is

a. gastrin
b. secretin
c. cholecystokinin
d. enterocrinin

_____ 8. An intestinal hormone that stimulates the gall bladder to release bile is

a. enterokinase
b. CCK
c. gastrin
d. secretin

_____ 9. An intestinal hormone that stimulates the release of insulin from the pancreatic islet cells is

a. secretin
b. cholecystokinin
c. enterokinase
d. enterocrinin

_____ 10. The intestinal hormone that stimulates parietal cells and chief cells in the stomach to secrete is

a. CCK
b. GIP
c. gastrin
d. enterokinase

Objective 8 Describe the structure and functions of the pancreas, liver, and gall bladder and explain how their activities are regulated.

_____ 1. The hormone which causes the pancreas to release its digestive enzymes into the small intestine is

a. cholecystokinin
b. secretin
c. gastrin
d. enterogastrone

2. Bile is stored in the _______________.

3. The basic functional unit of the liver is the _______________.

4. The liver cells which are arranged into a series of irregular plates like the spokes of a wheel are called _______________.

_____ 5. The primary function(s) of the liver includes

a. metabolic regulation
b. hemotological regulation
c. synthesis and secretion of bile
d. all of the above

_____ 6. The exocrine portion of the pancreas is composed of

a. islets of Langerhans
b. pancreatic crypts
c. pancreatic acini
d. pancreatic lobules

_____ 7. The pancreas produces and releases

a. lipases
b. carbohydrases
c. proteinases
d. all of the above

_____ 8. The hormone which causes the gall bladder to release bile is

a. cholecystokinin
b. secretin
c. gastrin
d. enterogastrone

Objective 9 Describe the structure of the large intestine, its movements, and its absorptive processes.

_____ 1. The large intestine

a. releases nutrients into the bloodstream
b. absorbs water from the bloodstream
c. releases water into the bloodstream
d. serves only as an organ of excretion

_____ 2. One of the functions of the large intestine is

a. bacteria in the large intestine manufacture vitamin K
b. the large intestine produces vitamins
c. the large intestine removes some vitamins from digested food
d. absorption of nutrients from the large intestine into the blood stream

_____ 3. Approximately 87% of the water that enters the large intestine is recycled daily into the bloodstream.

a. true
b. false

_____ 4. At the hepatic flexure, the colon becomes the

a. ascending colon
b. transverse colon
c. sigmoid colon
d. descending colon

_____ 5. Peristalic action occurs in the esophagus, small and large intestine.

a. true
b. false

6. The external pouches of the large intestine are called _____________.

7. The three longitudinal bonds of muscle located beneath the serosa of the colon are the _______________.

8. The sac-like structure that joins the ileum at the ileocecal valve is the _______________.

9. The small, fingerlike structure attached to the "blind" end of the cecum is the _______________.

10. Using the selections below, identify and label the structures of the large intestine. Place your answers in the spaces provided below the drawing.

Figure 17-6 The Large Intestine

cecum
aorta
transverse colon
ascending colon
ileum
vermiform appendix
rectum
ileocecal valve
sigmoid colon
taenia coli
inferior vena cava
descending colon
haustra
splenic flexure
hepatic portal vein
greater omentum
splenic vein

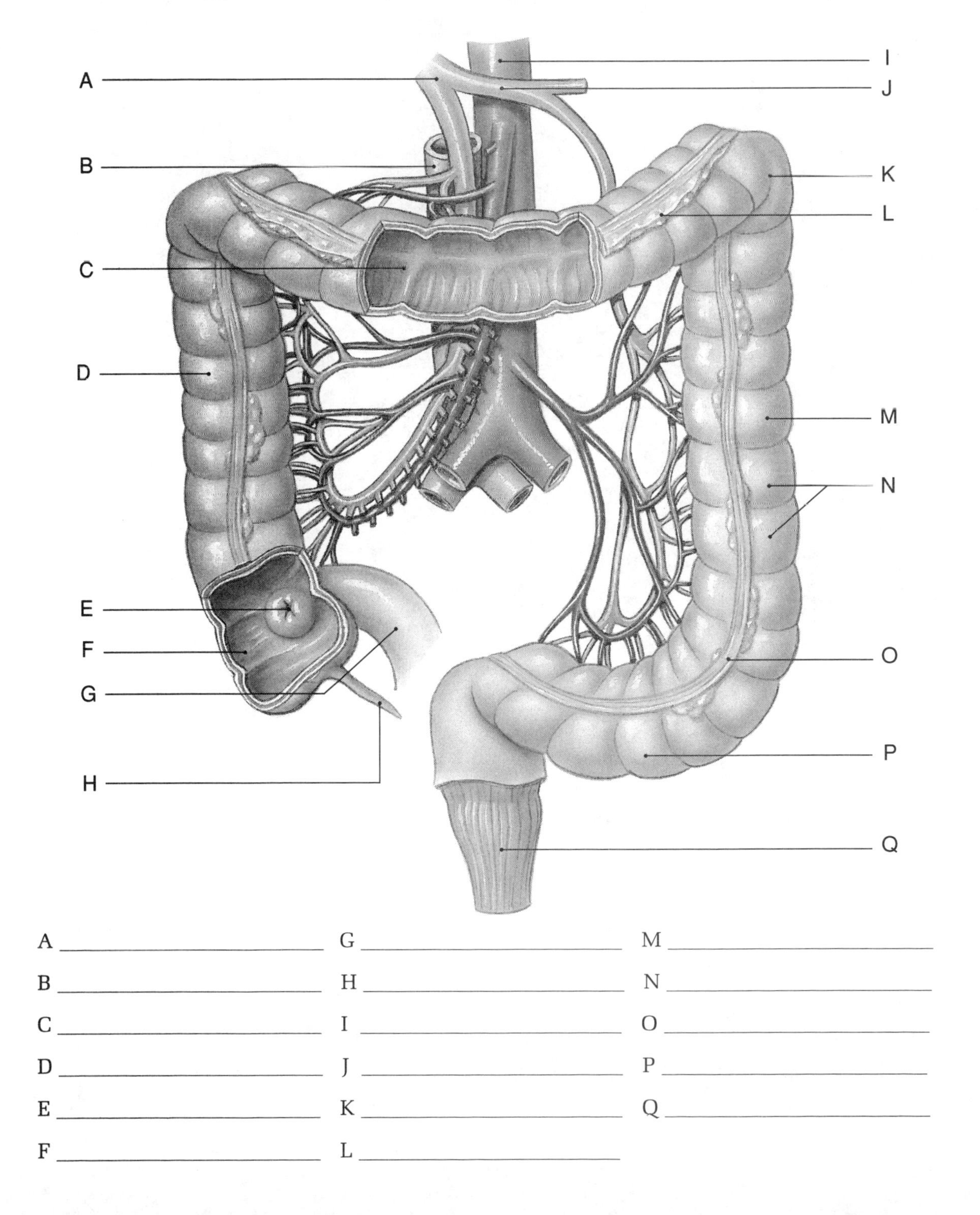

A ____________________ G ____________________ M ____________________

B ____________________ H ____________________ N ____________________

C ____________________ I ____________________ O ____________________

D ____________________ J ____________________ P ____________________

E ____________________ K ____________________ Q ____________________

F ____________________ L ____________________

Objective 10 Describe the digestion and absorption of carbohydrates, lipids and proteins.

_____ 1. Salivary amylase, which is secreted in the mouth, aids in the digestion of

a. protein
b. fat
c. carbohydrate
d. all of the above

_____ 2. An enzyme that will digest proteins into polypeptides is

a. lipase
b. amylase
c. nuclease
d. trypsin

_____ 3. Enzymes involved in fat digestion are called

a. pepsins
b. lipases
c. carboxypeptidases
d. amylases

_____ 4. Hydrochloric acid in the stomach functions primarily to

a. activate enzymes involved in protein digestion
b. hydrolyze peptide bonds
c. facilitate lipid digestion
d. facilitate carbohydrate digestion

_____ 5. Digested fats are absorbed into the _____ and then enter the _____.

a. bloodstream, lacteals
b. lacteals, villi
c. villi, bloodstream
d. villi, lacteals

Objective 11 Describe the changes in the digestive tract that occur with aging.

_____ 1. Weaker peristaltic contractions in the elderly results in more frequent

a. diarrhea
b. constipation
c. heartburn
d. all of the above

_____ 2. Dietary changes that affect the entire body in the elderly result from a

a. weakening of muscular sphincters
b. weaker peristaltic contractions
c. decline in sense of smell and taste
d. decreased movement within the G. I. tract

_____ 3. The types of cancers most common in the elderly are

a. oral and pharyngeal
b. lung and liver
c. colon and kidney
d. skin and lymphatic

Part II: Chapter Comprehensive Exercises

A. Word Elimination

Circle the term that does not belong in each of the following groupings.

1. ingestion digestion secretion circulation absorption
2. mucosa mesentery submucosa muscularis externa serosa
3. analysis excretion mechanical processing lubrication begin digestion
4. tongue parotid sublingual submandibular salivary glands
5. incisors cuspids dentin bicuspids molars
6. cardia fundus chyme body pylorus
7. gastric pits gastric glands parietal cells chief cells intrinsic factor
8. gastrin secretin CCK HCI GIP
9. trypsin amylase chymotrypsin charboxypeptidase proteolytic
10. haustra ascending sigmoid transverse descending

B. Matching

Match the terms in Column B with the terms in Column A. Place your answers in the spaces provided.

COLUMN A	COLUMN B
___ 1. mesenteries	A. soupy mixture
___ 2. bolus	B. secrete pepsinogen
___ 3. chyme	C. absorption of vitamin B12
___ 4. rugae	D. colon pouches
___ 5. intrinsic factor	E. small mass of food
___ 6. chief cells	F. terminal lymphatic
___ 7. parietal cells	G. serous membrane
___ 8. lacteal	H. bands of muscle
___ 9. haustra	I. ridges and folds
___ 10. taenia coli	J. secrete HCI

C. Concept Map I - The Digestive System

Using the following terms, fill in the numbered, blank spaces to complete the concept map. Follow the numbers that comply with the organization of the map.

Digestive tract movements
Intestinal mucosa
Large intestine
Hormones
Amylase
Hydrochloric acid
Stomach
Bile
Pancreas

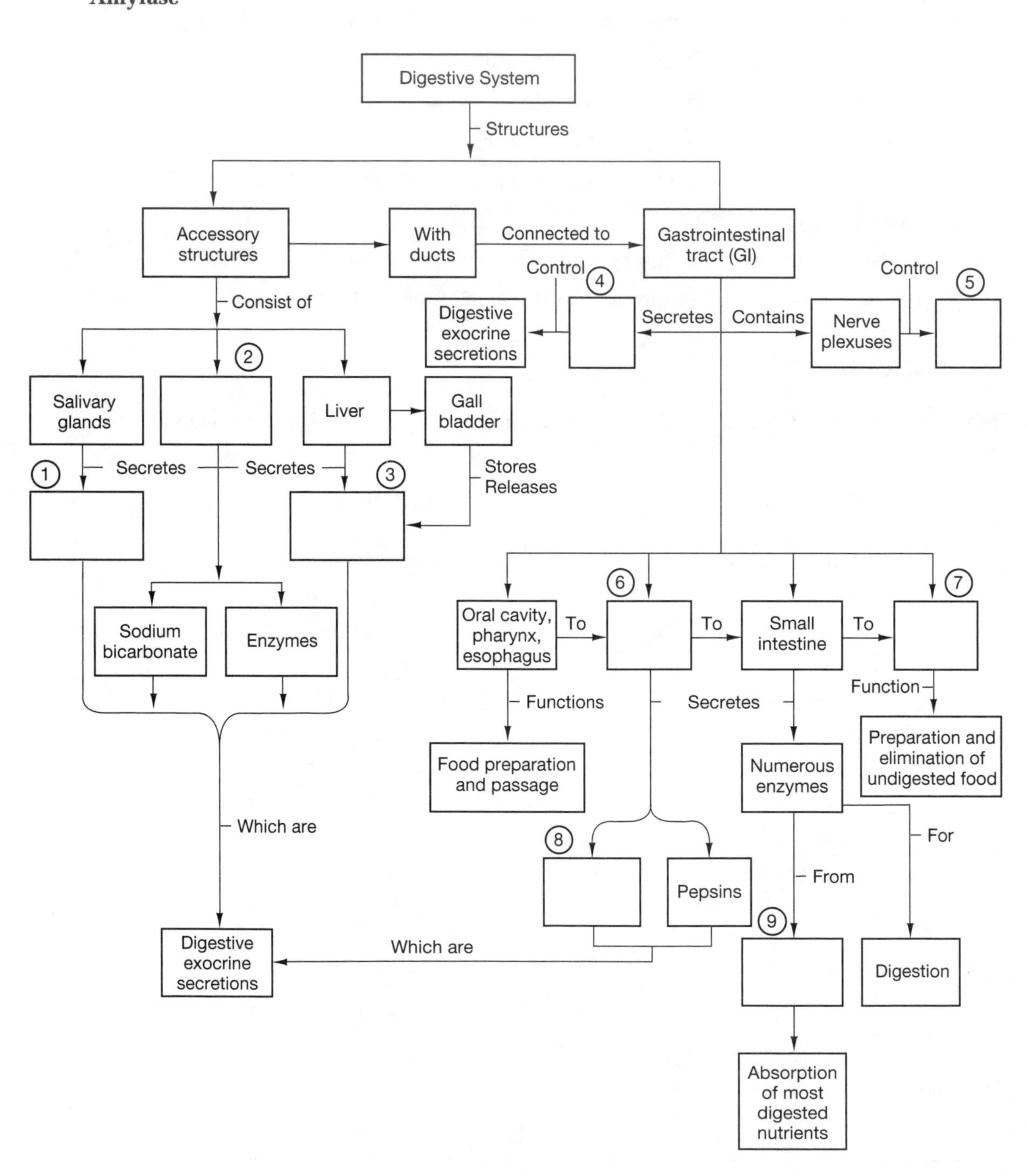

Concept Map II - Chemical Events in Digestion

Using the following terms, fill in the circled, numbered, blank spaces to complete the concept map. Follow the numbers that comply with the organization of the map.

Simple sugars
Esophagus
Triglycerides
Monoglycerides, fatty acids in micelles
Disaccharides, trisaccharides
Lacteal
Polypeptides
Small intestine
Complex sugars and starches
Amino acids

Chemical events in digestion

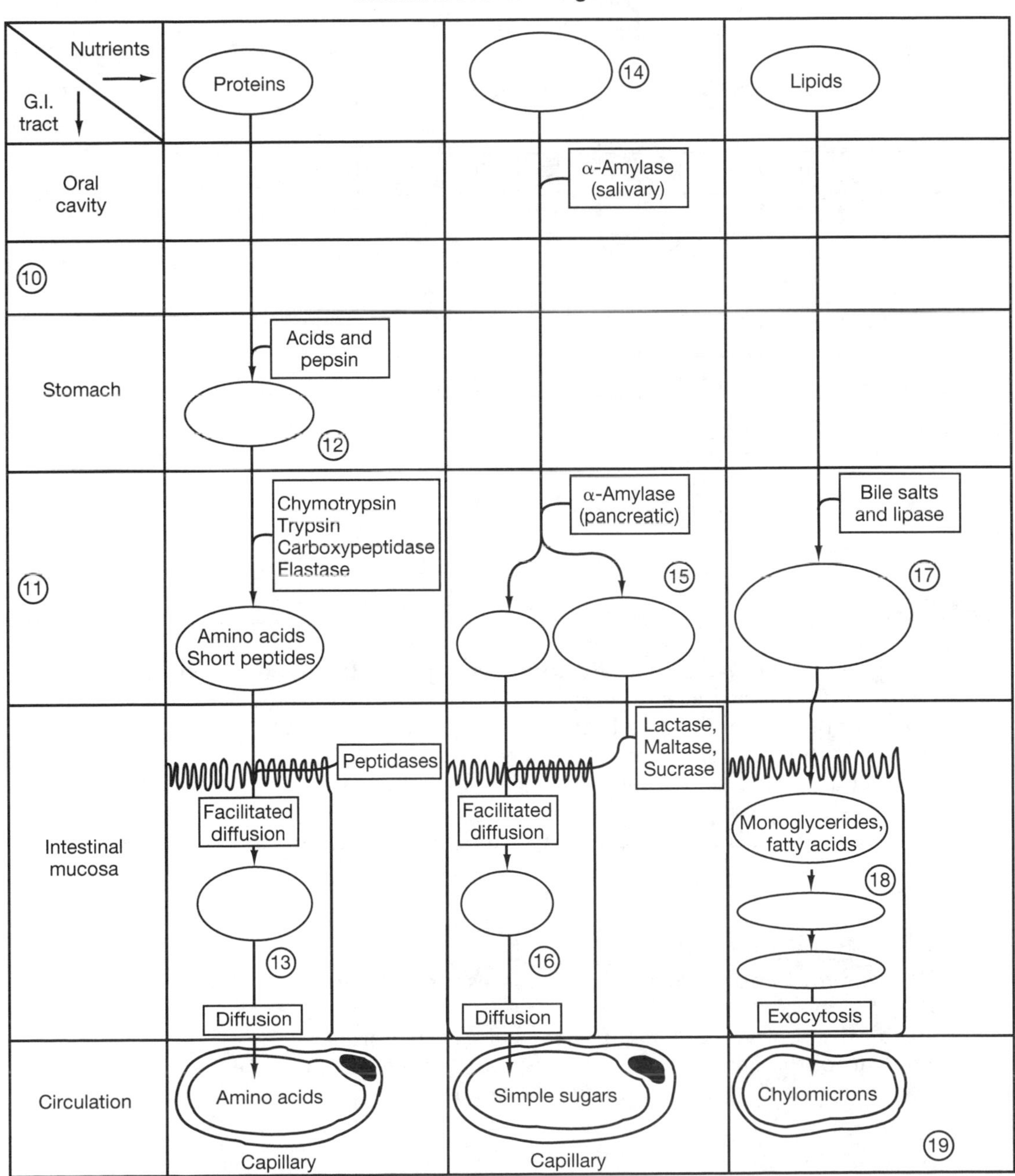

D. Crossword Puzzle

The following crossword puzzle reviews the material in Chapter 17. To complete the puzzle, you must know the answers to the clues given, and must be able to spell the terms correctly.

ACROSS

3. The process of moving nutrients from the digestive tract to the bloodstream.
4. The gland that produces salivary amylase.
5. The first portion of the small intestine.
6. Where most of the digestion in the digestive tract occurs.
8. The pancreas and gallbladder secrete their products into the _______ of the small intestine.
9. The pancreas and liver are considered to be _______ organs of the digestive system.

DOWN

1. Salivary amylase digests ________ to form glucose.
2. To reach to stomach the esophagus passes through the esophageal ______ of the diaphragm.
7. The fleshy protrusion hanging at the back of the throat.
10. The part of the large intestine to which the appendix is attached.
11. The muscular ridges found inside the stomach.

E. Short Answer Questions

Briefly answer the following questions in the spaces provided below.

1. What seven (7) integrated steps comprise the digestive functions?

2. What three (3) pairs of salivary glands secrete saliva into the oral cavity?

3. What four (4) types of teeth are found in the oral cavity and what is the function of each type?

4. What are the similarities and differences between parietal cells and chief cells in the wall of the stomach?

5. (a) What three (3) phases are involved in the regulation of gastric function?

   (b) What regulatory mechanism(s) dominate each phase?

6. What are the three (3) most important hormones that regulate intestinal activity?

7. What are the three (3) principal functions of the large intestine?

8. What three (3) basic categories describe the functions of the liver?

9. What two (2) distinct functions are performed by the pancreas?

CHAPTER

18

Nutrition and Metabolism

Overview

The knowledge of what happens to the food we eat and how the nutrients from food are utilized by the body continues to capture the interest of those who believe that "we are what we eat." Many people are usually more concerned with the taste of food or their "likes" than with the nutritional value of the food. Health claims about foods and food supplements have become a daily part of the news media, causing the "magic revolution" or "quick fix" for a healthy body to become believable in the minds of many people. Which claims are ridiculous, and which ones really work? A basic understanding of nutrition and knowledge of the fate of absorbed nutrients, can help us to answer these and other questions, so that we can make good food choices and develop adequate food plans that provide for happy, healthy living.

All cells require energy for the processes of metabolism. Provisions for these cellular activities are made by the food we eat; that is, it provides energy, promotes tissue growth and repair, and serve as regulatory mechanisms for the body's metabolic machinery.

The exercises in this chapter will provide some basic review of what you have learned in previous chapters and will be useful to you as you begin to understand how the integrated processes of the digestive and cardiovascular systems serve to advance the work of metabolism. They will also provide you with the basic principles of good nutrition and what the body requires to be healthy.

Review of Chapter Objectives

1. Define metabolism and explain why cells need to synthesize new organic components.
2. Describe the basic steps in glycolysis, the TCA cycle, and the electron transport system.
3. Describe the pathways involved in lipid metabolism.
4. Discuss protein metabolism and the use of proteins as an energy source.
5. Discuss nucleic acid metabolism.
6. Explain what constitutes a balanced diet, and why it is important.
7. Discuss the functions of vitamins, minerals and other important nutrients.
8. Describe the significance of the caloric value of foods.
9. Define metabolic rate and discuss the factors involved in determining an individual's metabolic rate.
10. Discuss the homeostatic mechanisms that maintain a constant body temperature.

Part I: Objective Based Questions

Objective 1 Define metabolism and explain why cells need to synthesize new organic components.

_____ 1. The sum of all the biochemical processes in the human body at any given time is called

a. anabolism
b. catabolism
c. oxidation
d. metabolism

_____ 2. The metabolic components of the body that interact to pressure homeostasis are

a. heart, lungs, kidneys, brain, and pancreas
b. blood, lymph, cerebrospinal fluid, hormones, and bone marrow
c. liver, skeletal muscle, adipose tissue, neural tissue, and other peripheral tissues
d. bones, muscles, integument, glands, and the heart

_____ 3. In resting skeletal muscles, a significant portion of the metabolic demand is met through the

a. catabolism of glucose
b. catabolism of fatty acids
c. catabolism of glycogen
d. anabolism of ADP to ATP

_____ 4. The process that breaks down organic substrates, releasing energy that can be used to synthesize ATP or other high energy compounds is

a. metabolism
b. anabolism
c. catabolism
d. oxidation

Objective 2 Describe the basic steps in glycolysis, the TCA cycle, and the electron transport system.

_____ 1. During glycolysis, carbon glucose molecules are broken down into 2 three-carbon molecules of

a. pyruvic acid
b. acetyl-CoA
c. citric acid
d. oxaloacetic acid

_____ 2. The first step in a sequence of enzymatic reactions in the TCA cycle is the formation of

a. acetyl-CoA
b. citric acid
c. oxaloacetic acid
d. pyruvic acid

_____ 3. The carbon dioxide released during cellular respiration is formed during

a. electron transport
b. glycolysis
c. TCA cycle
d. the formation of pyruvic acid

_____ 4. The TCA cycle must turn _______ times to completely metabolize the pyruvic acid produced from one glucose molecule.

a. 1
b. 2
c. 3
d. 4

_____ 5. During glycolysis, each molecule of glucose metabolized releases enough energy to form a net gain of ____ molecules of ATP.

a. 2
b. 4
c. 36
d. 38

_____ 6. In the process of cellular respiration, each molecule of glucose metabolized releases enough energy to form _____ molecules of ATP.

a. 2
b. 4
c. 36
d. 38

Objective 3 Describe the pathways involved in lipid metabolism.

_____ 1. During lipolysis

a. triglycerides are converted into molecules of acetyl-CoA
b. lipids are converted into glucose molecules
c. triglycerides are broken down into glycerol and fatty acids
d. lipids are formed from excess carbohydrates

_____ 2. Beta oxidation

a. occurs in the mitochondria
b. is the process that breaks down fatty acids into two-carbon fragments that can be metabolized by the TCA cycle
c. yields large amounts of ATP, while requiring coenzymes A, NAD, and FAD
d. all of the above

_____ 3. Lipids

a. release less energy than an equivalent amount of glucose
b. are difficult to store since they are not water soluble
c. are easily mobilized from their reserves
d. release more energy than an equivalent amount of glucose

4. Lipogenesis generally begins with __________________.

5. The largest metabolic reserves for the average adult are stored as ____________.

6. Lipoproteins that contain triglycerides manufactured in the liver which they transport to peripheral tissues are called ________________.

7. Lipoproteins that carry mostly cholesterol and phospholipids from peripheral tissues to the liver are called ______________.

Objective 4 Discuss protein metabolism and the use of proteins as an energy source.

_____ 1. The first step in amino acid catabolism is the removal of the

a. carboxyl group
b. amino group
c. keto acid
d. hydrogen from the central carbon

_____ 2. The process of deamination produces

a. keto acids
b. urea
c. ammonia
d. acetyl-CoA

_____ 3. Urea is formed in the

a. liver
b. kidneys
c. small intestine
d. pancreas

_____ 4. In transamination, the amino group of an amino acid is

a. converted to ammonia
b. converted to urea
c. transferred to acetyl-CoA
d. transferred to another carbon chain

_____ 5. The factor(s) that make protein metabolism an impractical source of quick energy is (are)

a. their energy yield is less than that of lipids
b. the byproduct ammonia is a toxin that can damage cells
c. proteins are important structural and functional cell components
d. a, b, and c are correct

Objective 5 Discuss nucleic acid metabolism.

_____ 1. All cells synthesize RNA, but DNA synthesis occurs only in

a. red blood cells
b. cells that are preparing for mitosis
c. lymph and cerebrospinal fluid
d. the bone marrow

_____ 2. The only parts of a nucleotide of RNA which can provide energy when broken down are the

a. sugars and phosphates
b. sugars and purines
c. sugars and pyrimidines
d. purines and pyrimidines

_____ 3. Nucleotides from RNA

a. are deaminated to form ammonia
b. can provide sugars for glycolysis
c. can be used to synthesize proteins
d. cannot be used as a source of energy for the production of ATP

_____ 4. During RNA catabolism the pyrimidines, cytosine and uracil, are converted to acetyl-CoA and metabolized via

a. glycolysis
b. the TCA cycle
c. gluconeogenesis
d. all of the above

Objective 6 Explain what constitutes a balanced diet, and why it is important.

_____ 1. A balanced diet contains all the ingredients necessary to

a. prevent starvation
b. prevent life-threatening illnesses
c. prevent deficiency diseases
d. maintain homeostasis

_____ 2. The foods that are deficient in dietary fiber are

a. vegetables and fruits
b. breads and cereals
c. milk and meat
d. rice and pastas

_____ 3. Foods that are low in fats, calories, and proteins are

a. vegetables and fruit
b. milk and cheese
c. meat, poultry, and fish
d. breads, cereals, and rice

_____ 4. Minerals, vitamins, and water are classified as *essential* nutrients because

a. they are used by the body in large quantities
b. the body cannot synthesize the nutrients in sufficient quantities
c. they are major providers of calories for the body
d. all of the above

_____ 5. The *trace* minerals found in extremely small quantities in the body include

a. sodium, potassium, chloride, calcium
b. phosphorus, magnesium, calcium, iron
c. iron, zinc, copper, manganese
d. phosphorus, zinc, copper, potassium

_____ 6. Hypervitaminosis involving water-soluble vitamins is relatively uncommon because excessive amounts are

a. stored in adipose tissue
b. readily excreted in the urine
c. stored in the bones
d. readily absorbed into skeletal muscle tissue

_____ 7. The nutrients that yield approximately 4 calories per gram when metabolized are

a. water and vitamins
b. carbohydrates and fats
c. fats and proteins
d. proteins and carbohydrates

8. Identify and label the foods which comprise the food pyramid. Place your answers in the spaces provided below the drawing. Select from the following:

Figure 18-1 The Food Pyramid

Fruit group	**Fats, oils, and sweets**
Bread, cereal, rice	**Vegetable group**
Meat, poultry, fish, nuts, beans	**Milk, yogurt, cheese**

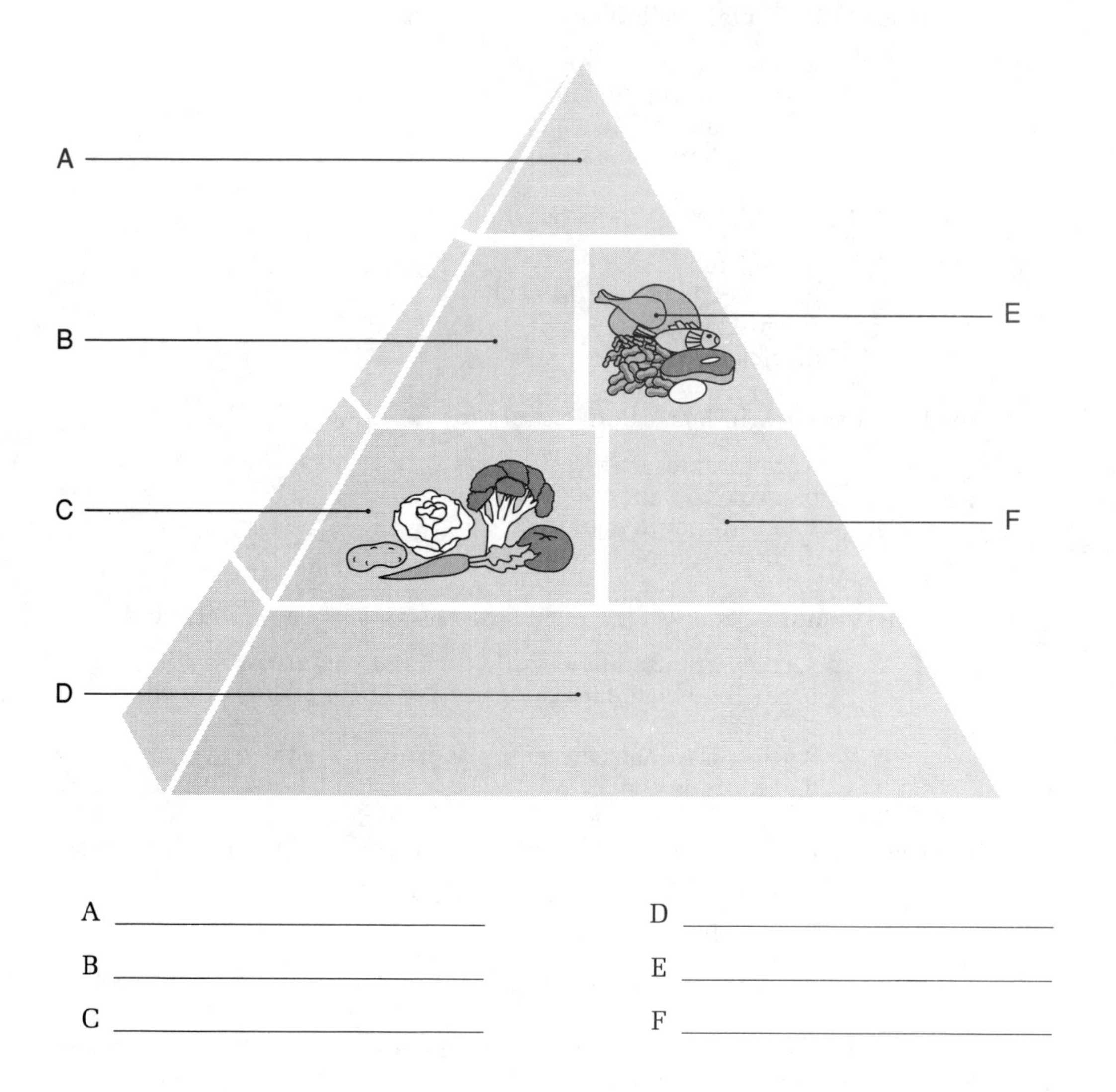

A ______________________ D ______________________

B ______________________ E ______________________

C ______________________ F ______________________

Objective 7 Discuss the functions of vitamins, minerals, and other important nutrients.

1. The major anion in body fluids is ______________.
2. A cation that is essential for muscle contraction, nerve function, and blood clotting is ___________.
3. An ion that is a necessary component of high-energy compounds and nucleic acids, and a structural component of bone is _________________.
4. A mineral that is a component of hemoglobin, myoglobin, and cytochromes is _____________.
5. A mineral that is a necessary cofactor for hemoglobin synthesis is _________________.
6. The vitamin required for the synthesis of visual pigments is __________________.
7. The vitamin required for proper bone growth and calcium absorption and retention is ___________________.
8. The vitamin essential for the production of several clotting factors is ______________.
9. The vitamin that is a constituent of the coenzymes FAD and FMN is ______________.
10. The vitamin that is a constituent of the coenzyme NAD is _______________.
11. The vitamin that plays the role of a coenzyme in amino acid and lipid metabolism is _____________.
12. The vitamin that is a coenzyme in amino acid and nucleic acid metabolism is _______________.

Objective 8 Describe the significance of the caloric value of foods.

1. The amount of energy needed to raise the temperature of 1 kilogram of water one degree centigrade is the _______________.

_____ 2. The catabolism of lipids releases approximately ____________ calories per gram of energy

a. 4.18
b. 4.32
c. 9.46
d. none of the above

_____ 3. The catabolism of carbohydrates releases ________________ calories per gram of energy

a. 4.18
b. 4.32
c. 9.46
d. 12.21

_____ 4. The catabolism of proteins yields approximately ____________ calories of energy.

a. 4.18
b. 4.32
c. 9.46
d. 2.61

Objective 9 Define metabolic rate and discuss the factors involved in determining an individual's metabolic rate.

_____ 1. The sum total of all the varied anabolic and catabolic processes occurring in the body represents and individual's

a. bioenergetic condition
b. thermoregulatory status
c. physiological condition
d. metabolic rate

_____ 2. The factor(s) that influence an individual's BMR is (are)

a. sex (gender)
b. age
c. body weight and genetics
d. all of the above

_____ 3. To examine the metabolic state of an individual, results may be expressed as

a. calories per hour
b. calories per day
c. calories per unit of body weight per day
d. a, b, and c are correct

_____ 4. An individual's basal metabolic rate ideally represents

a. the minimum, resting energy expenditure of an awake, alert person
b. genetic differences among ethnic groups
c. the amounts of circulating hormone levels in the body
d. a measurement of the daily energy expenditures for a given individual

Objective 10 Discuss the homeostatic mechanisms that maintain a constant body tem perature.

_____ 1. The greatest amount of the daily water intake is obtained by

a. consumption of food
b. drinking fluids
c. metabolic processes
d. decreased urination

_____ 2. The four processes involved in heat exchange with the environment are

a. physiological, behavioral, generational, acclimatization
b. radiation, conduction, convection, evaporation
c. sensible, insensible, heat loss, heat gain
d. thermogenesis, dynamic action, pyrexia, thermalphasic

_____ 3. The primary mechanisms for increasing heat loss in the body include

a. sensible and insensible
b. acclimatization and pyrexia
c. physiological mechanisms and behavioral modification
d. vasomotor and respiratory

4. The homeostatic process that keeps body temperatures within acceptable limits regardless of environmental conditions is called _______________.

5. The nonpathological term for an elevated body temperature is _______________.

Part II: Chapter Comprehensive Exercises

A. Word Elimination

Circle the term that does not belong in each of the following groupings.

1. Lipids Proteins Glycolysis Carbohydrates Water
2. Glycolysis TCA Electron transport Krebs cycle anaerobic
3. Glucose NAD Pyruvic acid Acetyl-CoA citric acid
4. ATP fatty acids glycerol glucose amino acids
5. linoleic acid arachidonic linolenic EFA EAA
6. leucine lysine trytophan linoleic phenylalanine
7. milk minerals meat vegetables bread
8. sodium potassium zinc chloride calcium
9. vitamin A vitamin D vitamin C vitamin E vitamin K
10. thermoregulation radiation conduction convection evaporation

B. Matching

Match the terms in Column B with the terms in Column A. Use letters for answers in the spaces provided.

COLUMN A	COLUMN B
___ 1. catabolism	A. lipid storage
___ 2. anabolism	B. N compounds
___ 3. adipose tissue	C. resting energy expenditure
___ 4. ketone bodies	D. formation of new chemical bonds
___ 5. purines and pyrimidines	E. B complex and C
___ 6. gluconeogenesis	F. increasing muscle tone
___ 7. water-soluble vitamins	G. energy release
___ 8. basal metabolic rate	H. glucose synthesis
___ 9. shivering termogenesis	I. increasing metabolism
___ 10. non-shivering thermogenesis	J. protein catabolism

C. Concept Map

Using the following terms, fill in the circled, numbered, blank spaces to complete the concept map. Follow the numbers that comply with the organization of the map.

Vitamins	**Meat**	**Proteins**
Metabolic regulators	**Tissue growth and repair**	**Carbohydrates**
Vegetables	**9 cal/gram**	

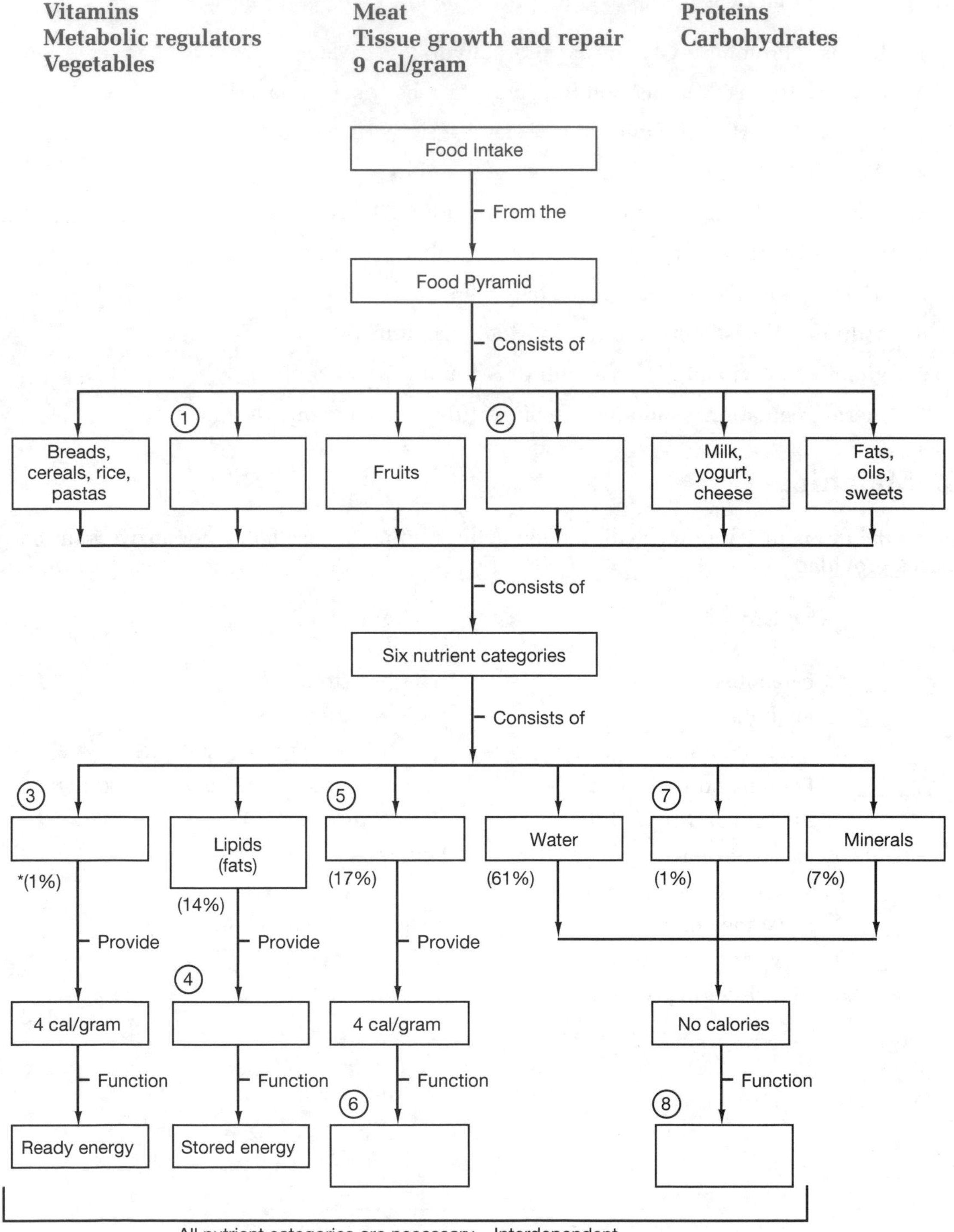

Concept Map II

Using the following terms, fill in the circled, numbered, blank spaces to complete the concept map. Follow the numbers that comply with the organization of the map.

Gluconeogenesis	**Amino acids**	**Lipogenesis**
Beta oxidation	**Lipolysis**	**Glycolysis**

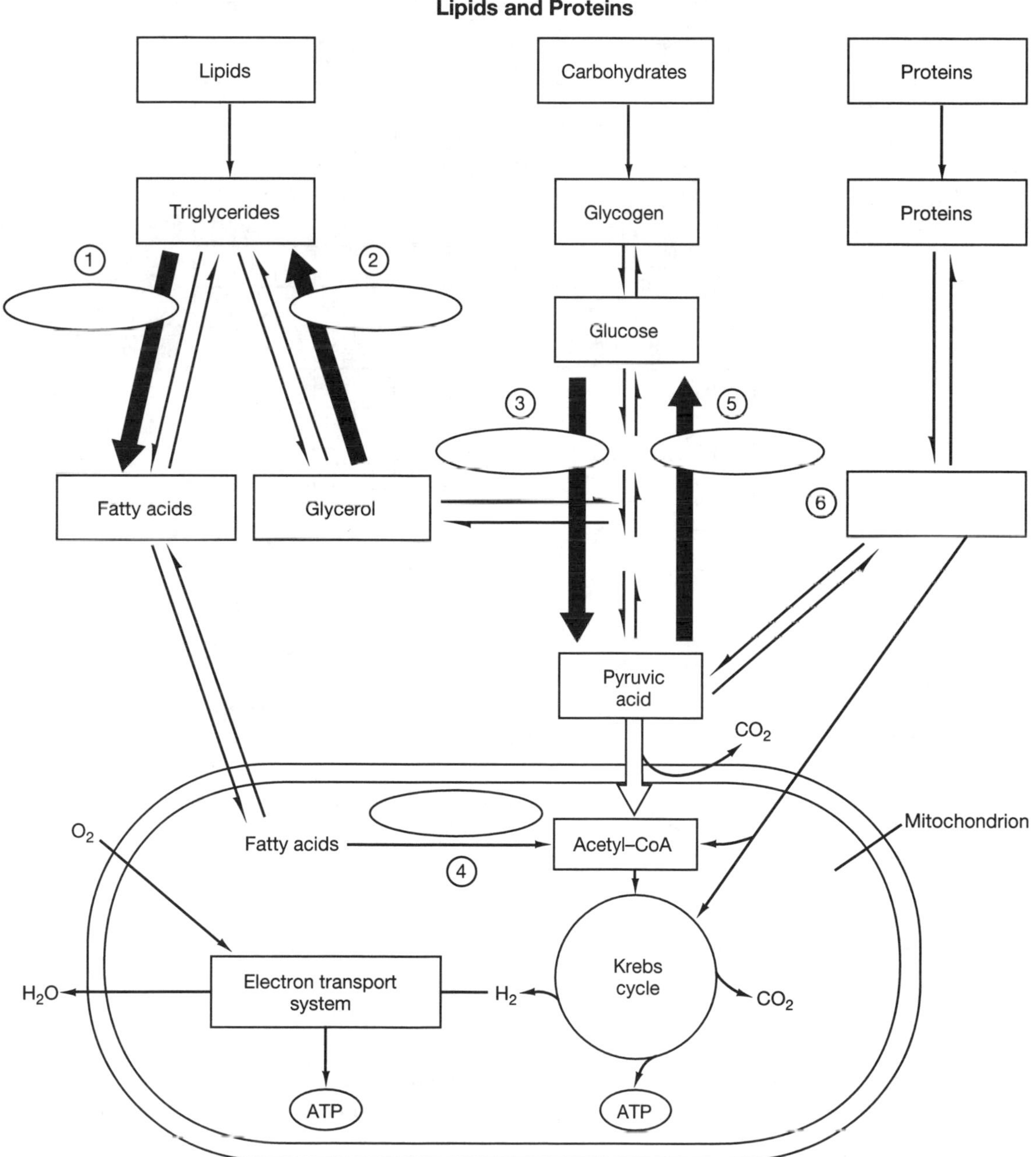

D. Crossword Puzzle

The following crossword puzzle is a review of the material in Chapter 18. To complete the puzzle, you must know the answers to the clues given, and must be able to spell the terms correctly.

ACROSS

3. A molecule required by the body but which the body cannot make.
4. Metabolic reactions that break down molecules.
5. One of the essential fatty acids.
8. One of the essential amino acids.
10. A unit of measure that refers to the amount of energy in a specific type of food.
11. The digestion of fats produces fatty acids and ________________.

DOWN

1. The digesting of protein produces ___________.
2. The sum of all the chemical reactions in the body.
6. A food substance that has all the amino acids the body requires is known as a(n) __________ protein.
7. The digestion of carbohydrates produces ____________.
9. The molecule from which mitochondria ultimately produce ATP is ____________ acid.

E. Short Answer Questions

Briefly answer the following questions in the spaces provided below.

1. What factors make protein catabolism an impractical source of quick energy?

2. Why are nucleic acids insignificant contributors to the total energy reserves of the cell?

3. From a metabolic standpoint, what five (5) distinctive components are found in the human body?

4. What six (6) food groups in the Food Pyramid provide the basis for a balanced diet?

5. Even though minerals do not contain calories, why are they important in good nutrition?

6. (a) List the fat-soluble vitamins.

   (b) List the water-soluble vitamins.

CHAPTER

19

The Urinary System

Overview

Although several organ systems are involved in the body's excretory processes, the urinary system is primarily responsible for the removal of nitrogenous wastes from the blood. The kidneys play a crucial homeostatic role by cleansing the blood of waste products and excess water by forming urine, which is then transferred to the urinary bladder and eventually excreted from the body during urination. In addition to the kidneys, the urinary system consists of the ureters, urinary bladder, and urethra, components that are responsible for transport, storage, and conduction and elimination of urine to the exterior. Few people are aware of the many functions the kidneys perform. In addition to the preparation and excretion of urine they serve to: (1) regulate blood pressure; (2) conserve valuable nutrients; (3) regulate blood ions; and (4) regulate blood pH. The kidneys receive more blood from the heart than any other body organ. One and one-half quarts of blood every minute circulate through a complex renal pathway housed in each kidney. The blood filtering system of the kidneys consists of millions of tubules and specialized cells that are designed to filter, reabsorb, and secrete a filtrate, ultimately leaving about 1% of the original filtrate volume as urine to be excreted.

Chapter 19 focuses on the structural and functional organization of the urinary system, the regulatory mechanisms that control urine formation and modification, and urine transport, storage, and elimination.

Review of Chapter Objectives

1. Identify the components of the urinary system and describe their functions.
2. Describe the structural features of the kidney.
3. Describe the structure of the nephron and the processes involved in urine formation.
4. Trace the path of blood flow through the kidney.
5. List and describe the factors that influence filtration pressure and the rate of filtrate formation.
6. Describe the changes that occur in the filtrate as it moves through the nephrons and exits as urine.
7. Describe the structures and functions of the ureters, urinary bladder and urethra.
8. Discuss the process of urination and how it is controlled.

9. Explain how the urinary system interacts with other body systems to maintain homeostasis in body fluids.
10. Describe how water and electrolytes are distributed within the body.
11. Explain the basic concepts involved in the control of fluid and electrolyte regulation.
12. Explain the buffering systems that balance the pH of the intracellular and extracellular fluid.
13. Describe the effects of aging on the urinary system.

Part I: Objective Based Questions

Objective 1 Identify the components of the urinary system and describe their functions.

_____ 1. Urine production occurs in the

a. ureters
b. kidney
c. urethra
d. urinary bladder

_____ 2. Urine leaves the kidneys on its way to the urinary bladder via the

a. urethra
b. nephrons
c. ureters
d. trigone

_____ 3. The organ or structure that does not belong to the urinary system is the

a. gall bladder
b. urethra
c. kidneys
d. ureters

_____ 4. Urine exits from the body via the

a. ureters
b. urinary bladder
c. penis
d. urethra

_____ 5. The initial factors which determines if urine production occurs is

a. secretion
b. absorption
c. sympathetic activation
d. filtration

_____ 6. Along with the urinary system, the other systems of the body that affect the composition of body fluids are

a. nervous, endocrine, and cardiovascular
b. lymphatic, cardiovascular, and respiratory
c. integumentary, respiratory, and digestive
d. muscular, digestive, and lymphatic

7. Identify and label the structures comprising the Urinary System. Place your answers in the spaces provided below the drawing.

Figure 19-1 Components of the Urinary System

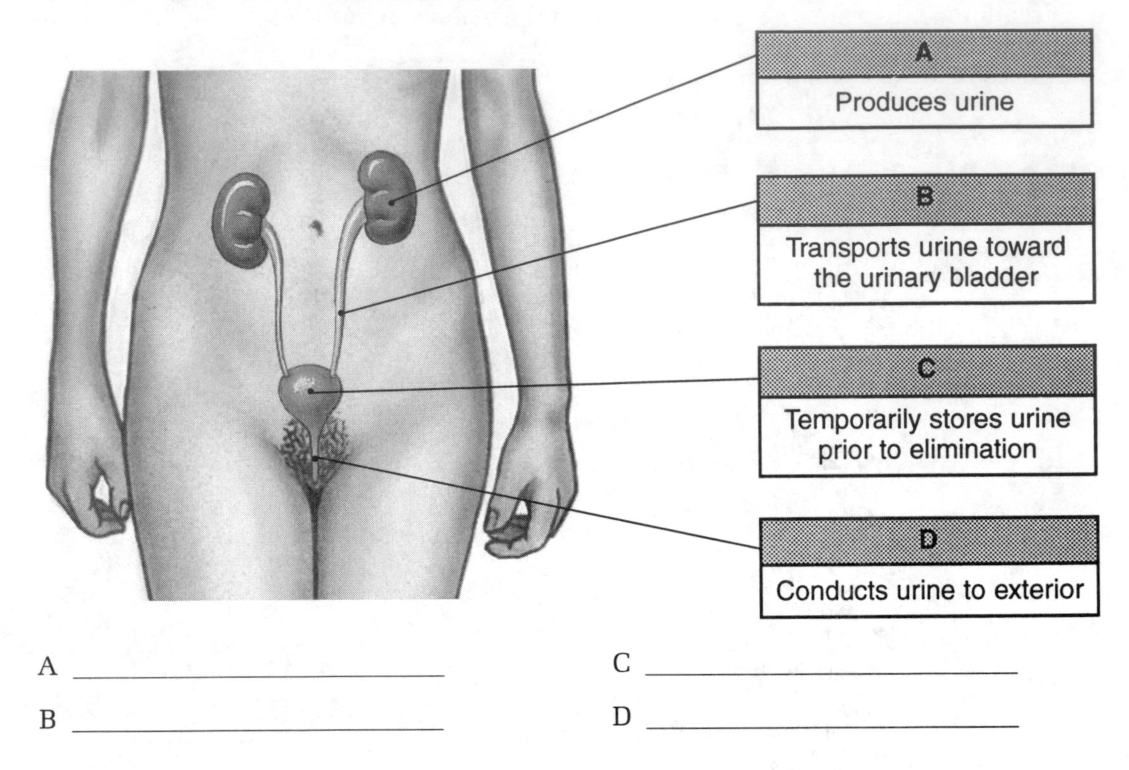

A ______________________ C ______________________

B ______________________ D ______________________

Objective 2 Describe the structural features of the kidney.

_____ 1. Seen in sections, the kidney is divided into

a. renal columns and renal pelvis
b. an outer cortex and an inner medulla
c. major and minor calyces
d. a renal tubule and renal corpuscle

_____ 2. The basic functional unit in the kidney is the

a. glomerulus
b. loop of Henle
c. Bowman's capsule
d. nephron

_____ 3. The three consecutive layers of connective tissue that protect and anchor the kidneys are the

a. hilus, renal sinus, renal corpuscle
b. cortex, medulla, papillae
c. renal capsule, adipose capsule, renal fascia
d. major calyces, minor calyces, renal pyramids

_____ 4. The hilus is the pint where

a. a ureter exits the kidney
b. blood vessels enter and exit the kidney
c. the urethra exits the kidney
d. a and b are correct

_____ 5. The renal pyramids are located in the

a. calyces
b. cortex
c. medulla
d. renal pelvis

_____ 6. Once formed, urine will follow a sequential pathway through the kidney which includes

a. renal pyramid, major calyx, minor calyx, renal pelvis, ureter
b. renal pyramid, minor calyx, major calyx, renal pelvis, ureter
c. renal pyramid, minor calyx, major calyx, renal pelvis, urethra
d. renal pyramid, renal cortex, medulla, renal pelvis, ureter

7. Identify and label the structures in the kidney. Place your answers in the spaces provided below the drawing. Select your answers from the following choices.

Figure 19-2 Sectional Anatomy of the Kidney

Ureter	**Cortex**	**Renal vein**
Minor calyx	**Renal capsule**	**Renal column**
Renal pelvis	**Major calyx**	**Renal pyramid**

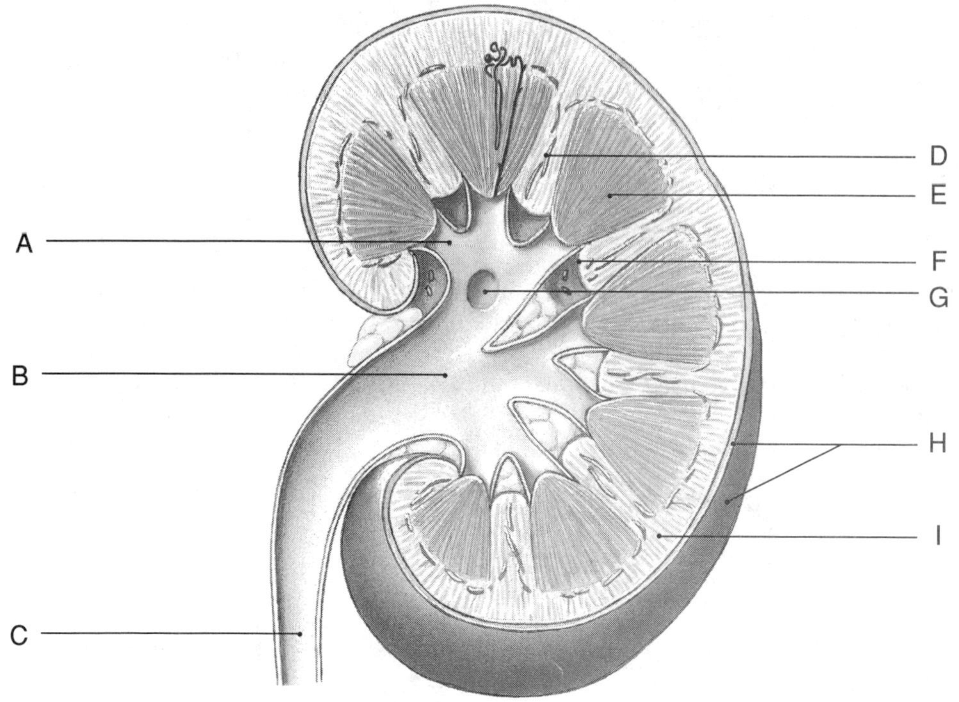

A ______________________ F ______________________

B ______________________ G ______________________

C ______________________ H ______________________

D ______________________ I ______________________

E ______________________

Objective 3 Describe the structure of the nephron and the processes involved in urine formation.

_____ 1. The "tuft" of capillaries that lies within the renal corpuscle is the

a. glomerulus
b. loop of Henle
c. juxtaglomerulas apparatus
d. major calyces

_____ 2. In a nephron, the tubular passageway through which the filtrate passes includes

a. collecting tubule, collecting duct, papillary duct
b. renal corpuscle, renal tubule, renal pelvis
c. proximal and distal convoluted tubules and loop of Henle
d. loop of Henle, collecting and papillary ducts

_____ 3. The primary site of regulating water, sodium, and potassium ion loss in the nephron is the

a. distal convoluted tubule
b. loop of Henle
c. proximal convoluted tubule
d. glomerulus

_____ 4. The three processes involved in urine formation are

a. diffusion, osmosis, and filtration
b. co-transport, countertransport, facilitated diffusion
c. regulation, elimination, micturition
d. filtration, reabsorption, secretion

_____ 5. The primary site for secretion of substances into the filtrate is the

a. renal corpuscle
b. loop of Henle
c. distal convoluted tubule
d. proximal convoluted tubule

_____ 6. The outflow across the walls of the glomerulus produces a protein-free solution known as the

a. urine
b. plasma
c. filtrate
d. lymph

7. Identify and label the following selections the structures comprising a nephron. Place your answers in the spaces provided below the drawing.

Figure 19-3 Diagrammatic View of a Nephron

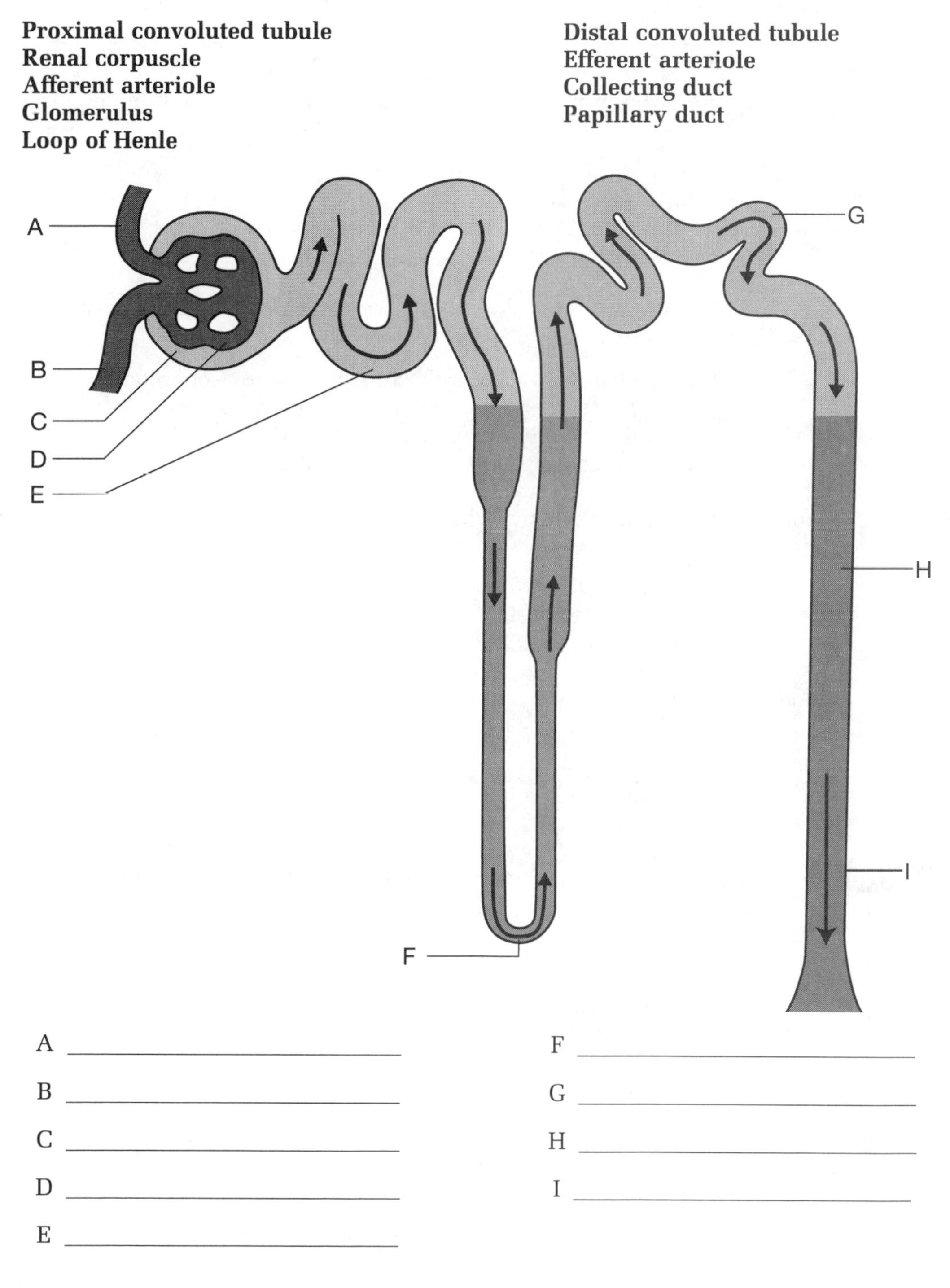

A ______________________

B ______________________

C ______________________

D ______________________

E ______________________

F ______________________

G ______________________

H ______________________

I ______________________

Objective 4 Trace the path of blood flow through the kidney.

_____ 1. Each kidney receives blood from a

a. afferent arteriole
b. arcuate artery
c. renal artery
d. interlobular artery

_____ 2. The slender capillary network that accompanies the loop of Henle into the medulla are known as the

a. macula densa
b. vasa recta
c. juxtaglomerular apparatus
d. papillary capillaries

_____ 3. Blood supply to the proximal and distal convoluted tubules of the nephron is provided by the

a. peritubular capillaries
b. afferent arterioles
c. segmental veins
d. interlobular veins

_____ 4. Dilation of the afferent arteriole and glomerular capillaries and construction of the efferent arteriole causes

a. elevation of glomerular blood pressure to normal levels
b. a decrease in glomerular blood pressure
c. a decrease in the glomerular filtration rate
d. an increase in the secretion of renin and erythropoietin

_____ 5. Blood enters the glomerulus via the

a. renal artery
b. interlobar artery
c. efferent arteriole
d. afferent arteriole

_____ 6. Blood leaves the kidney via the

a. efferent arteriole
b. renal vein
c. interlobar vein
d. arcuate vein

Objective 5 List and describe the factors that influence filtration pressure and the rate of filtrate formation.

_____ 1. The glomerular filtration rate is regulated by

a. autoregulation
b. hormonal regulation
c. autonomic regulation
d. all of the above

_____ 2. The pressure that represents the resistance to flow along the nephron and conducting system is the

a. blood colloid osmotic pressure (BCOP)
b. glomerular hydrostatic pressure (GHP)
c. capsular hydrostatic pressure (CHP)
d. capsular colloid osmotic pressure (CCOP)

_____ 3. The process of filtration occurs at the

a. proximal convoluted tubule
b. Bowman's capsule
c. loop of Henle
d. distal convoluted tubule

_____ 4. The most selective pores in the filtration membrane are located in the

a. podocytes
b. capillary endothelium
c. capsular space
d. lamina densa

_____ 5. The process of filtration is driven by

a. capsular hydrostatic pressure
b. active transport
c. blood osmotic pressure
d. blood hydrostatic pressure

_____ 6. The ability to form a concentrated urine depends on the functions of the

a. Bowman's capsule
b. collecting duct
c. loop of Henle
d. proximal convoluted tubule

Objective 6 Describe the changes that occur in the filtrate as it moves through the nephrons and exits as urine.

_____ 1. Approximately 60% of the volume of the filtrate produced in the renal corpuscle is reabsorbed in the

a. loop of Henle
b. proximal convoluted tubule
c. distal convoluted tubule
d. collecting duct

_____ 2. The distal convoluted tubule secretes

a. hydrogen ions
b. creatinine
c. potassium ions
d. all of the above

_____ 3. When the level of antidiuretic hormone (ADH) increases

a. more urine is produced
b. less water is reabsorbed by the nephron
c. less urine is produced
d. more salts are secreted by the nephron

_____ 4. Increased levels of aldosterone cause the kidneys to produce

a. a larger volume of urine
b. urine with a lower specific gravity
c. urine with a lower concentration of sodium ions
d. urine with a higher concentration of potassium ions

_____ 5. An abnormal characteristic of a urine sample is

a. cloudy
b. acidic pH
c. specific cavity greater than 1.0
d. odor of ammonia

_____ 6. From the following selections, identify the *abnormal* constituent of a urine sample

a. urea
b. creatinine
c. large proteins
d. hydrogen ions

Objective 7 Describe the structures and functions of the ureters, urinary bladder, and urethra.

_____ 1. When urine leaves the kidneys it travels to the urinary bladder via the

a. urethra
b. ureters
c. renal hilus
d. renal calyces

_____ 2. The expanded, funnel-shaped upper end of the ureter in the kidney is the

a. renal pelvis
b. urethra
c. renal hilus
d. renal calyces

_____ 3. Contraction of the muscular bladder forces the urine out of the body through the

a. ureter
b. penis
c. urethra
d. all of the above

4. The muscular ring that provides involuntary control over the discharge of urine from the urinary bladder is the ____________________.

5. In a relaxed condition the epithelium of the urinary bladder forms a series of prominent folds called ______________.

6. The area surrounding the urethral entrance of the urinary bladder is the __________________.

7. Identify and label the following structures of the male urinary bladder. Place your answers in the spaces provided below the drawing.

Figure 19-4 The Male Urinary Bladder

Detrusor muscle
Prostate gland
Ureter
Urethral openings
Internal sphincter
Trigone
External sphincter
Urethra

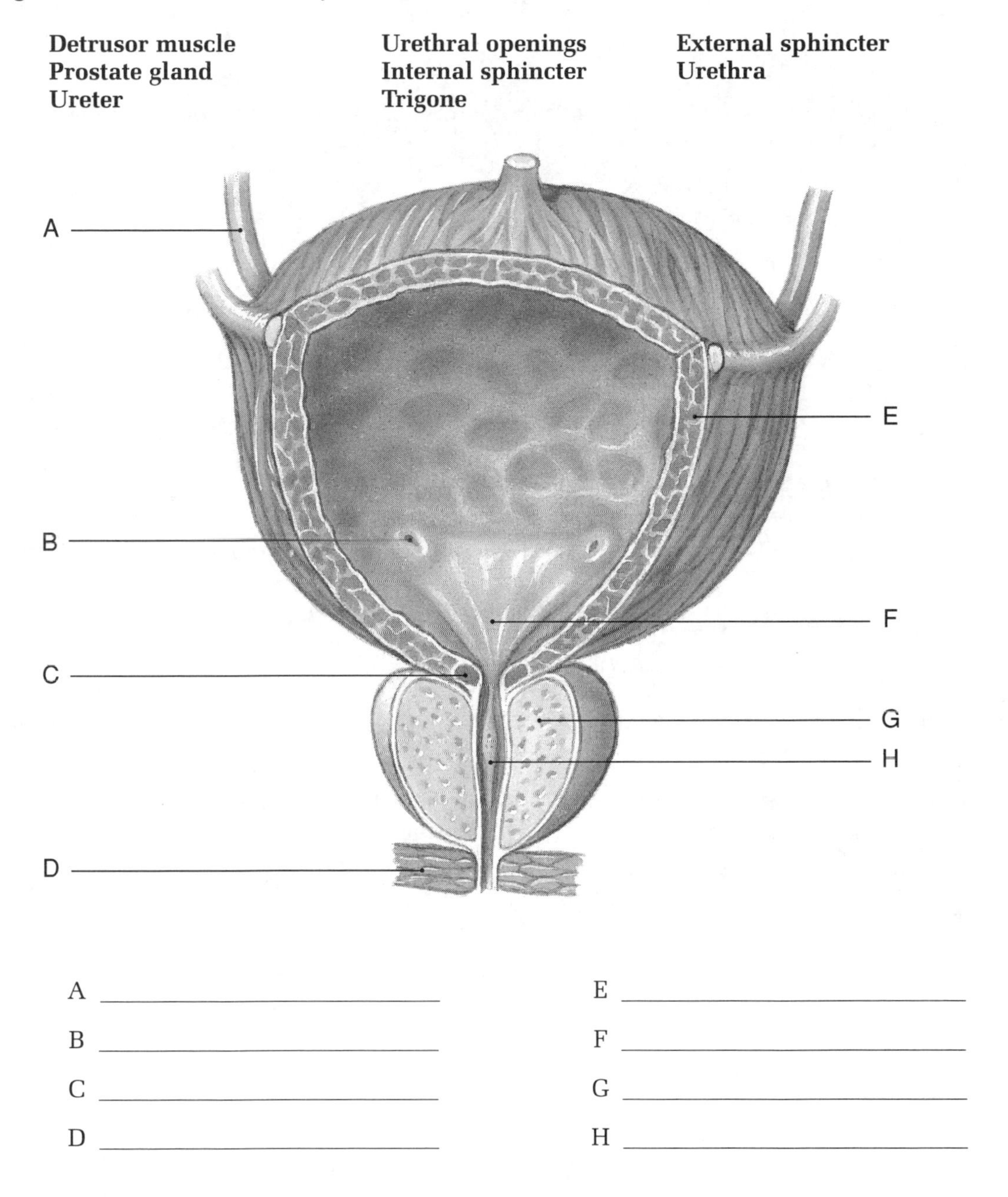

A ______________________ E ______________________

B ______________________ F ______________________

C ______________________ G ______________________

D ______________________ H ______________________

Objective 8 Discuss the process of urination and how it is controlled.

_____ 1. The first signal that triggers urination is when the urinary bladder contains about _______ ml of urine.

a. 10
b. 100
c. 200
d. 500

_____ 2. The involuntary internal sphincter opens when the urinary bladder contains about _______ ml of urine.

a. 100
b. 200
c. 500
d. 1000

_____ 3. The male urethra is used as a passageway for urine and seminal fluid, whereas the female is for urine only.

a. true
b. false

_____ 4. The muscle that compresses the urinary bladder and expels urine through the urethra is the

a. internal sphincter
b. external sphincter
c. rectus abdominis
d. detrusor

_____ 5. During the micturition reflex

a. stretch receptors in the bladder are stimulated
b. parasympathetic stimulation initiates smooth muscle activity in the bladder
c. the internal sphincter is consciously relaxed
d. the external sphincter relaxes

Objective 9 Explain how the urinary system interacts with other body systems to maintain homeostasis in body fluids.

1. The system involved when water and electrolyte losses in perspiration affect plasma volume and composition is the ____________________ system.

2. The system involved when compounds such as acetone and water evaporate into the alveoli and are eliminated during exhalation is the __________________ system.

3. The system involved when metabolic wastes are excreted in liver bile and variable amounts of water are lost in the feces is the ____________________ system.

Objective 10 Describe how water and electrolytes are distributed within the body.

_____ 1. The intracellular fluid (ICF) is found in

a. blood vessels
b. lymph
c. the cells of the body
d. the interstitial spaces

_____ 2. The extracellular fluid (ECF) of the body includes

a. blood plasma
b. cerebral spinal fluid
c. interstitial fluid
d. a, b, and c are correct

_____ 3. The principal cation in the ICF is

a. sodium
b. potassium
c. calcium
d. chloride

_____ 4. If the ECF solute concentration rises, water will

a. leave the ICF
b. enter the ICF
c. move into the ICF and back out again to maintain homeostasis
d. leave the ECF

_____ 5. When salt is consumed, water will

a. enter the urinary system and raise the blood pressure
b. leave the ECF and enter the ICF, causing a rise in the blood pressure
c. leave the ICF and enter the ECF causing the blood pressure to rise
d. cause a rise in the sodium concentration causing a rise in blood pressure

Objective 11 Explain the basic concepts involved in the control of fluid and electrolytes regulation.

_____ 1. When water is lost but electrolytes are retained

a. the osmolarity of the ECF falls
b. water moves by osmosis from the ICF to the ECF
c. the ICF and the ECF become diluted
d. there is an increase in the volume of the ICF

_____ 2. When large amounts of water are consumed the

a. ICF becomes hypertonic to the ICF
b. volume of the ECF will decrease
c. volume of the ICF will decrease
d. osmolarities of the ICF and ECF will be slightly lower

_____ 3. Consuming a high intake of salt will

a. result in a temporary increase in blood volume
b. decrease thirst
c. cause hypotension
d. activate the renin-angiotensin mechanism

_____ 4. When the level of sodium ions in the ECF decreases

a. osmoreceptors are stimulated
b. a person experiences increased thirst
c. more ADH is released
d. there is an increase in the level of aldosterone

_____ 5. Large amounts of chloride ions are lost each day in the urine.

a. true
b. false

6. The hormone that stimulates water conservation in the kidneys is _______________.

7. The hormone that promotes sodium retention in the kidneys is ________________.

8. Excessive potassium ions are eliminated from the body by the ________________.

9. The amount of potassium secreted by the kidneys is regulated by ________________.

10. Calcium reabsorption by the kidneys is promoted by the hormone ________________.

Objective 12 Explain the buffering systems that balance the pH of the intracellular and extracellular fluid.

___ 1. Buffers stabilize the pH of body fluids by

a. adding or removing bicarbonate ions from the fluid
b. adding or removing hydrogen ions from the fluid
c. adding or removing phosphate ions from the fluid
d. adding or removing protein from the fluid

_____ 2. Hemoglobin acts as a buffer because it

a. adds or removes hydrogen ions from the blood
b. adds or removes protein from the blood
c. transports oxygen and carbon dioxide
d. all of the above are correct

_____ 3. When exhaling, there is a decrease in the blood levels of

a. carbon dioxide and oxygen
b. carbon dioxide and ultimately hydrogen ions
c. oxygen and ultimately hydrogen ions
d. carbon dioxide only

_____ 4. When exhaling rapidly and excessively, the blood pH

a. drops
b. fluctuates
c. does not change
d. rises

_____ 5. A rise in carbon dioxide in the blood causes the blood pH to

a. drop
b. fluctuate
c. not change
d. rise

_____ 6. A rise in carbon dioxide in the blood will ultimately cause the breathing rate to

a. decrease
b. fluctuate
c. increase
d. not change

_____ 7. The part of the kidney involved in removing or adding hydrogen ions to the blood is the

a. cortex
b. medulla
c. nephrons
d. renal pyramids

_____ 8. When a person holds their breath the

a. blood CO_2 rises and the blood pH drops
b. blood CO_2 drops and the blood pH rises
c. blood CO_2 rises and the blood pH rises
d. blood CO_2 drops and the blood pH drops

Objective 13 Describe the effects of aging on the urinary system.

_____ 1. In the elderly

a. there is about a 40% decline in nephron function
b. the nephrons become less responsive to ADH
c. the filtration rate slows down
d. a, b, and c are correct

_____ 2. As the aging process proceeds, the elderly may lose control of the

a. internal urethral sphincter
b. external urethral sphincter
c. prostate gland
d. both a and b

_____ 3. Urination in the elderly may occur more frequently because

a. the nephrons are not responding to ADH
b. incontinence
c. the lack of ADH
d. of excessive ADH production and secretion

_____ 4. Elderly men may experience difficulty with urination due to

a. excessive secretion of ADH
b. an increase in the size of the glomerulus
c. an enlarged prostate
d. all of the above

Part II: Chapter Comprehensive Exercises

A. Word Elimination

Circle the term that does not belong in each of the following groupings.

1. kidney ureter prostate urinary bladder urethra
2. cortex nephron medulla pyramids columns
3. calyces PCT loop of Henle DCT collecting duct
4. interlobar arteries arcuate arteries interlobular arteries afferent arteriole vasa recta
5. urea creatinine ammonia urine uric acid
6. sodium potassium glucose chloride bicarbonate
7. specific gravity pH osmolarity color ADH
8. adrenalin angiotensin II aldosterone ADH ANP
9. interstitial fluid blood plasma urine lymph synovial fluid
10. trigone neck urethral sphincter ureter detrusor

B. Matching

Match the terms in Column B with the terms in Column A. Use letters for answers in the spaces provided.

COLUMN A	COLUMN B
___ 1. renal artery	A. dominant cation - ICF
___ 2. Loop of Henle	B. concentration of dissolved solutes
___ 3. glycosuria	C. kidney failure
___ 4. uremia	D. ICF and ECF
___ 5. anuria	E. pH increases
___ 6. fluid compartments	F. blood to kidney
___ 7. potassium	G. dominant cation - ECF
___ 8. sodium	H. countercurrent multiplication
___ 9. osmolarity	I. glucose in the urine
___ 10. decreased pCO2	J. no urine production

C. Concept Map I - Urinary System

Using the following terms, fill in the circled, numbered, blank spaces to complete the concept map. Follow the numbers that comply with the organization of the map.

Renal sinus	**Ureters**	**Glomerulus**
Urinary bladder	**Minor calyces**	**Proximal convoluted tubule**
Medulla	**Nephrons**	**Collecting tubules**

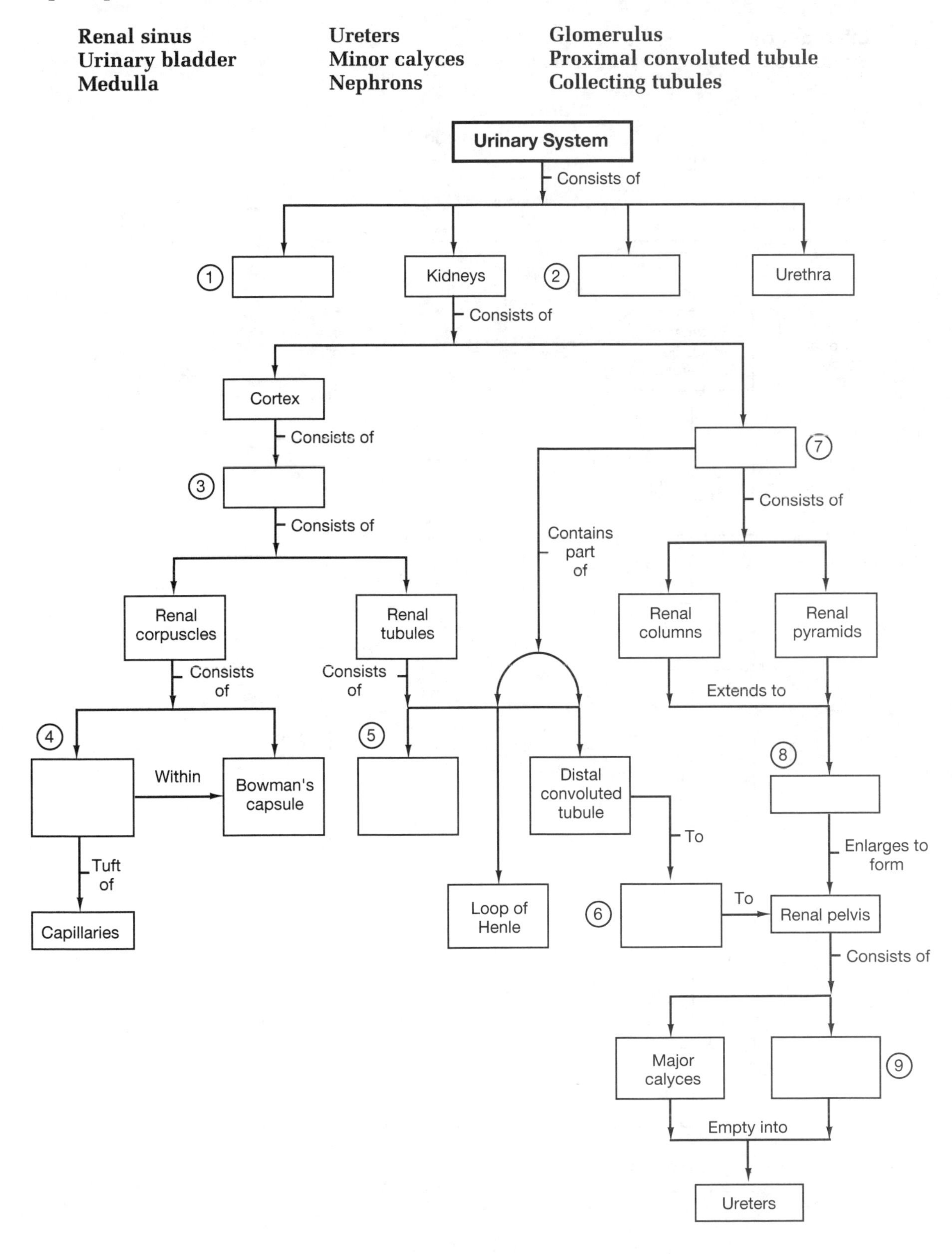

Concept Map II - Kidney Circulation

Using the following terms, fill in the circled, numbered, blank spaces to complete the concept map. Follow the numbers that comply with the organization of the map.

Efferent artery	**Interlobular vein**	**Afferent artery**
Renal artery	**Interlobar vein**	**Arcuate artery**

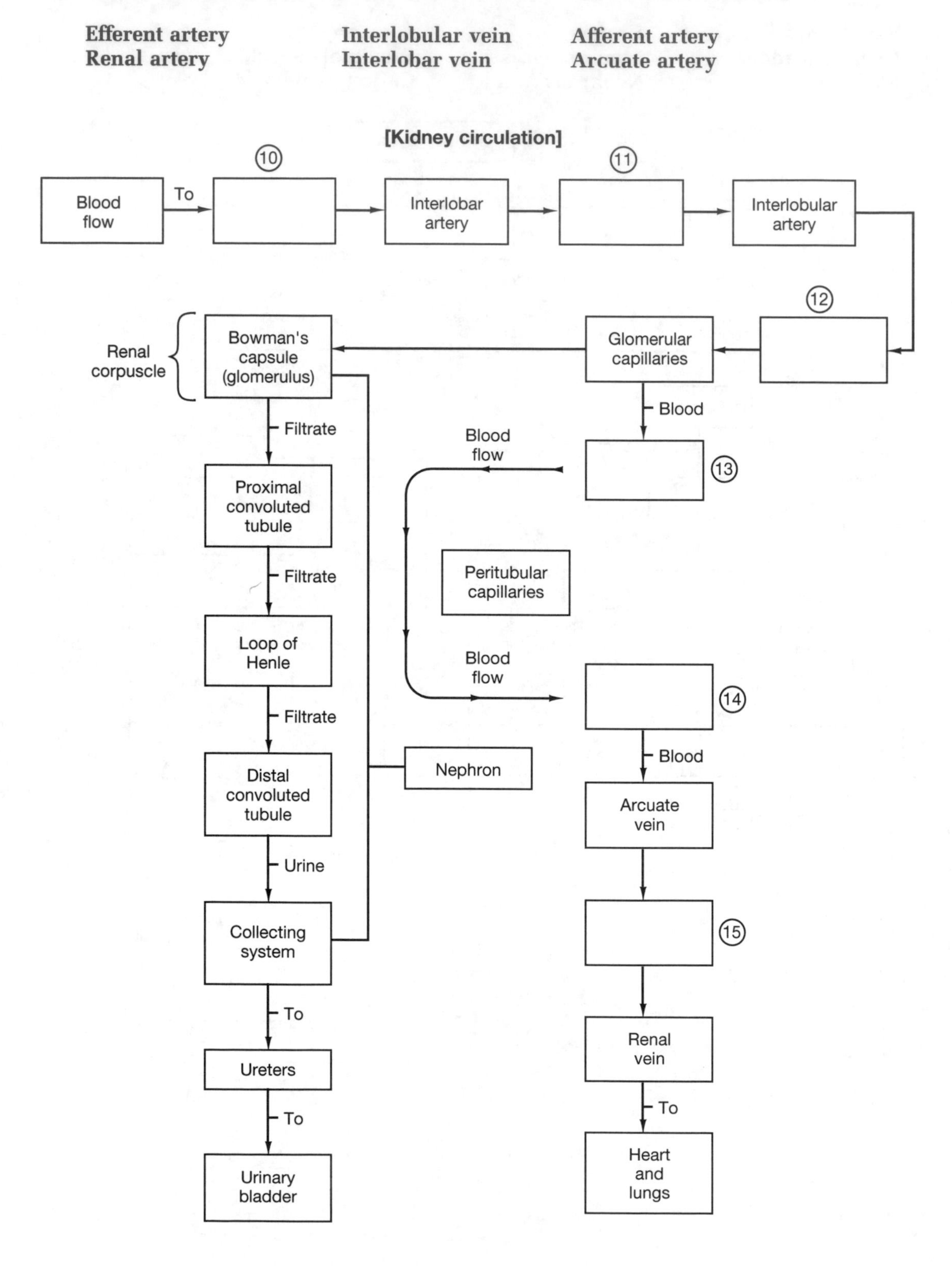

D. Crossword Puzzle

The following crossword puzzle reviews the material in Chapter 19. To complete the map, you must know the answers to the clues given, and must be able to spell the terms correctly.

ACROSS

3. The urethral sphincter that is under voluntary control.
4. An increase in carbon dioxide will increase the _______ ions in the blood.
5. Chemicals that resist changes in pH are known as _______.
9. The arteriole that takes material to the glomerulus.
10. The area where blood vessels enter and exit the kidneys.
12. When the prostate gland swells, it constricts the _________, making urination difficult.
13. The ______ ducts pass through the renal pyramid area.

DOWN

1. The capillaries inside the Bowman's capsule.
2. At about 500 ml, the ____ sphincter automatically opens.
6. This arteriole contains "cleaner" blood because it has been filtered by the glomerulus.
7. The chemical that activates agiotensin II.
8. The main functioning unit of the kidney.
11. A tube that leads from the kidneys to the urinary bladder.

E. Short Answer Questions

Briefly answer the following questions in the spaces provided below.

1. What are the essential functions of the urinary system?

2. What known functions result from sympathetic innervation into the kidneys?

3. What control mechanisms are involved with regulation of the glomerular filtration rate (GFR)?

4. What four (4) hormones affect urine production? Describe the role of each.

5. What two (2) major effects does ADH have on maintaining homeostatic volumes of water in the body?

6. How do pulmonary and renal mechanisms support the chemical buffer system?

CHAPTER

20

The Reproductive System

Overview

The structures and functions of the reproductive system are notably different from those of any other organ system in the human body. Most organ systems of the body show little difference between males and females. The structural differences between males and females and the important roles these differences play in human behavior emphasize the significance of the reproductive system. Even though major differences exist between the reproductive organs of the male and female, both are primarily involved with propagation of the species and passing genetic material from one generation to another. In addition, they produce hormones that control development of male and female sex characteristics.

The reproductive system is the only system that is not essential to the life of the individual. It is specialized to ensure survival, not of the individual, but the species. The other organ systems in the body are functional at birth or shortly thereafter; however, the reproductive system does not become functional until hormonal influences are induced during puberty.

This chapter provides a series of exercises that will help you to review and reinforce the principles of anatomy and physiology of the male and female reproductive tracts, hormonal influences, and changes that occur during the maturation and aging processes.

Review of Chapter Objectives

1. Summarize the functions of the human reproductive system and its principal components.
2. Describe the components of the male reproductive system.
3. Describe the process of spermatogenesis.
4. Describe the roles the male reproductive tract and the accessory glands play in the maturation and transport of spermatozoa.
5. Describe the hormonal mechanisms that regulate male reproductive functions.
6. Describe the components of the female reproductive system.
7. Describe the process of oogenesis in the ovary.
8. Detail the physiological processes involved in the ovarian and menstrual cycles.
9. Discuss the physiology of sexual intercourse as it affects the reproductive systems of males and females.
10. Describe the changes in the reproductive system that occur with aging.
11. Explain how the reproductive system interacts with other body systems.

Part I: Objective Based Questions

Objective 1 Summarize the functions of the human reproductive system and its principal components.

_____ 1. The reproductive system

a. produces and transports gametes
b. produces FSH and LH
c. stores and nourishes gametes
d. only a and c are correct

_____ 2. The reproductive system produces gametes via a process called

a. fertilization
b. meiosis
c. ejaculation
d. coitus

_____ 3. The testes produce

a. hormones
b. semen
c. sperm
d. all of the above

_____ 4. Embryo development takes place in the

a. ovaries
b. uterine tubes
c. uterus
d. vagina

_____ 5. The reproductive system consists of

a. gonads
b. ducts that receive and transport the gametes
c. accessory glands and organs that secrete fluids
d. all of the above

_____ 6. The reproductive organs that produce gametes and hormones are the

a. accessory glands
b. gonads
c. vagina and penis
d. all of the above

_____ 7. The systems involved in an adequate sperm count, correct pH and nutrients, and erection and ejaculation are

a. reproductive and digestive
b. endocrine and nervous
c. cardiovascular and urinary
d. a, b and c are correct

Objective 2 Describe the components of the male reproductive system.

_____ 1. The structures inside the testes responsible for sperm production are the

a. seminiferous tubules
b. interstitial cells
c. sustentacular cells
d. seminal vesicles

_____ 2. The correct sequence of travel for sperm through the male's reproductive tract is

a. ductus deferens, epididymis, ejaculatory duct
b. ejaculatory duct, ductus deferens, epididymis
c. epididymis, ductus deferens, ejaculatory duct
d. ejaculatory duct, epididymis, ductus deferens

_____ 3. The seminal fluid exits the penis by travelling through the

a. ureter
b. urethra
c. ductus deferens
d. ejaculatory duct

_____ 4. A bundle of tissue that contains the ductus deferens, blood vessels, nerves, and lymphatics that serve the testes is the

a. epididymis
b. rete testes
c. corpora cavernosa
d. spermatic coral

_____ 5. Androgens are produced in the

a. Sertoli cells
b. seminal vesicles
c. interstitial cells
d. sustentacular cells

_____ 6. Sperm storage and maturation occurs in the

a. epididymis
b. seminal vesicles
c. interstitial cells
d. ductus deferens

7. Identify and label the structures comprising the male reproductive tract from the following selections. Place your labels in the spaces provided below the drawing.

Figure 20-1 Male Reproductive System

Urinary bladder	**Bulbourethral gland**	**Ductus deferens**
Penile urethra	**Urethral meatus**	**Ureter**
Penis	**Rectum**	**Scrotum**
Testis	**Seminal vesicle**	**Ejaculatory duct**
Prostate gland	**Pubic symphysis**	**Epididymis**

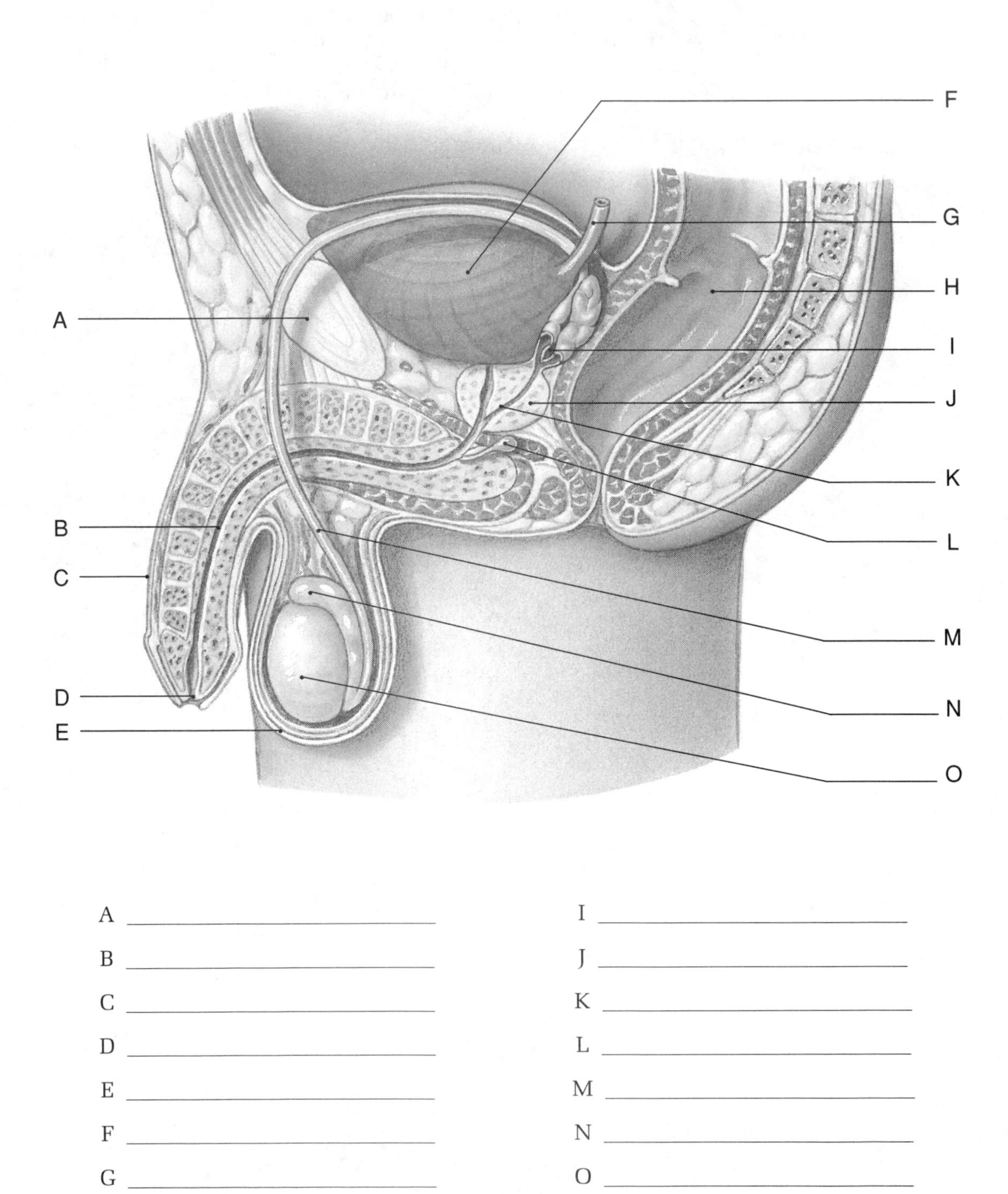

A ______________________

B ______________________

C ______________________

D ______________________

E ______________________

F ______________________

G ______________________

H ______________________

I ______________________

J ______________________

K ______________________

L ______________________

M ______________________

N ______________________

O ______________________

Objective 3 Describe the process of spermatogenesis.

_____ 1. Sperm production occurs in the

a. seminal vesicles
b. seminiferous tubules
c. rete testis
d. epididymis

_____ 2. Spermatogenesis ultimately produces

a. male gametes
b. gonads
c. sustentacular cells
d. both a and b

_____ 3. Sperm develop from "primordial cells" called

a. spermatids
b. primary spermatocytes
c. spermatogonia
d. secondary spermatocytes

_____ 4. When the process of meiosis begins developing sperm cells become

a. spermatids
b. spermatocytes
c. spermatozoans
d. sustentacular cells

_____ 5. At the end of meiosis the immature sperm are called

a. spermatozoa
b. secondary spermatocytes
c. spermatogonia
d. spermatids

_____ 6. The process of spermiogenesis produces

a. spermatazoa
b. spermatids
c. secondary spermatocytes
d. Sertoli cells

Objective 4 Describe the roles the male reproductive tract and the accessory glands play in the maturation and transport of spermatozoa.

_____ 1. Before sperm cells leave the body of the male they travel from the testes to the

a. ductus deferens → epididymis → urethra → ejaculatory duct
b. ejaculatory duct → epididymis → ductus deferens → urethra
c. epididymis → ductus deferens → ejaculatory duct → urethra
d. epididymis → ejaculatory duct → ductus deferens → urethra

_____ 2. The accessory organs in the male that secrete into the ejaculatory ducts and the urethra are

a. epididymis, seminal vesicles, vas deferens
b. prostate gland, inguinal canals, raphe
c. adrenal glands, bulbourethral glands, seminal glands
d. seminal vesicles, prostate gland, bulbourethral glands

_____ 3. Sperm move from the epididymis to the urethra via the

a. ductus deferens
b. rete testis
c. ejaculatory duct
d. corpora cavernosum

_____ 4. The fleshy pouch suspended below the base of the penis and anterior to the anus is the

a. corpus cavernosum
b. scrotum
c. corpus spongiosum
d. prepuce

_____ 5. The gland that contributes about 60% to the volume of semen and produces a secretion that contains fructose and is slightly alkaline is the

a. prostate gland
b. seminal vesicle
c. bulbourethral glands
d. Bartholin's gland

_____ 6. The gland that surrounds the urethra at the base of the urinary bladder and produces an alkaline secretion is the

a. bulbourethral gland
b. seminal vesicle
c. prostate gland
d. Bartholin's gland

_____ 7. The small paired glands at the base of the penis that produce a lubricating secretion are the

a. bulbourethral glands
b. Bartholin's glands
c. seminal vesicles
d. prostate glands

8. The male organ of copulation is the _________________.

9. The fold of skin that covers the tip of the penis is the ___________________.

10. The erectile tissue that surrounds the urethra is the ___________________.

11. The erectile tissue located on the ventral surface of the penis is the _________________.

Objective 5 Describe the hormonal mechanisms that regulate male reproductive functions.

_____ 1. The hormone synthesized in the hypothalamus which initiates release of pituitary hormones is

a. FSH - Follicle-Stimulating Hormone
b. ICSH - Interstitial Cell-Stimulating Hormone
c. LH - Lutenizing Hormone
d. GnRH - Gonadotropin-Releasing Hormone

_____ 2. The hormone that promotes spermatogenesis along the seminiferous tubules is

a. ICSH
b. FSH
c. GnRH
d. LH

_____ 3. Testosterone plays a secondary role in sperm production because it

a. causes the production of FSH
b. (at high concentrations) stimulates GnRH to activate the release of FSH
c. (at low concentrations) stimulates the release of FSH
d. directly stimulates and initiates spermatogeneis

_____ 4. The hormone responsible for secondary sex characteristics in the male is

a. testosterone
b. FSH
c. LH
d. ICSH

_____ 5. The pituitary hormone that stimulates the interstitial cells to secrete testosterone is

a. FSH
b. LH
c. ACTH
d. ADH

Objective 6 Describe the components of the female reproductive system.

_____ 1. The ovum is transported from the ovary to the uterus via the

a. infundibulum
b. uterine tube
c. vagina
d. myometrium

_____ 2. A developing embryo receives mechanical protection and nutritional support in the

a. cervix
b. ovary
c. uterus
d. uterine tube

_____ 3. The mass of erectile tissue located at the anterior margin of the labia minora is the

a. hymen
b. fornix
c. zona pellucida
d. clitoris

_____ 4. The muscular region extending between the uterus and the external genitalia is the

a. mons pubis
b. vagina
c. ampullae
d. fornix

_____ 5. The muscular layer of the uterus is the

a. endometrium
b. zona pellucida
c. mons pubis
d. myometrium

6. The inferior portion of the uterus that projects into the vagina is the ____________________.

7. In the mammary glands, milk production occurs in the ____________________.

8. The central space bounded by the labia minora is the ____________________.

9. Fleshy folds that encircle and partially conceal the labia minora and vestibular structures are the ____________________.

10. The generally dark, pigmented skin that surrounds the mammary nipple is called the ____________________.

11. Identify and label the structures comprising the female reproductive tract from the following selections. Place your labels in the spaces provided below the drawing.

Figure 20-2 The Female Reproductive System

Sigmoid colon
Uterine tube
Urethra
Fornix
Uterus
Greater vestibular gland

Pubic symphysis
Urinary bladder
Ovary
Cervix
Vagina
Labium majus

Labium minus
Ovarian follicle
Clitoris
Endometrium
Anus

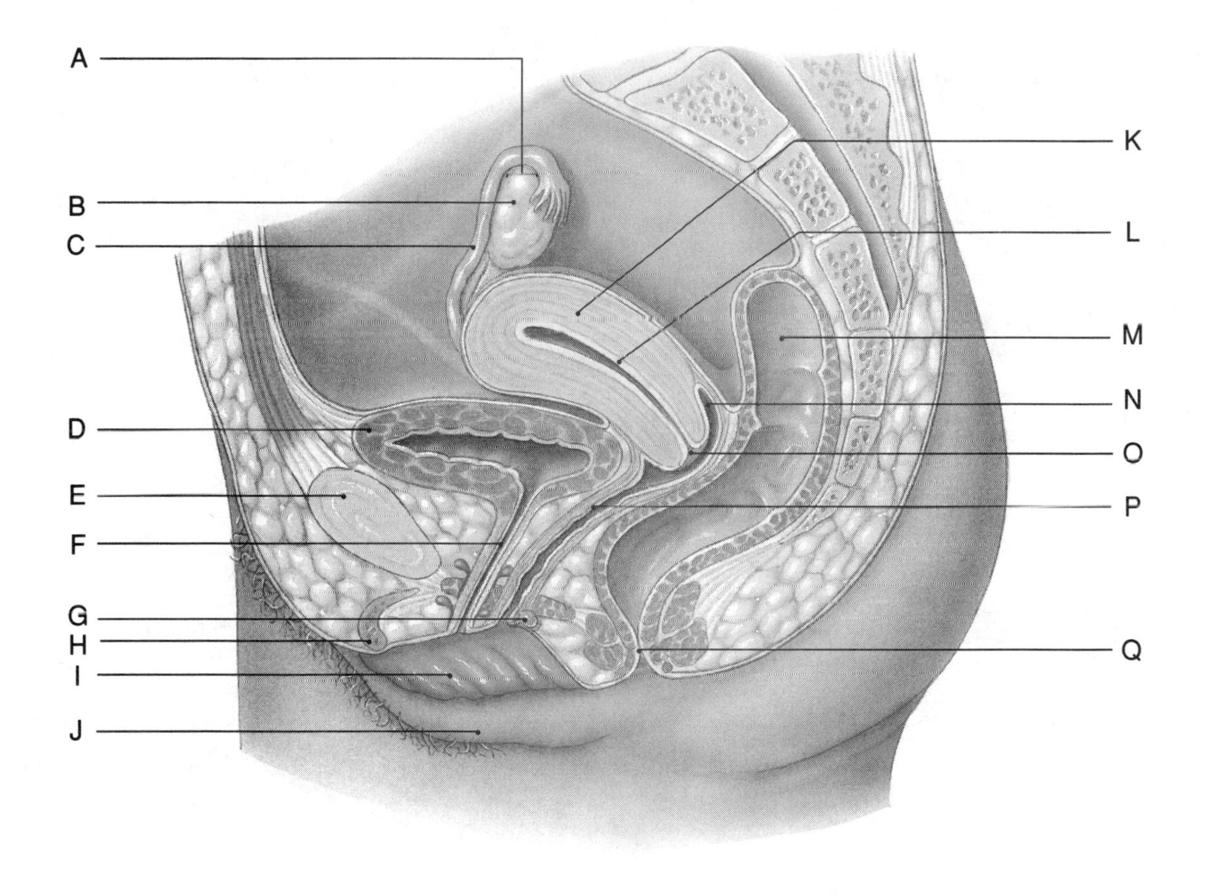

A ______________________

B ______________________

C ______________________

D ______________________

E ______________________

F ______________________

G ______________________

H ______________________

I ______________________

J ______________________

K ______________________

L ______________________

M ______________________

N ______________________

O ______________________

P ______________________

Q ______________________

Objective 7 Describe the process of oogenesis in the ovary.

_____ 1. The process of oogenesis produces three non-functional polar bodies and a

a. primordial follicle
b. granulosa cell
c. functional ovum
d. zona pellucida

_____ 2. Ova develop from stem cells called

a. primary oocytes
b. oogenia
c. secondary oocytes
d. polar bodies

_____ 3. An ovum will only complete meiosis if it is fertilized.

a. true
b. false

_____ 4. The first meiotic division is completed just prior to ovulation.

a. true
b. false

_____ 5. The process of oogenesis produces many viable ova.

a. true
b. false

_____ 6. Oogenesis occurs continuously from puberty until menopause.

a. true
b. false

Objective 8 Detail the physiological processes involved in the ovarian and menstrual cycles.

_____ 1. The proper sequence that describes the ovarian cycle involves the formation of

a. primary follicles, secondary follicles, tertiary follicles, ovulation, and formation and destruction of the corpus luteum.
b. primary follicles, secondary follicles, tertiary follicles, corpus luteum, and ovulation
c. corpus luteum, primary, secondary, and tertiary follicles, and ovulation
d. primary and tertiary follicles, secondary follicles, ovulation, and formation and destruction of the corpus luteum

_____ 2. Under normal circumstances, in a 28 day cycle, ovulation occurs on day _____, and the menses begins on day _____.

a. 1; 4
b. 28; 14
c. 14; 1
d. 6; 14

_____ 3. During the *proliferative* phase of the menstrual cycle

a. ovulation occurs
b. a new functional layer is formed in the uterus
c. secretory glands and blood vessels develop in the endometrium
d. the old functional layer is sloughed off

_____ 4. During the *secretory* phase of the menstrual cycle

a. a new uterine lining is formed
b. the corpus luteum is formed
c. the old functional layer is sloughed off
d. glands enlarge and their rate of secretion is accelerated

_____ 5. During the *menses*

a. the old functional layer is sloughed off
b. ovulation occurs
c. a new uterine lining is formed
d. the corpus luteum is formed

_____ 6. At *puberty* in the male and the female

a. levels of FSH and LH increase
b. gametogenesis begins
c. secondary sex characteristics begin to appear
d. all of the above

7. The hormone that initiates oogenesis is ________________.

8. The hormone that initiates ovulation is ________________.

9. The principal hormone secreted by the corpus luteum is ________________.

10. The hormone which causes a thickening of the endometrium to prepare the body for pregnancy is ________________.

11. During pregnancy, the levels of estrogen and progesterone remain quite high to inhibit the release of ________________.

Objective 9 Discuss the physiology of sexual intercourse as it affects the reproductive systems of males and females.

_____ 1. The functional result of sexual intercourse, or coitus is to

a. satisfy the needs associated with arousal
b. introduce semen into the female reproductive tract
c. relieve pressure within the reproductive tract
d. provide viable eggs and sperm

_____ 2. Male sexual function is coordinated by reflex pathways involving both divisions of the ANS.

a. true
b. false

_____ 3. During arousal, erotic thoughts or stimulation of sensory nerves in the genital region increase the parasympathetic outflow causing

a. emission
b. ejaculation
c. erection
d. detumescence

_____ 4. Emission occurs under

a. sympathetic stimulation
b. parasympathetic stimulation
c. both divisions of the ANS
d. none of the above

_____ 5. Orgasm is an intense, pleasurable sensation associated with

a. emission
b. arousal
c. ejaculation
d. detumescence

_____ 6. Engorgement of blood vessels at the nipple results from

a. sympathetic stimulation
b. parasympathetic stimulation
c. both divisions of the ANS
d. none of the above

Objective 10 Describe the changes in the reproductive system that occur with aging.

1. In females, the time that ovulation and menstruation cease is referred to as _________________.

2. Changes that occur in the male reproductive system over a period of time are known as the male _____________________.

_____ 3. Menopause is accompanied by a sharp and sustained rise in the production of ____________, while circulating concentrations of _________________ decline.

a. estrogen and progesterone; GnRH, FSH, LH
b. GnRH, FSH, LH; estrogen and progesterone
c. LH, estrogen; progesterone, GnRH, and FSH
d. FSH, LH, progesterone; estrogen and GnRH

_____ 4. In the male between the age of 50 and 60, circulating ______________ levels begin to decline, coupled with increases in circulating levels of ______________.

a. FSH and LH; testosterone
b. FSH; testosterone and LH
c. testosterone; FSH and LH
d. LH; testosterone and FSH

Objective 11 Explain how the reproductive system reacts with other body systems.

1. For all other body systems, the system that provides secretion of hormones with effects on growth and metabolism is the ___________________ system.

2. The system that distributes reproductive hormones, provides nutrients and oxygen, and facilitates waste removal for a fetus is the __________________ system.

3. The system that secretes and releases pituitary hormones that regulate sexual development and function is the ___________________________ system.

4. The system that controls sexual behavior and sexual function is the _____________________ system.

Part II: Chapter Comprehensive Exercises

A. Word Elimination

Circle the term that does not belong in each of the following groupings.

1. ovaries uterine tubes cremaster muscle uterus vagina
2. scrotum epididymis ductus deferens ejaculatory duct urethra
3. primary spermatocyte secondary spermatocyte spermatids mitosis sperm
4. head neck middle piece tail epididymis
5. seminal vesicles penis prostate gland Cowper's glands bulbourethral glands
6. primary oocyte polar body secondary oocyte ovum corpus luteum
7. areola vulva vestibule labia minora clitoris
8. nipple lactiferous duct vestibular glands lactiferous sinus areola
9. FSH LH progestin androgens estradiol
10. arousal emission coitus ejaculation orgasm

B. Matching

Match the terms in Column B with the terms in Column A. Use letters for answers in the spaces provided.

COLUMN A	COLUMN B
___ 1. gametes	A. sperm production
___ 2. sustentacular cells	B. egg production
___ 3. seminiferous tubules	C. hormone - secondary sex characteristics
___ 4. puberty in male	D. follicular degeneration
___ 5. testosterone	E. sertoli cells
___ 6. ovaries	F. indicates pregnancy
___ 7. puberty in female	G. reproductive cells
___ 8. atresia	H. milk ejection
___ 9. human chorionic hormone	I. menarche
___ 10. oxytocin	J. spermatogenesis begins

C. Concept Map I - Male Reproductive Tract

Using the following terms, fill in the numbered, blank spaces to complete the concept map. Follow the numbers that comply with the organization of the map.

Urethra	**Produce testosterone**	**Ductus deferens**
Seminiferous tubules	**FSH**	**Bulbourethral glands**
Penis	**Seminal vesicles**	

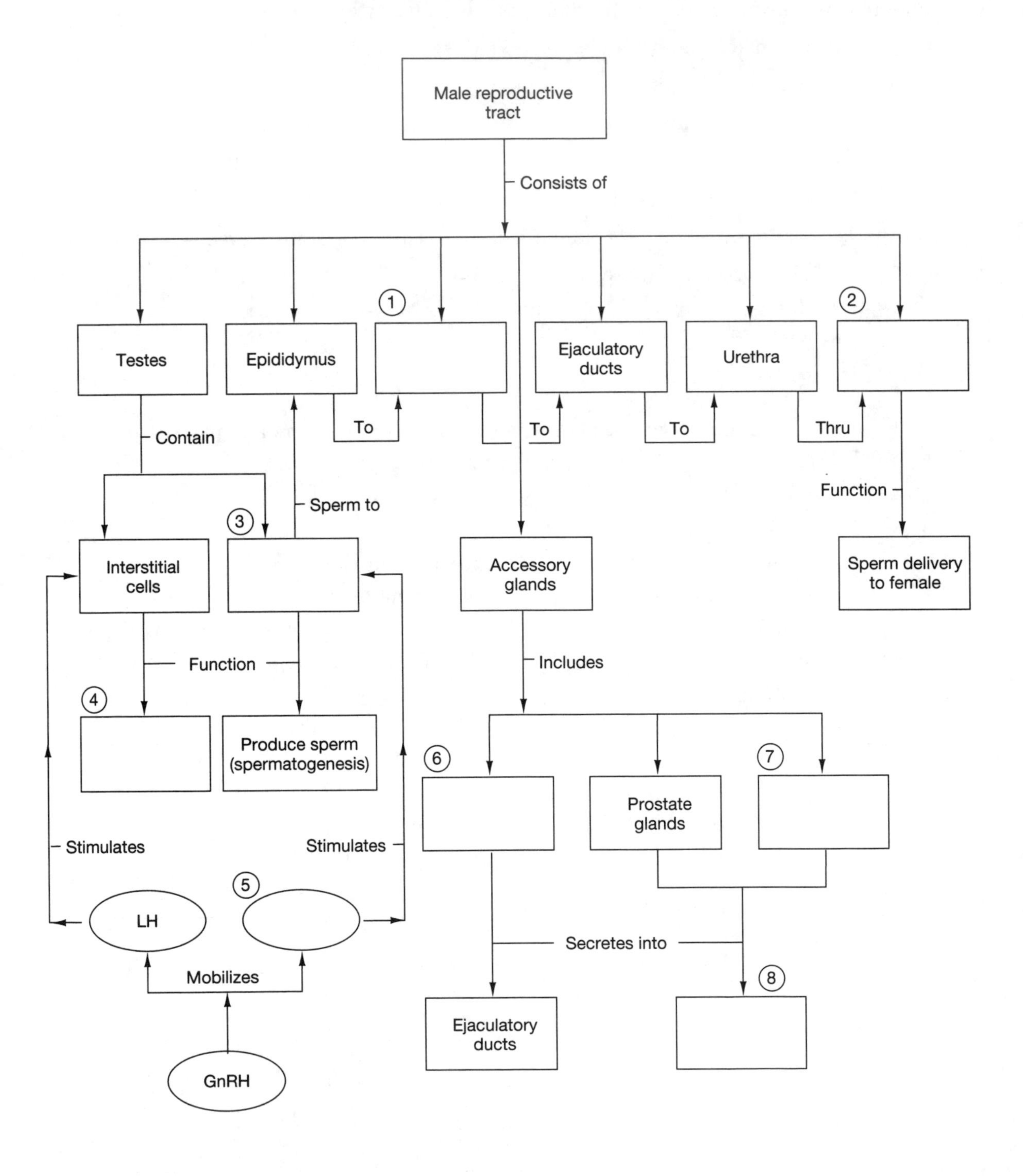

Concept Map II - Female Reproductive Tract

Using the following terms, fill in the circled, numbered, blank spaces to complete the concept map. Follow the numbers that comply with the organization of the map.

nutrients	**vulva**	**vagina**
follicles	**supports fetal development**	**granulosa and thecal cells**
uterine tubes	**endometrium**	**clitoris**
labia majora and minora		

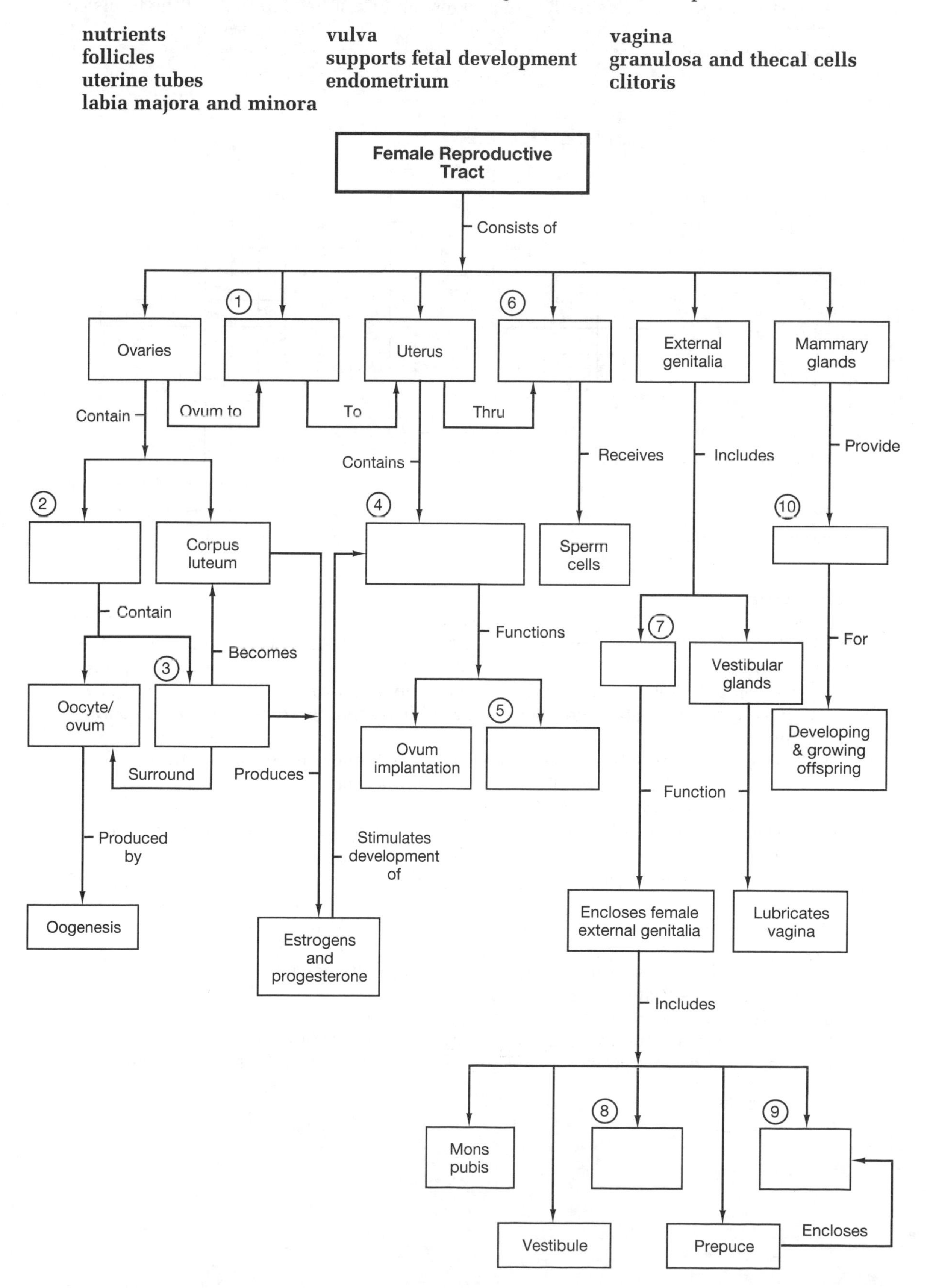

D. Crossword Puzzle

The following crossword puzzle reviews the material in Chapter 20. To complete the puzzle, you must know the answers to the clues given, and must be able to spell the terms correctly.

ACROSS

1. A type of cellular reproduction that reduces the number of chromosomes in each gamete.
5. If a zygote is not present, progesterone levels will drop, and _______ will begin.
7. _______ degrees Fahrenheit below normal body temperature is ideal for sperm production.
9. The tubules inside the testes that are responsible for sperm production.
10. Sperm cells and urine both travel through this tube.
11. Successful fertilization occurs in the _________ tube.
12. A chemical source of energy for the sperm cells that comes from the seminal vesicle
13. The presence of this hormone indicates pregnancy.
14. A genetic name of sperm cells and egg cells.

DOWN

2. A zygote implants itself in the ___________ of the uterus.
3. The hormone that causes ovulation.
4. One of the glands that produces alkaline semen.
6. Upon ovulation the ovarian follicle ruptures and becomes a ___________.
8. The hormone that targets the testes and causes them to release testosterone.

E. Short Answer Questions

Briefly answer the following questions in the spaces provided below.

1. (a) What three (3) glands secrete their products into the male reproductive tract?

   (b) What are the four (4) primary functions of these glands?

2. What is the difference between *seminal fluid* and *semen*?

3. What is the difference between *emission* and *ejaculation*?

4. What are the five (5) primary function of testosterone in the male?

5. What are the three (3) reproductive functions of the vagina?

6. What are the three (3) phases of female sexual function and what occurs in each phase?

7. What are the five (5) steps involved in the ovarian cycle?

8. What are the three (3) stages of the menstrual cycle?

9. What hormones are secreted by the placenta?

10. What is colostrum and what are its major contributions to the infant?

F. Formation of Gametes - Gametogenesis

Identify the stages in each of the following processes. Place your answers in the spaces provided below each drawing.

1. Name of process ______________________

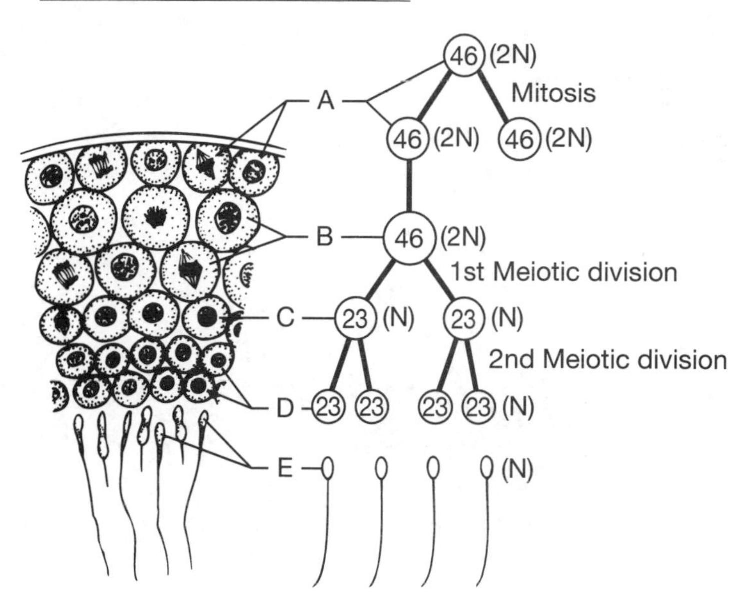

A ______________________

B ______________________

C ______________________

D ______________________

E ______________________

2. Name of process ______________________

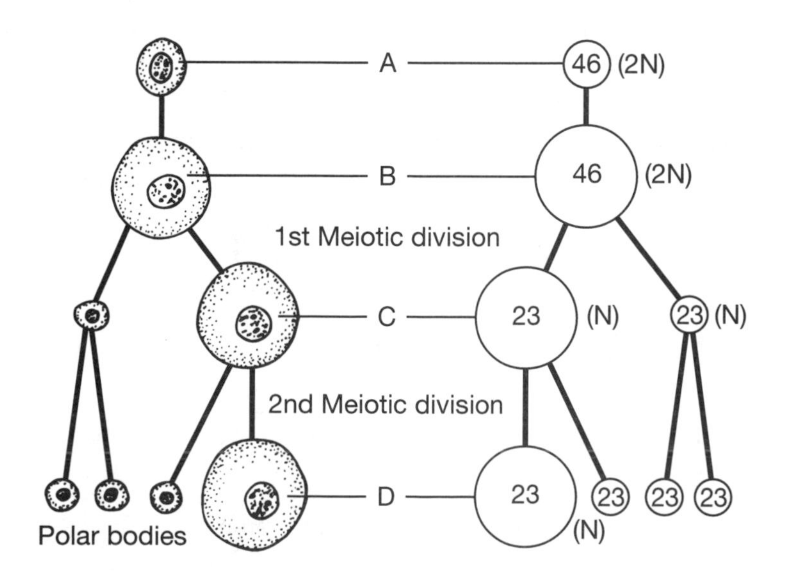

A ______________________

B ______________________

C ______________________

D ______________________

CHAPTER

21

Development and Inheritance

Overview

The events of development, differentiation and inheritance occur as a result of a complex and orderly sequence of progressive changes that begin at conception and have a profound effect on an individual for a lifetime. Few topics in anatomy and physiology are as fascinating as the "miracle of life" and the events that occur during the nine months before birth. When a new life begins in the female reproductive system, new genetic combinations similar to the parents, yet different, are made and nourished until they emerge from the female tract to take up life on their own - at first highly dependent individuals but eventually growing independent as continued development and maturation occurs. What was once a single cell has developed and grown into a complex organism consisting of trillions of cells which are organized into tissues, organs, and organ systems - the human body - a complete living organism - carrying on all the functions of life!

Chapter 21 highlights the major events which occur during development and how our genetic inheritance profoundly effects everything about us, including our physical appearance, the way we behave, how long we live, and the likelihood of developing certain diseases. The chapter concludes with how developmental patterns can be modified for the good or ill of the individual.

Review of Chapter Objectives

1. Describe the process of fertilization.
2. List the three prenatal periods and describe the major events associated with each period.
3. Describe the origin of the three primary germ layers and their participation in the formation of the extraembryonic membranes.
4. Describe the interplay between the maternal organ systems and the developing fetus.
5. List and describe the events that occur during labor and delivery.
6. Discuss the major stages of life after delivery.
7. Relate basic principles of genetics to the inheritance of human traits.

Part I: Objective Based Questions

Objective 1 Describe the process of fertilization.

_____ 1. The normal male genotype is __________, and the normal female genotype is __________.

a. xx; xy
b. x; y
c. xy; xx
d. y; x

_____ 2. Normal fertilization occurs in the

a. lower part of the uterine tube
b. upper one-third of the uterine tube
c. upper part of the uterus
d. antrum of a tertiary follicle

_____ 3. Fertilization is completed with the

a. formation of a gamete with 23 chromosomes
b. formation of the male and female pronuclei
c. completion of the meiotic process
d. formation of a zygote containing 46 chromosomes

4. Sperm cannot fertilize an egg until they have undergone an activation in the vagina called ____________.

5. The fusion of the male and female pronuclei is called ______________________.

6. One chromosome in each pair is contributed by the sperm and the other by the egg at ________________.

Objective 2 List the three prenatal periods and describe the major events associated with each period.

_____ 1. The period of gestation that is characterized by *rapid fetal growth* is the ____________ trimester.

a. first
b. second
c. third

_____ 2. The period of gestation when the rudiments of all major *organ systems appear* is the _________ trimester.

a. first
b. second
c. third

_____ 3. The period of gestation when organs and organ systems complete most of their development and the *fetus looks distinctly human* is the ____________ trimester.

a. first
b. second
c. third

_____ 4. A blastocyst is

a. an extraembryonic membrane that forms blood vessels
b. a solid ball of cells
c. a hollow ball of cells
d. a part of the placenta

_____ 5. The inner cell mass of the blastocyst will form

a. the placenta
b. the morula
c. the embryo
d. blood vessels of the placenta

_____ 6. During implantation

a. the syncytial trophoblast erodes a path through the endometrium
b. the inner cell mass begins to form the placenta
c. maternal blood vessels in the endometrium are walled-off from the blastocyst
d. the inner cell mass is temporarily deprived of nutrients

_____ 7. During gastrulation

a. the blastodisc is formed
b. the placenta is formed
c. endodermal cells migrate to the ectoderm
d. germ layers are formed

_____ 8. The chorionic villi

a. increase the surface area for exchange between the placenta and maternal blood
b. form the umbilical cord
c. form the umbilical vein and arteries
d. form the major part of the placenta

9. The process of cell division that occurs after fertilization is called _______________.

10. The penetration of the endometrium by the blastocyst is referred to as _______________.

11. Identical cells produced by early cleavage are called _______________.

12. The solid ball of cells formed after several rounds of cell division following fertilization is called a _______________.

13. Neural tissues develop from the _______________ . (germ layer)

14. Muscle tissue develops from the _______________. (germ layer)

15. The digestive lining develops from the _______________ . (germ layer)

Objective 3 Describe the origin of the three primary germ layers and their participation in the formation of the extraembryonic membranes.

_____ 1. Germ-layer formation results from the process of

a. embryogenesis
b. fertilization
c. gastrulation
d. parturition

_____ 2. The extraembryonic membranes that develop from the endoderm and mesoderm are

a. amnion and chorion
b. yolk sac and allantois
c. allantois and chorion
d. yolk sac and amnion

_____ 3. The chorion development from the

a. endoderm and mesoderm
b. ectoderm and mesoderm
c. trophoblast and endoderm
d. mesoderm and trophoblast

_____ 4. The extraembryonic membrane that forms a fluid-filled sac is the

a. amnion
b. chorion
c. allantois
d. yolk sac

_____ 5. The extraembryonic membrane that forms the fetal portion of the placenta is the

a. yolk sac
b. chorion
c. amnion
d. allantois

Objective 4 Describe the interplay between the maternal organ systems and the developing fetus.

_____ 1. The vital link between maternal and embryonic systems that support the fetus during development is the

a. chorion
b. amnionic sac
c. placenta
d. yolk sac

_____ 2. The umbilical cord or umbilical stalk contains

a. the amnion, allantois, and chorion
b. paired umbilical arteries and the amnion
c. a single umbilical vein and the chorion
d. the allantois, blood vessels, and yolk stalk

_____ 3. The hormone that is the basis for a pregnancy test is

a. estrogen
b. progesterone
c. human chorionic gonadotropin (hCG)
d. human placental lactogen (hPL)

_____ 4. The developing fetus is totally dependent on maternal organ systems for nourishment, respiration, and waste removal.

a. true
b. false

_____ 5. By the end of gestation, the maternal blood volume decreases by almost 50%.

a. true
b. false

_____ 6. The reason pregnant women need to urinate frequently is because the volume of urine produced increases and the weight of the uterus presses down on the urinary bladder.

a. true
b. false

Objective 5 List and describe the events that occur during labor and delivery.

_____ 1. The correct sequence which describes labor and delivery is

a. dilation, expulsion, placental
b. dilation, placental, expulsion
c. expulsion, dilation, placental
d. placental, dilation, expulsion

_____ 2. The dilation stage involves dilation of the

a. cervix
b. uterus
c. vagina
d. uterine tubes

_____ 3. When the "water breaks" the

a. hymen is ruptured and a watery fluid is released
b. urinary bladder releases fluid due to excessive abdominal pressure
c. amniotic sac bursts and releases amniotic fluid
d. chorionic villi release excessive fluid

_____ 4. The "afterbirth" is the expulsion of

a. fetal waste
b. maternal waste
c. amniotic sac
d. the placenta

Objective 6 Discuss the major stages after delivery.

_____ 1. The sequential stages that identify the features and functions associated with the human experience are

a. neonatal, childhood, infancy, maturity
b. neonatal, postnatal, childbirth, adolescence
c. infancy, childhood, adolescence, maturity

_____ 2. The systems that were relatively nonfunctional during the fetus's prenatal period that must become functional at birth are the

a. circulatory, muscular, skeletal
b. integumentary, reproductive, nervous
c. endocrine, nervous, circulatory
d. respiratory, digestive, excretory

_____ 3. The stage at which levels of sex hormones rise and gametogenesis begins is referred to as

a. puberty
b. maturity
c. senescence
d. neonatal

4. The stage associated with the end of growth in the late teens or early twenties is referred to as ____________.

5. The process of aging is referred to as ________________.

Objective 7 Relate basic principles of genetics to the inheritance of human traits.

_____ 1. The normal chromosome complement of a typical somatic, or body, cell is:

a. 23
b. N or haploid
c. 46
d. 92

_____ 2. Gametes are different from ordinary somatic cells because:

a. they contain only half the normal number of chromosomes
b. they contain the full complement of chromosomes
c. the chromosome number doubles in gametes
d. gametes are diploid, or 2N

_____ 3. During gamete formation, meiosis splits the chromosome pairs, producing:

a. diploid gametes
b. haploid gametes
c. gametes with a full chromosome complement
d. duplicate gametes

_____ 4. The first meiotic division:

a. results in the separation of the duplicate chromosomes
b. yields four functional spermatids in the male
c. produces one functional ovum in the female
d. reduces the number of chromosomes from 46 to 23

_____ 5. Spermatogenesis produces:

a. four functional spermatids for every primary spermatocyte undergoing meiosis
b. functional spermatozoan with the diploid number of chromosomes
c. secondary spermatocytes with the 2N number of chromosomes
d. a, b, and c are correct

_____ 6. Oogenesis produces:

a. an oogonium with the haploid number of chromosomes
b. one functional ovum and three nonfunctional polar bodies
c. a secondary oocyte with the diploid number of chromosomes
d. a, b, and c are correct

_____ 7. If an allele is dominant it will be expressed in the phenotype:

a. if both alleles agree on the outcome of the phenotype
b. by the use of lower-case abbreviations
c. regardless of any conflicting instructions carried by the other allele
d. by the use of capitalized abbreviations

_____ 8. If a female X chromosome of an allelic pair contains sex-linked character for color blindness, the individual would be:

a. normal
b. color blind
c. color blind in one eye
d. a, b, or c could occur

9. The special form of cell division leading to the production of sperm or eggs is ____________.

10. The formation of gametes is called ____________________.

11. Chromosomes with genes that affect only somatic characteristics are referred to as ______________ chromosomes.

12. If both chromosomes of a homologus pair carry the same allele of a particular gene, the individual is ________________ for that trait.

13. When an individual has two different alleles carrying different instructions, the individual is ______________ for that trait.

Development and Inheritance Questions

Using principles and concepts learned about development and inheritance, answer the following questions. Write your answers on a separate sheet of paper.

14. A common form of color blindness is associated with the presence of a dominant or recessive gene on the X chromosome. Normal color vision is determined by the presence of a dominant gene (C), and color blindness results from the presence of the recessive gene (c). Suppose a heterozygous normal female marries a normal male. Is it possible for any of their children to be color blind? Show the possibilities by using a Punnett square.

15. Albinism (aa) is inherited as a homozygous recessive trait. If a homozygous-recessive mother and a heterozygous father decide to have children, what are the possibilities of their offspring inheriting albinism? Use a Punnett square to show the possibilities.

16. Tongue rolling is inherited as a dominant trait. Even though a mother and father are tongue rollers (T), show how it would be possible to bear children who do not have the ability to roll the tongue. Use a Punnett square to show the possibilities.

Part II: Chapter Comprehensive Exercises

A. Word Elimination

Circle the term that does not belong in each of the following groupings.

1. conception birth development differentiation inheritance
2. fertilization embryology pronucleus spindle formation cleavage
3. gestation blastomere morula blastocyst trophoblast
4. yolk sac amnion allantois placenta chorion
5. hCG colostrum prolactin hPL relaxin
6. childhood infancy fetus maturity adolescence
7. deafness albunism phenylketonuria Tay-Sacks Marfan's syndrome
8. curly hair albinism blond hair red hair Type O blood
9. parturition dilation gastrulation expulsion placental
10. endoderm amnion germ layers ectoderm mesoderm

B. Matching

Match the terms in Column B with the terms in Column A. Use letters for answers in the spaces provided.

COLUMN A	COLUMN B
___ 1. fertilization	A. stimulates smooth muscle contractions
___ 2. gestation	B. newborn infant
___ 3. prostaglandin	C. reductional division
___ 4. true labor	D. visible characteristics
___ 5. neonate	E. zygote formation
___ 6. adolescence	F. equational division
___ 7. meiosis I	G. chromosomes and component genes
___ 8. meiosis II	H. prenatal development
___ 9. phenotype	I. begins at puberty
___ 10. genotype	J. positive feedback

C. Concept Map I - Fertilization and Development

This concept map summarizes the information in Chapter 21. Use the following terms to complete the map by filling in the boxes identified by the circled numbers, 1-8.

Germ layer
allantois
muscle
relaxin
endoderm
progesterone
zygote
yolk sac

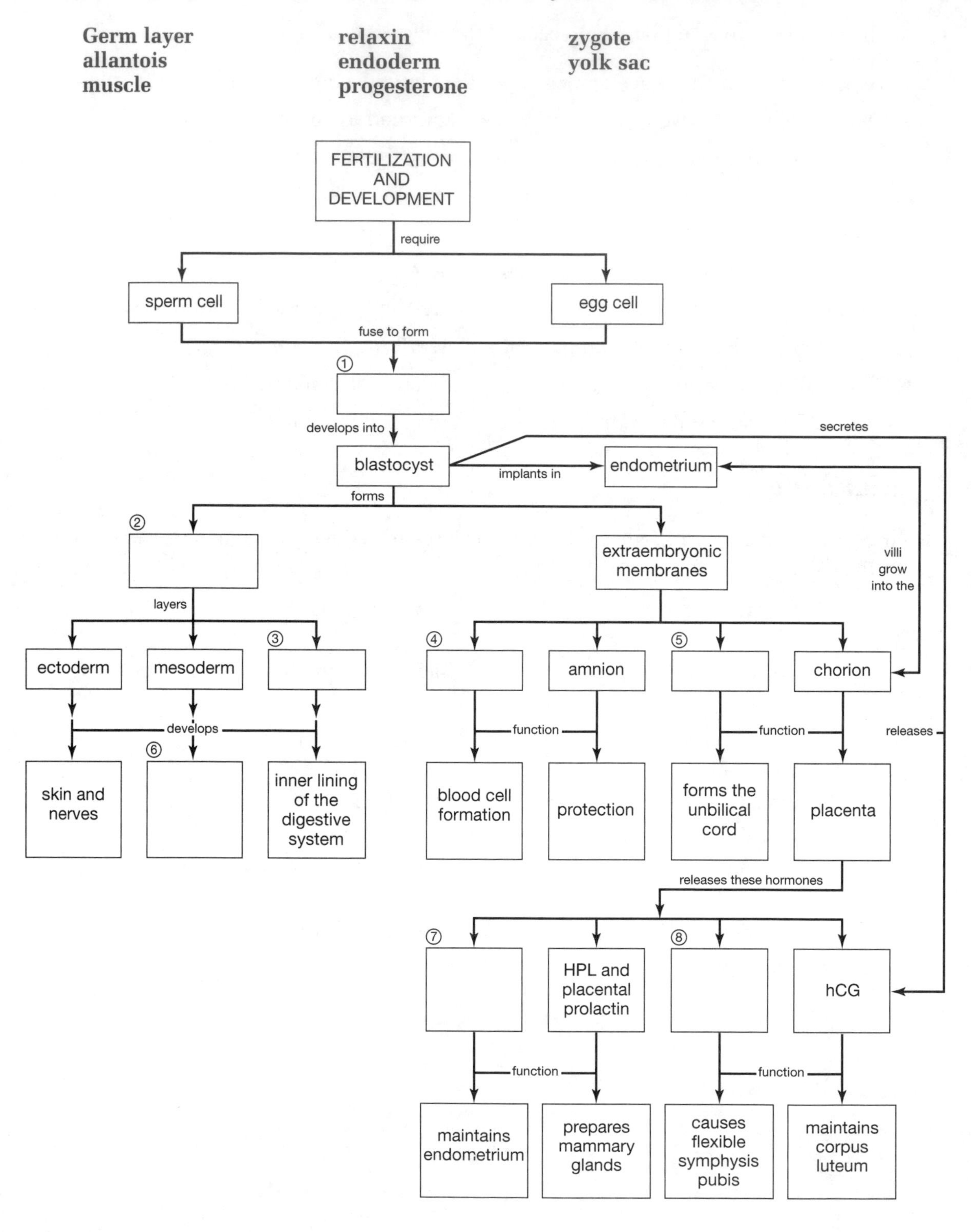

Concept Map II - Cleavage and Blastocyst Formation

Using the terms below, identify the structures or processes at the numbered locations in the drawing/illustration through the female reproductive tract. Record your answers in the spaces below.

early blastocyst	**fertilization**	**secondary oocyte**
morula	**implantation**	**zygote**
2-cell stage	**8-cell stage**	

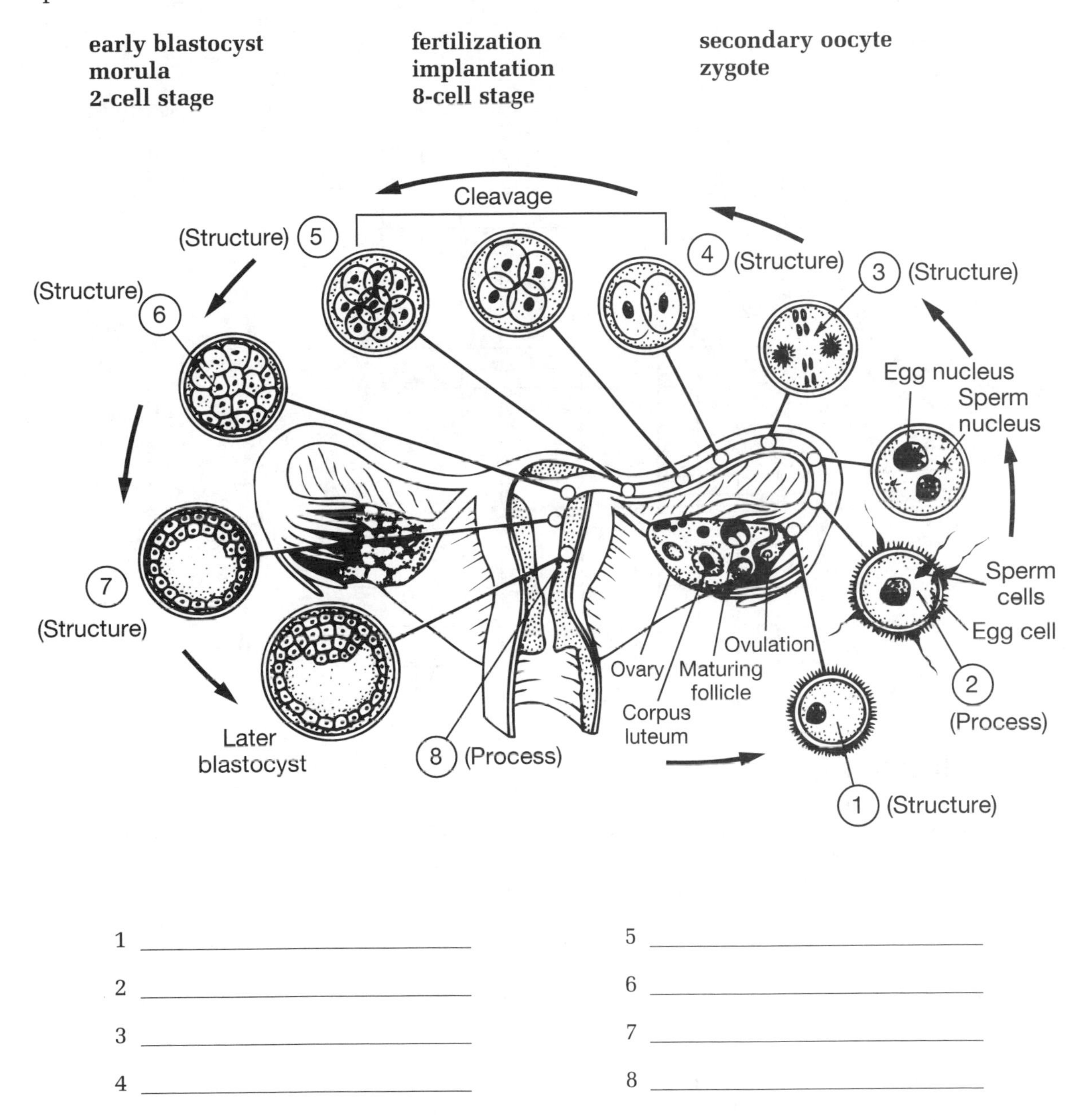

1 ____________________

2 ____________________

3 ____________________

4 ____________________

5 ____________________

6 ____________________

7 ____________________

8 ____________________

D. Crossword Puzzle

The following crossword puzzle reviews the material in Chapter 21. To complete the puzzle, you must know the answers to the clues given, and must be able to spell the terms correctly.

ACROSS

3. A female that has a genetic trait that is not expressed is called a ______ of that genetic trait.
6. The expulsion of the ____ is referred to as the "afterbirth."
8. The hormone that causes milk release.
9. The gestation period is divided into __________.

DOWN

1. The membranous structure that forms the placenta.
2. The hormone that causes milk production.
4. The membranous structure that eventually becomes a part of the umbilical cord.
5. Human _________ gonadotropin hormone targets the corpus luteum.
7. A fertilized egg.

E. Short Answer Questions

Briefly answer the following questions in the spaces provided below.

1. (a) What are the four (4) extraembryonic membranes that are formed from the three (3) germ layers?

   (b) From which given layer(s) does each membrane originate?

2. What are the major compensatory adjustments necessary in the maternal systems to support the developing fetus?

3. What primary factors interact to produce labor contractions in the uterine wall?

4. What three (3) events interact to promote increased hormone production and sexual maturation at adolescence?

5. What four (4) processes are involved with aging that influence the genetic programming of individual cells?

6. What are the three (3) stages of labor?

7. What are the identifiable life stages that comprise postnatal development of distinctive characteristics and abilities?

CHAPTER 1: *An Introduction to Anatomy and Physiology*

Part I: Objective Based Questions

Objective 1

1. B
2. C
3. A
4. D

Objective 2

1. C
2. B
3. A

Objective 3

1. C
2. A
3. D
4. B

Objective 4

1. D
2. C
3. C
4. skeletal
5. muscular
6. brain
7. spinal cord
8. peripheral nerve
9. nervous
10. pineal gland
11. pituitary
12. thyroid
13. thymus
14. pancreas
15. adrenal gland
16. ovaries
17. testes
18. endocrine
19. heart
20. veins
21. arteries
22. capillaries
23. cardiovascular
24. thymus
25. lymph nodes
26. spleen
27. lymphatic vessels
28. lymphatic system
29. salivary gland
30. pharynx
31. esophagus
32. liver
33. gall bladder
34. stomach
35. large intestine
36. small intestine
37. anus
38. digestive system
39. nasal cavity
40. larynx
41. trachea
42. bronchi
43. lungs
44. respiratory
45. kidney
46. ureters
47. urinary bladder
48. urethra
49. excretory
50. prostate
51. urethra
52. penis
53. scrotum (testes)
54. mammary glands
55. uterine tubes
56. ovary
57. uterus
58. vagina
59. reproductive

Objective 5

1. B
2. C
3. A
4. B

Objective 6

1. A
2. B
3. C
4. D
5. B
6. nervous
7. endocrine

Objective 7

1. D
2. A
3. C
4. B
5. coronal or frontal
6. inguinal
7. gluteal
8. distal
9. **Figure 1-2**
   A. R. hypochondriac
   B. R. lumbar
   C. R. iliac
   D. epigastric
   E. umbilical
   F. hypogastric
   G. L. hypochondriac
   H. L. lumbar
   I. L. iliac
10. **Figure 1-3**
   A. orbital
   B. buccal
   C. thorax
   D. axillary
   E. brachial
   F. cubital
   G. abdominal
   H. umbilical
   I. pubic
   J. palmar
   K. femoral
   L. patellar
   M. occipital
   N. deltoid
   O. scapular
   P. lumbar
   Q. gluteal
   R. popliteal
   S. calf

Objective 8

1. C
2. B
3. A
4. B
5. mesenteries
6. diaphragm
7. **Figure 1-4**
   A. cranial cavity
   B. dorsal body cavity
   C. spinal cavity
   D. pleural cavity
   E. pericardial cavity
   F. diaphragm
   G. abdominal cavity
   H. abdominopelvic cavity
   I. pelvic cavity
   J. ventral body cavity

Part II: Chapter Comprehensive Exercises

A. Word Elimination

1. biology
2. organism
3. digestion
4. organism
5. temperature
6. supine
7. prone
8. coronal
9. caudal
10. mediastinum

B. Matching

1. G
2. J
3. A
4. H
5. C
6. B
7. I
8. E
9. D
10. F

C. Concept Map

1. cranial cavity
2. spinal cord
3. two pleural cavities
4. heart
5. abdominopelvic cavity
6. pelvic cavity

D. Crossword Puzzle

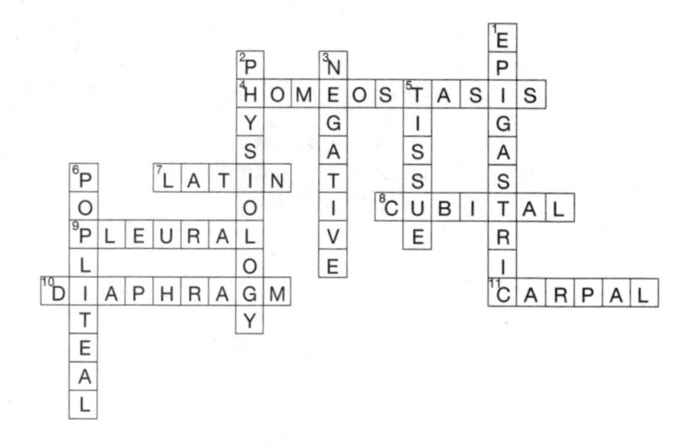

E. Short Answer Questions (Answers)

1. Any one of the following might be listed: responsiveness, adaptability, growth, reproduction, movement, absorption, respiration, excretion, digestion, circulation.

2. Subatomic particles - atoms - molecules - organelle - cell(s) - tissue(s) - organ(s) - system(s)

3. In negative feedback a variation outside of normal limits triggers an automatic response that corrects the situation. In positive feedback the initial stimulus produces a response that exaggerates the stimulus.

4. In anatomical position, the body is erect, feet are parallel and flat on the floor, eyes are directed forward, and the arms are at the sides of the body with the palm of the hands turned forward.

5. ventral, dorsal, cranial (cephalic), caudal

6. A sagittal section separates right and left positions. A transverse or horizontal section separates superior and inferior portions of the body.

7. (a) They protect delicate organs from accidental shocks, and cushion them from thumps and bumps that occur during walking, jumping, and running.
   (b) They permit significant changes in the size and shape of visceral organs.

CHAPTER 2: *The Chemical Level of Organization*

Part I: Objective Based Questions

Objective 1

1. D
2. B
3. C
4. B
5. A
6. A
7. C
8. element
9. nucleus
10. energy
11. (Atomic structure)

6P+
6N

12. eight
13. sixteen
14. two electrons
15. chemically active
16. C

Objective 2

1. C
2. A
3. D
4. C
5. A
6. B
7. B
8. ionic
9. covalent
10. covalent bonds
11. ionic bond

Objective 3

1. D
2. C
3. B
4. D
5. B

Objective 4

1. A
2. D
3. B
4. exergonic
5. equilibrium

Objective 5

1. C
2. B
3. A
4. D
5. C
6. B
7. buffers
8. acidic
9. neutral
10. alkaline

Objective 6

1. B
2. D
3. organic
4. inorganic

Objective 7

1. C
2. B
3. B
4. C
5. solution
6. electrolytes

Objective 8

1. D
2. B
3. C
4. D
5. B
6. D

Objective 9

1. C
2. A
3. C
4. B
5. D
6. D
7. B
8. C
9. D
10. A
11. B
12. B
13. C
14. D
15. peptide bond
16. deoxyribose
17. uracil
18. deoxyribose
19. adenine
20. thymine
21. guanine
22. phosphate
23. cytosine
24. hydrogen bond

25. **Table 2.1**

Table 2–1 Classification of Various Organic Molecules

Organic molecules	Carbohydrate	Lipid	Protein	Nucleic acid	High-energy compound
Amino Acid			X		
ATP					X
Cholesterol		X			
Cytosine				X	
Disaccharide	X				
DNA				X	
Fatty acid		X			
Glucose	X				
Glycerol		X			
Glycogen	X				
Guanine				X	
Monosaccharide	X				
Nucleotide				X	
Phospholipid		X			
Polysaccharide	X				
RNA				X	
Starch	X				

Objective 10

1. A
2. C
3. B
4. D
5. B

Part II: Chapter Comprehensive Exercises

A. Word Elimination

1. isotope
2. compound
3. buffer
4. glucose
5. carbonic acid
6. glycogen
7. glycogen
8. uracil
9. carbon dioxide
10. monosaccharide

B. Matching

1. F
2. L
3. H
4. J
5. D
6. A
7. B
8. K
9. C
10. E
11. G
12. I

C. Crossword Puzzle

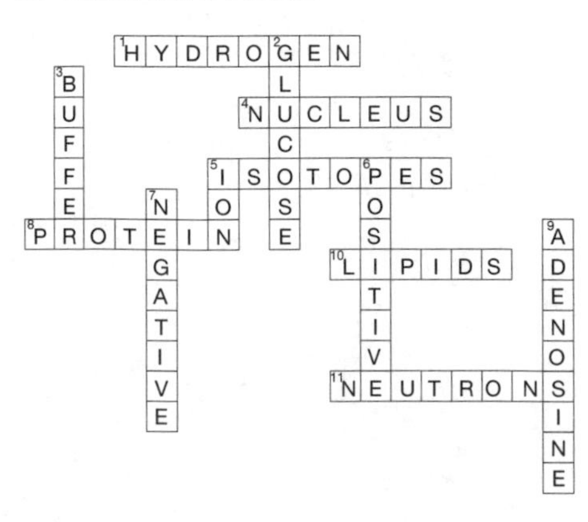

D. Short Answer Questions (Answers)

1.

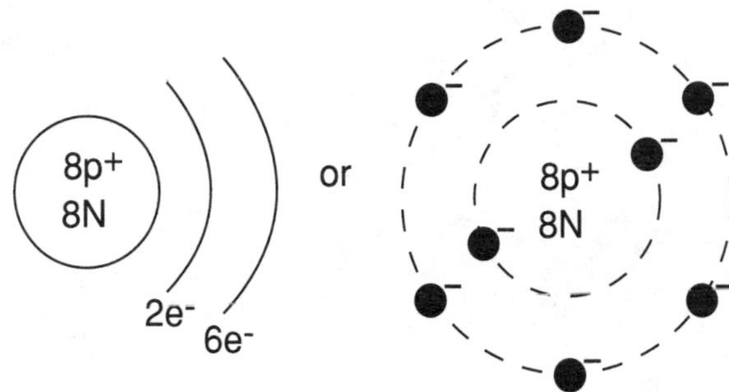

2. Their outer energy levels contain the maximum number of electrons and they will not react with one another nor combine with atoms of other elements.

3. The oxygen atom has a much stronger attraction for the shared electrons than do the hydrogen atoms, so the electrons spend most of this time in the vicinity of the oxygen nucleus. Because the oxygen atom has two extra electrons part of the time, it develops a slight negative charge. The hydrogens develop a slight positive charge because their electrons are away part of the time.

4. (1) freezing point 0 degrees C; boiling point 100 degrees C, (2) capacity to absorb and distribute heat, (3) heat absorbed during evaporation, (4) solvent properties,

5. (a) carbohydrates, ex. glucose; (b) lipids, ex. steroids; (c) proteins, ex. enzymes; (d) nucleic acid, ex. DNA

6. In a saturated fatty acid each carbon atom in the hydrocarbon tail has four single covalent bonds. If some of the carbon-to-carbon bonds are double covalent bonds, the fatty acid is unsaturated.

7. (1) Adenine nucleotide; (2) thymine nucleotide; (3) cytosine nucleotide; (4) guanine nucleotide

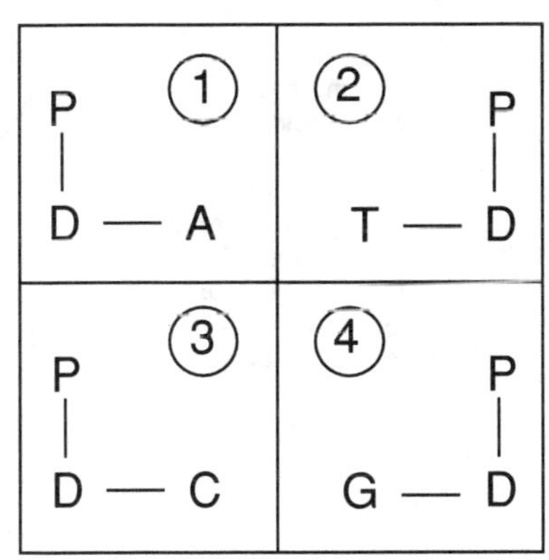

8. (a) adenine; (b) ribose; (c) phosphates

CHAPTER 3: *Cell Structure and Function*

Part I: Objective Based Questions

Objective 1

1. B
2. D
3. A
4. homeostasis

Objective 2

1. C
2. C
3. D
4. A
5. phospholipid bilayer
6. cell membrane
7. **Figure 3-1**
   A. heads
   B. phospholipid bilayer
   C. protein with channels
   D. carbohydrate chain
   E. tails
   F. cholesterol
   G. proteins
   H. cytoskeleton
   I. cell membrane

Objective 3

1. A
2. C
3. B
4. D
5. B
6. D
7. B
8. C
9. diffusion
10. phagocytosis
11. isotonic
12. **Figure 3-2**
    A. hypertonic
    B. hypotonic
    C. isotonic

Objective 4

1. A
2. C
3. B
4. D
5. B
6. C
7. fixed ribosomes
8. ribosomes
9. **Figure 3-3**
   A. cilia
   B. secretory vesicles
   C. centrioles
   D. Golgi apparatus
   E. mitochondrion
   F. rough E. R.
   G. nuclear envelope
   H. nuclear pores
   I. fixed ribosomes
   J. microvilli
   K. cytosol
   L. lysosome
   M. cytoskeleton
   N. cell membrane
   O. smooth E. R.
   P. free ribosomes
   Q. nucleolus
   R. chromatin

Objective 5

1. C
2. B
3. D
4. C
5. A

Objective 6

1. A
2. C
3. A
4. D
5. A
6. D
7. C
8. amino acids
9. **Figure 3-4**
   A. tRNA
   B. translation
   C. ribosomes
   D. mRNA
   E. transcription
   F. DNA
   G. nuclear pore

Objective 7

1. C
2. D
3. A
4. B
5. B
6. **Figure 3-5**
   A. nucleus
   B. nucleolus
   C. interphase
   D. centrioles
   E. spindle fibers
   F. chromatin
   G. early prophase
   H. centromere
   I. chromosome
   J. late prophase
   K. metaphase plate
   L. metaphase
   M. daughter chromosomes
   N. anaphase
   O. cytokinesis
   P. daughter cell
   Q. telophase

Objective 8

1. C
2. A
3. B
4. D

Part II: Chapter Comprehensive Exercises

A. Word Elimination

1. organelle
2. lysomes
3. glycocalyx
4. distance
5. saturation
6. diffusion
7. ribosomes
8. mitochondria
9. DNA
10. interphase

B. Matching

1. B
2. I
3. C
4. A
5. J
6. E
7. D
8. G
9. F
10. H

C. Concept Map

1. lipid bilayer
2. proteins
3. organelles
4. membranes
5. nucleolus
6. lysosomes
7. centrioles
8. ribosomes
9. fluid component

D. Crossword Puzzle

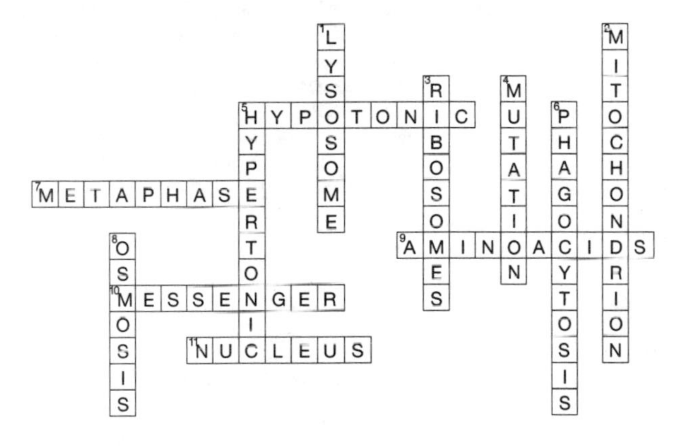

E. Short Answer Questions (Answers)

1. (a) Circulatory – RBC; (b) Muscular – muscle cell; (c) Reproductive – sperm cell; (d) Skeletal – bone cell; (e) Nervous – neuron (Other systems and cells could be listed for this answer.)

2. (a) Physical isolation; (b) Regulation of exchange with the environment; (c) Sensitivity; (d) Structural support

3. Cytosol – high concentration of K^+; (Extra Cellular Fluid E.C.F.) – high concentration of Na^+
   Cytosol – higher concentration of dissolved proteins
   Cytosol – smaller quantities of carbohydrates and lipids

4. Cytoskeleton, centrioles, ribosomes, mitochondria, nucleus, nucleolus, ER, Golgi apparatus, lysosomes

5. Centrioles – move DNA during cell division
   Cilia – move fluids or solids across cell surfaces
   Flagella – move cell through fluids

6. (a) Lipid solubility; (b) Channel size; (c) Electrical interactions

CHAPTER 4: *The Tissue Level of Organization*

Part I: Objective Based Questions

Objective 1

1. D
2. B
3. C

Objective 2

1. C
2. D
3. B
4. C
5. epithelial
6. protection

Objective 3

1. B
2. C
3. D
4. A
5. B
6. D
7. C
8. E
9. A
10. F
11. H
12. L
13. I
14. J
15. G
16. K

Objective 4

1. C
2. A
3. A
4. D
5. B
6. B
7. D
8. C
9. B
10. A
11. connective
12. collagen
13. areolar
14. reticular
15. **Figure 4-2**
    A. adipose
    B. loose connective
    C. collagen fibers
    D. central canal
    E. canaliculi
    F. matrix
    G. bone tissue
    H. hyaline cartilage
    I. dense connective

Objective 5

1. A
2. C
3. B
4. C
5. A
6. D
7. lamina propria
8. synovial
9. **Figure 4-3**
    A. cutaneous
    B. mucous
    C. serous
    D. synovial

Objective 6

1. D
2. A
3. B
4. D
5. D
6. contraction
7. skeletal
8. **Figure 4-4**
    A. muscle fiber (cell)
    B. striations
    C. skeletal muscle
    D. nucleus
    E. smooth muscle cell
    F. smooth muscle
    G. nucleus
    H. intercalated disc
    I. striations
    J. cardiac muscle

Objective 7

1. B
2. D
3. B
4. A
5. B
6. **Figure 4-5**
    A. dendrites
    B. nucleus
    C. soma
    D. axon

Objective 8

1. C
2. A
3. D
4. B

Objective 9

1. C
2. D
3. D

Part II: Chapter Comprehensive Exercises

A. Word Elimination

1. adipose
2. storage
3. microvilli
4. flagella
5. connective
6. covering
7. matrix
8. lymph
9. visceral
10. neuroglia

B. Matching

1. I
2. F
3. A
4. J
5. B
6. D
7. C
8. F
9. G
10. H

C. Concept Map

1. connective
2. columnar
3. skeletal
4. neuron
5. loose
6. adipose
7. ligaments
8. cartilage
9. blood
10. immunity

D. Crossword Puzzle

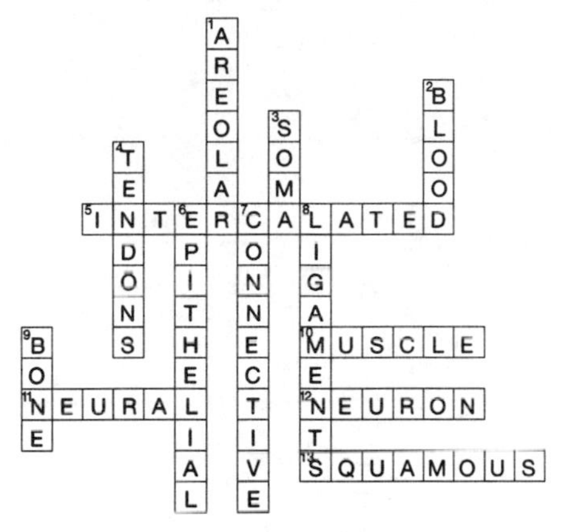

1. Epithelial, connective, muscle, and neural tissue

2. Provides physical protection, controls permeability, provides sensations, and provides specialized secretion.

3. Microvilli are abundant on epithelial surfaces where absorption and secretion take place. A cell with microvilli has at least twenty times the surface area of a cell without them. A typical ciliated cell contains about 250 cilia that beat in coordinated fashion. Materials are moved over the epithelial surface by the synchronized beating of cilia.

4. Merocrine secretion – the product is released through exocytosis.
   Apocrine secretion – loss of cytoplasm as well as the secretory produce.
   Holocrine secretion – product is released, cell is destroyed.
   Merocrine and apocrine secretions leave the cell intact and are able to continue secreting; holocrine does not.

5. Serous – watery solution containing enzymes.
   Mucous – viscous mucous.
   Mixed glands – serous and mucous secretions.

6. Specialized cells, extracellular protein fibers, and ground substance.

7. Connective tissue proper, fluid connective tissues, supporting connective tissues.

8. Collagen, reticular, elastic.

9. Mucous membranes, serous membranes, cutaneous membranes, and synovial membranes.

10. Skeletal, cardiac, smooth.

11. Neurons – transmission of nerve impulses from one region of the body to another.
    Neuroglia – support framework for neural tissue.

CHAPTER 5: *The Integumentary System*

Part I: Objective Based Questions

Objective 1

1. C
2. A
3. D
4. B
5. D

Objective 2

1. D
2. A
3. A
4. D
5. stratum corneum
6. stratum lucidum
7. **Figure 5-1**
   A. epidermis
   B. dermis
   C. hypodermis
   D. hair shaft
   E. sebaceous gland
   F. arrector pili muscle
   G. hair follicle
   H. touch pressure receptor
   I. nerve fiber
   J. sweat gland
   K. blood vessels
   L. fat (adipose)

Objective 3

1. C
2. D
3. A
4. B
5. melanin
6. melanin
7. B

Objective 4

1. D
2. C
3. A
4. B
5. D

Objective 5

1. D
2. D
3. D
4. DNA
5. ultraviolet light

Objective 6

1. C
2. D
3. C
4. A
5. B
6. B
7. nails
8. follicle
9. eyelashes
10. goosebumps
11. **Figure 5-2**
    A. free edge
    B. hyponychium (underneath)
    C. nail bed (underneath)
    D. lateral nail groove
    E. lunular (cuticle)
    F. eponychium
    G. nail root
    H. eponychium
    I. lunula
    J. nail body
    K. hyponychium
    L. phalanx (bone of fingertips)

Objective 7

1. D
2. C
3. D
4. B
5. C
6. A

Objective 8

1. A
2. D
3. B
4. C
5. D

Objective 9

1. D
2. D
3. C
4. D
5. melanocyte
6. glandular

Part II: Chapter Comprehensive Exercises

A. Word Elimination

1. cutaneous
2. dermis
3. secretion
4. melanocytes
5. stabilize
6. apocrine
7. sebaceous
8. arrector pili
9. increased immunity
10. epidermis

B. Matching

1. E
2. H
3. G
4. C
5. A
6. B
7. J
8. F
9. I
10. D

C. Concept Map

1. dermis
2. glands
3. vitamin D synthesis
4. lubrication
5. produce secretions
6. sensory reception

D. Crossword Puzzle

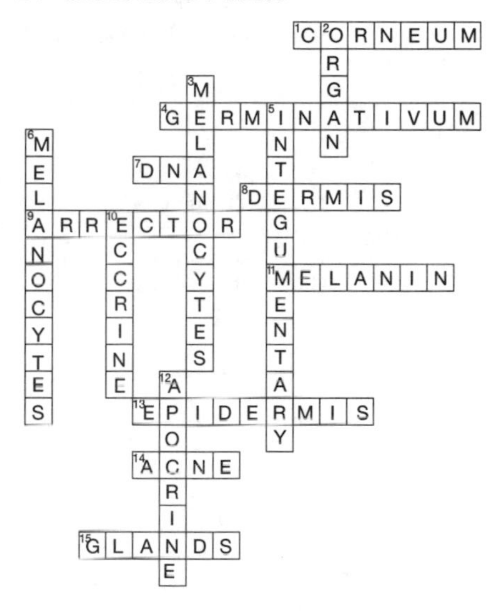

E. Short Answer Questions (Answers)

1. Palms of hand; soles of feet. These areas are covered by 30 or more layers of cornified cells. The epidermis in these locations may be six times thicker than the epidermis covering the general body surface.

2. Bacteria break down the organic secretions of the apocrine glands.

3. Stratum corneum, stratum granulosum, stratum spinosium, stratum germinativum, papillary layer, reticular layer.

4. Long term damage can result from chronic exposure, and an individual attempting to acquire a deep tan places severe stress on the skin. Alterations in underlying connective tissue lead to premature wrinkling and skin cancer can result from chromosomal damage or breakage.

5. Hairs are dead, keratinized structures, and no amount of oiling or shampooing with added ingredients will influence either the exposed hair or the follicles buried in the dermis.

6. "Whiteheads" contain accumulated, stagnant secretions. "Blackheads" contain more solid material that has been invaded by bacteria.

7. Bedsores are a result of circulatory restrictions. They can be prevented by frequent changes in body position that vary the pressure applied to specific blood vessels.

8. Special smooth muscles in the dermis, the arrector pili muscles, pull on the hair follicles and elevate the hair, producing "goose bumps."

CHAPTER 6: *The Skeletal System*

Part I: Objective Based Questions

Objective 1

1. D
2. B
3. A
4. C

Objective 2

1. A
2. D
3. C
4. B
5. osteon
6. epiphysis
7. **Figure 6-1**
A. proximal epiphysis
B. diaphysis
C. distal epiphysis
D. spongy bone
E. articular cartilage
F. periosteum
G. compact bone
H. marrow cavity
I. endosteum
J. blood vessel
8. **Figure 6-2**
A. osteons
B. artery
C. vein
D. compact bone
E. trabeculae (spongy bone)

Objective 3

1. A
2. B
3. C
4. C
5. A
6. D
7. A
8. **Figure 6-3**
A. disintegrating chondrocytes
B. enlarging chondrocytes
C. epiphysis
D. diaphysis
E. marrow cavity
F. blood vessels
G. epiphyseal plate

Objective 4

1. C
2. B
3. D
4. A
5. A
6. A
7. D
8. D
9. B
10. B

Objective 5

1. B
2. C
3. A
4. D
5. C
6. A
7. C
8. muscles
9. extremities
10. clavicle
11. **Figure 6-4**
A. parietal
B. frontal bone
C. clavicle
D. scapula
E. humerus
F. vertebrae
G. os coxae
H. femur
I. patella
J. talus
K. metatarsals
L. sphenoid bone
M. maxilla
N. mandible
O. sternum
P. rib
Q. radius
R. sacrum
S. ulna
T. carpals
U. metacarpals
V. phalanges
W. tibia
X. fibula
Y. tarsals
Z. phalanges

Objective 6

1. B
2. A
3. B
4. D
5. C
6. D
7. C
8. D
9. B
10. **Figure 6-5**
A. frontal bone
B. sphenoid bone
C. ethmoid bone
D. lacrimal bone
E. nasal bone
F. zygomatic bone
G. maxilla
H. mandible
I. parietal bone
J. temporal bone
K. occipital bone
L. zygomatic bone
M. mastoid process
N. styloid process
11. **Figure 6-6**
A. parietal bone
B. frontal bone
C. sphenoid bone
D. temporal bone
E. zygomatic bone
F. maxilla
G. nasal bone
H. ethmoid bone
I. lacrimal bone
J. nasal concha
K. vomer bone
L. mandible
12. **Figure 6-7**
A. zygomatic bone
B. zygomatic arch
C. vomer
D. styloid process
E. mastoid process
F. occipatal bone
G. maxilla
H. palatine bone
I. sphenoid bone
J. temporal bone
K. occipital condyle
L. foramen magnum
13. **Figure 6-8**
A. coronal suture
B. frontal bone
C. sphenoidal fontanel
D. maxilla
E. mandible
F. squamosal suture
G. mastoid fontanel
H. occipital bone
I. lambdoidal suture
J. frontal fontanel
K. coronal suture
L. sagittal suture
M. parietal bone
N. occipital fontanel

Objective 7

1. B
2. C
3. C
4. B
5. C
6. A
7. B
8. C
9. A
10. C
11. coccyx
12. centrum
13. cervical
14. **Figure 6-9**
    A. cervical
    B. thoracic
    C. lumbar
    D. sacral
    E. coccygeal
15. **Figure 6-10**
    A. lamina
    B. pedicle
    C. spinous process
    D. transverse process
    E. vertebral foramen
    F. body
16. **Figure 6-11**
    A. sternum
    B. manubrium
    C. body of sternum
    D. xiphoid process
    E. floating ribs
    F. true ribs
    G. false ribs
    H. costal cartilage

Objective 8

1. A
2. C
3. B
4. A
5. D
6. B
7. D
8. A
9. C
10. B
11. C
12. D
13. A
14. D
15. B
16. C
17. clavicle
18. coxae
19. styloid
20. malleolus
21. acetabulum
22. knee
23. childbirth
24. clavicle
25. femur
26. **Figure 6-12**
    A. superior border
    B. spine
    C. medial border
    D. coracoid process
    E. acromion process
    F. glenoid fossa
    G. body
    H. lateral border
27. **Figure 6-13**
    A. ileum
    B. sacrum
    C. pubis
    D. symphysis pubis
    E. ischium
    F. iliac crest
    G. coccyx
    H. acetabulum
    I. obturator foramen
28. **Figure 6-14**
    A. ulna
    B. radius
    C. carpals
    D. metacarpals
    E. phalanges

Objective 9

1. A
2. C
3. D
4. B
5. B
6. D
7. D
8. suture
9. synostosis
10. symphysis
11. synovial
12. hip
13. knee
14. elbow
15. **Figure 6-15**
    A. extensor muscle
    B. bursa
    C. femur
    D. tendon
    E. patella
    F. fat pad
    G. joint capsule
    H. meniscus
    I. joint cavity
    J. intracapsular ligament
    K. patellar ligament
    L. tibia

Objective 10

1. A
2. C
3. D
4. A
5. D
6. B
7. C
8. C
9. D
10. A
11. **Figure 6-16**
    A. extension
    B. flexion
    C. hyperextension
    D. flexion
    E. extension
    F. flexion
    G. extension
    H. hyperextension
    I. abduction
    J. adduction
    K. supination
    L. pronation
    M. head rotation
    N. limb rotation
12. **Figure 6-17**
    A. inversion
    B. eversion
    C. dorsiflexion
    D. plantar flexion
    E. opposition
    F. retraction
    G. protraction
    H. elevation
    I. depression

Objective 11

1. D
2. B
3. A
4. A
5. B
6. D
7. **Figure 6-17**
    A. gliding joint
    B. hinge joint
    C. pivot joint
    D. ellipsoidal joint
    E. saddle joint
    F. ball-and-socket joint

Objective 12

1. D
2. D
3. B
4. D

Part II: Chapter Comprehensive Exercises

A. Word Elimination

1. secretion
2. vitamin D_3
3. ostopenia
4. pelvis
5. hyoid bone
6. occipital
7. maxilla
8. scoliosis
9. olecranon
10. pronation

B. Matching

1. D
2. H
3. A
4. J
5. D
6. I
7. E
8. B
9. G
10. F

C. Concept Map I - Skeletal System

1. support
2. osteons
3. osteocytes
4. lacunae
5. axial
6. sternum
7. pectoral girdle
8. lumbar
9. true
10. tibia

Concept Map II - Joints

1. no movement
2. sutures
3. cartilagimous
4. amphiarthrosis
5. fibrous
6. symphysis
7. synovial
8. monoaxial
9. wrist

D. Crossword Puzzle

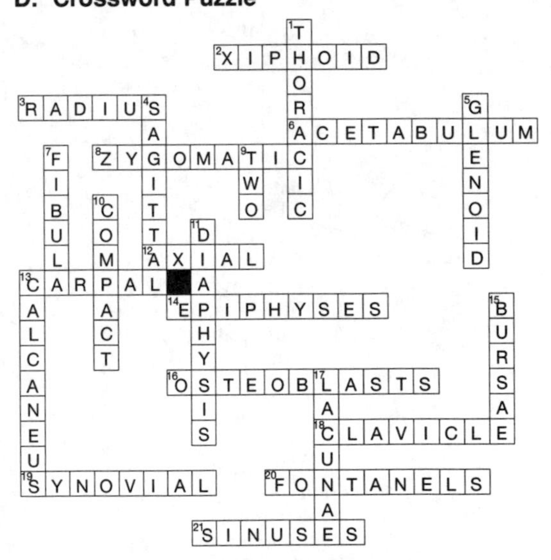

E. Short Answer Questions (Answers)

1. (1) Create a framework that supports and protects organ systems in the dorsal and ventral body cavities. (2) Provide a surface area for attachment of muscles that adjust the positions of the head, neck , and trunk. (3) Performs respiratory movements. (4) Stabilize or position elements of the appendicular system.

2. The auditory ossicles consist of three (3) tiny bones on each side of the skull that are enclosed by the temporal bone. The hyoid bone lies below the skull suspended by the stylohyoid ligaments.

3. Craniostenosis is premature closure of one or more fontanels, which results in unusual distortions of the skull.

4. The thoracic and sacral curves are called primary curves because they begin to appear late in fetal development. They accomodate the thoracic and abdominopelvic viscera.

   The lumbar and cervical curves are called scondary curves because they do not appear until several months after birth. They help position the body weight over the legs.

5. *Kyphosis*: normal thoracic curvature becomes exaggerated, producing "roundback" appearance.
   Lordosis: exaggerated lumbar curvature produces "swayback" appearance.
   Scoliosis: abnormal lateral curvature.

6. The *true* ribs reach the anterior body wall and are connected to the sternum by separate cartilaginous extensions. The *false* ribs do not attach directly to the sternum.

7. Provides control over the immediate environment; changes your position in space, and makes you an active, mobile person.

8. Pectoral girdle: scapula, clavicle
   Pelvic girdle: ilium, ischium, pubis

9. A tendon attaches muscle to bone. Ligaments attach bones to bones.

10. (a) Bursa are small, synovial-fillied pockets in connective tissue that form where a tendon or ligament rubs against other tissues. Their function is to reduce friction and act as a shock absorber.
    (b) Menisci are articular discs that: (1) subdivide a synovial cavity (2) channel the flow of synovial fluid, and (3) allow variations in the shape of the articular surfaces.

11. (1) Provides lubrication
    (2) Acts as a shock absorber
    (3) Nourishes the chondrocytes

12. The knee joint and elbow joint are both hinge joints.

13. The shoulder joint and hip joint are both ball-and-socket joints.

14. Intervertebral discs are not found between the first and second cervical vertebrae (atlas and axis), the sacrum and the coccyx. C, the atlas, sits on top of C, the axis; the den of the axis provides for rotation of the first cervical vertebrae which supports the head. An intervertebral head would prohibit rotation. The sacrum and the coccyx are fused bones.

15. (1) inversion
    (2) opposition
    (3) plantar flexion
    (4) protraction
    (5) depression
    (6) elevation

CHAPTER 7: *The Muscular System*

Part I: Objective Based Questions

Objective 1

1. B
2. D
3. C
4. A

Objective 2

1. B
2. A
3. D
4. D
5. C
6. A
7. **Figure 7-1**
   A. tendon
   B. skeletal muscle
   C. epimysium
   D. blood vessels and nerves
   E. perimysium
   F. muscle fascicle
   G. endomysium
   H. muscle fiber

Objective 3

1. A
2. B
3. D
4. A
5. B
6. **Figure 7-2**
   A. actin
   B. A-band
   C. I-band
   D. M-line
   E. myosin
   F. Z-line

Objective 4

1. D
2. A
3. C
4. B
5. C
6. **Figure 7-3**
   A. resting sarcomere
   B. active site exposure
   C. cross-bridge attachment
   D. pivoting of myosin head
   E. cross-bridge detachment
   F. myosin reactivation

Objective 5

1. C
2. A
3. C
4. A
5. treppe
6. incomplete

Objective 6

1. D
2. C
3. A
4. C
5. A
6. B

Objective 7

1. B
2. D
3. B

Objective 8

1. B
2. B
3. D
4. A
5. C

Objective 9

1. B
2. D
3. D

Objective 10

1. D
2. D
3. B
4. B
5. A
6. C
7. D
8. A
9. B
10. C

Objective 11

1. C
2. A
3. D
4. B
5. C
6. C
7. A
8. B
9. D
10. B
11. **Figure 7-4**
    A. temporalis
    B. orbicularis oculi
    C. zygomaticus
    D. orbicularis oris
    E. pectoralis major
    F. deltoid
    G. biceps brachii
    H. rectus abdominis
    I. external oblique
    J. adductor muscles
    K. gracilis
    L. sartorius
    M. tibialis anterior
    N. frontalis
    O. masseter
    P. sternocleidmastoid
    Q. transversus abdominis
    R. tensor fascia lata
    S. rectus femoris
    T. vastas medialis
    U. vastus lateralis
    V. perineus longus
12. **Figure 7-5**
    A. gluteus maximus
    B. semimembranosus
    C. biceps femoris
    D. semitendinosus
    E. gastrocnemius
    F. occipitalis
    G. trapezius
    H. deltoid
    I. triceps brachii
    J. latissimus dorsi
    K. external oblique
    L. gluteus medius
    M. soleus

Objective 12

1. D
2. A
3. B
4. C

Objective 13

1. skeletal
2. cardiovascular
3. lymphatic
4. endocrine
5. nervous

Part II: Chapter Comprehensive Exercises

A. Word Elimination

1. support
2. sarcomere
3. contraction
4. myogram
5. DNA
6. myoglobin
7. jogging
8. anaerobic
9. sartorius
10. heart

B. Matching

1. E
2. G
3. A
4. C
5. B
6. D
7. F
8. J
9. H
10. I

C. Concept Maps - I and II

I. Muscle Tissue

1. heart
2. striated
3. smooth
4. involuntary
5. non-striated
6. bones
7. multinucleated

II. Muscle Structure

8. fascicles (muscle bandles)
9. myofibrils
10. sarcomeres
11. actin
12. Z-lines
13. thick filaments
14. H-zone

D. Crossword Puzzle

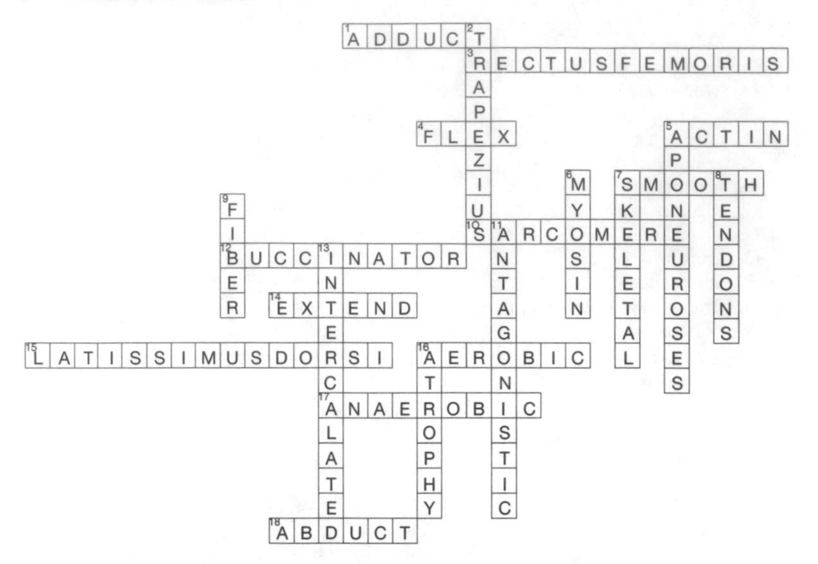

E. Short Answer Questions (Answers)

1. (a) produce skeletal movement
   (b) maintain posture and body position
   (c) support soft tissues
   (d) guard entrances and exits
   (e) maintain body temperature

2. (a) an outer *epimysium*, (b) a central *perimysium*, and (c) an inner *endomysium*

3. (a) active site exposure
   (b) cross-bridge attachment
   (c) pivoting
   (d) cross-bridge detachment
   (e) myosin activation

4. *Isometric* contraction – tension rises to maximum but the length of the muscle remains constant.

   Isotonic contraction – tension in the muscle builds until it exceeds the amount of resistance and the muscle shortens. As the muscle shortens, the tension in the muscle remains constant, at a value that just exceeds the applied resistance.

5. When fatigue occurs and the oxygen supply to muscles is depleted, aerobic respiration ceases owing to the decreased oxygen supply. Anaerobic glycolysis supplies the needed energy for a short period of time. The amount of oxygen needed to restore normal pre-exertion conditions is the oxygen debt.

6. Fast fiber muscles produce powerful contractions, which use ATP in massive amounts. Prolonged activity is primarily supported by anaerobic glycolysis, and fast fibers fatigue rapidly. Slow fibers are specialized to enable them to continue contracting for extended periods. The specializations include an extensive network of capillaries so supplies of O_2 are available, and the presence of myoglobin, which binds O_2 molecules and which results in the buildup of O_2 reserves. These factors improve mitochondrial performance.

7. The origin remains stationary while the insertion moves.

8. Prime mover or agonist, synergist, antagonists

9. (a) muscles of the head and neck
   (b) muscles of the spine
   (c) oblique and rectus muscles
   (d) muscles of the pelvic floor

10. (a) muscles of the shoulders and arms
    (b) muscles of the pelvic girdle and legs

CHAPTER 8: *Neural Tissue and the Central Nervous System*

Part I: Objective Based Questions

Objective 1

1. A
2. C
3. D
4. C
5. D
6. A

Objective 2

1. B
2. C
3. D
4. B
5. C
6. A
7. D
8. B
9. A
10. C
11. **Figure 8-1**
    A. dendrites
    B. soma
    C. nucleus
    D. axon hillock
    E. axon
    F. neurilemma
    G. axon terminals
12. multipolar
13. **Figure 8-2**
    A. neuron
    B. astrocyte
    C. myelinated axon
    D. oligodendrocyte
    E. microglial cell
    F. capillaries

Objective 3

1. B
2. C
3. A
4. B
5. D
6. D
7. B

Objective 4

1. continuous
2. saltatory

Objective 5

1. A
2. C
3. B
4. A
5. B
6. D

Objective 6

1. A
2. D
3. D
4. B
5. C
6. **Figure 8-3**
    A. white matter
    B. gray matter
    C. interneuron
    D. sensory neuron
    E. stimulus
    F. receptor
    G. motor neuron
    H. effector
    I. synapse

Objective 7

1. C
2. A
3. A
4. C
5. B
6. A

Objective 8

1. D
2. C
3. C
4. B
5. A
6. D
7. **Figure 8-4**
    A. posterior median sulcus
    B. white matter
    C. pia mater
    D. central canal
    E. anterior horn
    F. anterior median fissure
    G. spinal nerve
    H. dorsal root
    I. ventral root
    J. dura mater
    K. posterior horn
    L. subarachnoid space
    M. gray commissure

Objective 9

1. D
2. B
3. A
4. B
5. D
6. B
7. A
8. D
9. C
10. B
11. **Figure 8-5**
    A. cental sulcus
    B. parietal lobe
    C. parieto-occipital fissure
    D. occipital lobe
    E. cerebellum
    F. medulla oblongata
    G. postcentral gyrus
    H. precentral gyrus
    I. frontal lobe
    J. lateral fissure
    K. temporal lobe
    L. pons
12. **Figure 8-6**
    A. choroid plexus
    B. cerebral hemispheres
    C. corpus callosum
    D. pineal body
    E. cerebral peduncle
    F. cerebral aqueduct
    G. fourth ventricle
    H. cerebellum
    I. thalamus
    J. fornix
    K. third ventricle
    L. corpora quadrigemina
    M. optic chiasma
    N. pituitary gland
    O. mammilary gland
    P. pons
    Q. medulla oblongata

Objective 10

1. B
2. C
3. B
4. A
5. C
6. A
7. A
8. C
9. D
10. B
11. **Figure 8-7**
    A. precentral gyrus
    B. premotor cortex
    C. frontal lobe
    D. temporal lobe
    E. central sulcus
    F. postcentral gyrus
    G. parietal lobe
    H. occipital lobe

Part II: Chapter Comprehensive Exercises

A. Word Elimination

1. CNS
2. neuron
3. neuroglia
4. neuron
5. conduction
6. adrenergic
7. ganglia
8. sensory
9. gyrus
10. meninges

B. Matching

1. F
2. H
3. G
4. C
5. B
6. A
7. D
8. J
9. E
10. I

C. Concept Map I - Neural Tissue

1. Schwann cells
2. neuroglia
3. transmit nerve impulses
4. surround peripheral ganglia
5. central nervous system

Concept Map II - Major regions of the Brain

1. diencephalon
2. hypothalamus
3. corpora quadrigemina
4. 2 cerebellar hemispheres
5. pons
6. medulla oblongata

D. Crossword Puzzle

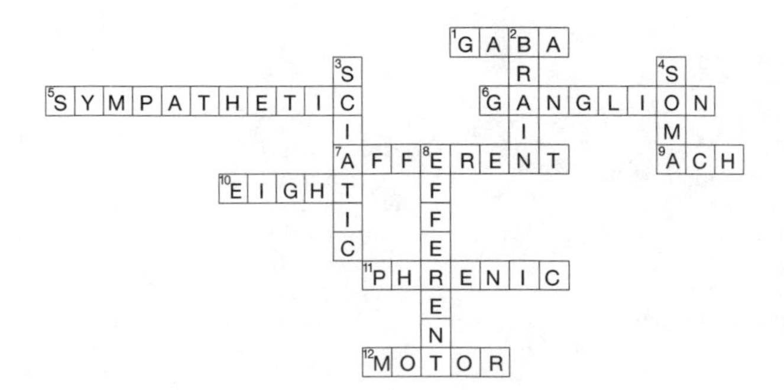

E. Short Answer Questions (Answers)

1. (a) providing sensation of the internal and external environments
   (b) integrating sensory information
   (c) coordinating voluntary and involuntary activities
   (d) regulating or controlling peripheral structures and systems

2. CNS consists of the brain and the spinal cord
   PNS consists of the somatic nervous system and the autonomic nervous system

3. astrocytes, oligodendrocytes, microglia, ependymal cells

4. *Neurons* are responsible for information transfer and processing in the nervous sytem.
   Neuroglia are specialized cells that provide support throughout the nervous system.

5. *Divergence* is the spread of information from one neuron to several neurons, or from one neuronal pool to multiple pools.
   Convergence – several neurons synapse on the same postsynaptic neuron.

6. The brain's versatility results from (a) the tremendous number of neurons and neuronal pools in the brain, and (b) the complexity of the interconnections between the neurons and neuronal pools.

7. (a) cerebrum; (b) diencephalon; (c) mesencephalon; (d) cerebellum; (e) pons; (f) medulla oblongata

8. "Higher centers" refers to nuclei, centers, and cortical areas of the cerebrum, cerebellum, diencephalon, and mesencephalon.

9. The limbic system includes nuclei and tracts along the border between the cerebrum and diencephalon. It is involved in the processing of memories, creation of emotional states, drives, and associated behaviors.

10. Most endocrine organs are under direct or indirect hypothalamic control. Releasing hormones and inhibiting hormones secreted by nuclei in the tuberal area of the hypothalamus promote or inhibit secretion of hormones by the anterior pituitary gland. The hypothalamus also secretes the hormones ADH (antidiuretic hormone) and oxytocin.

11. (a) The cerebellum oversees the postural muscles of the body, making rapid adjustments to maintain balance and equilibrium. (b) The cerebellum programs and times voluntary and involuntary movements.

12. (a) Sensory and motor nuclei for four of the cranial nerves; (b) nuclei concerned with the involuntary control of respiration; (c) tracts that link the cerebellum with the brain stem, cerebrum, and spinal cord; (d) ascending and descending tracts.

CHAPTER 9: *The Peripheral Nervous System and Integrated Neural Function*

Part I: Objective Based Questions

Objective 1

1. C
2. B
3. D
4. Olfactory
5. optic
6. oculomotor
7. opthalamic
8. VII
9. vagus
10. tongue
11. **Figure 9-1**
    A. olfactory CN I
    B. oculomotor CN III
    C. trigeminal CN V
    D. vestibulocochlear CN VIII
    E. glossopharyngeal CN IX
    F. hypoglossal CN XII
    G. optic CN II
    H. trochlear CN IV
    I. abducens CN VI
    J. facial CN VII
    K. vagus CN X
    L. accessory CN XI

Objective 2

1. B
2. A
3. D
4. C

Objective 3

1. C
2. B
3. D
4. C
5. C

Objective 4

1. D
2. B
3. A
4. C
5. A

Objective 5

1. C
2. B
3. A
4. D
5. B
6. C
7. C
8. D
9. A
10. **Figure 9-2**
    A. lateral corticospinal
    B. rubrospinal
    C. vestibulospinal
    D. reticulospinal
    E. tectospinal
    F. anterior corticospinal
    G. fasciculus gracilis
    H. fasciculus cuneatus
    I. posterior spinocerebellar
    J. lateral spinothalamic
    K. anterior spinocerebellar
    L. anterior spinothalamic

Objective 6

1. B
2. D
3. B
4. D
5. C

Objective 7

1. C
2. D
3. B
4. synapse
5. involuntary

Objective 8

1. B
2. D
3. D
4. C
5. B
6. C
7. B
8. C
9. C
10. B
11. D
12. A
13. A
14. fight or flight
15. rest and repose

Objective 9

1. B
2. C
3. C
4. A
5. B
6. A
7. D

Objective 10

1. B
2. C
3. D

Objective 11

1. reproduction
2. skeletal
3. endocrine
4. digestive
5. cardiovascular

Part II: Chapter Comprehensive Exercises

A. Word Elimination

1. CNS
2. N I
3. N IV
4. peripheral
5. muscle
6. pyramidal
7. ganglia
8. splanchnic
9. lymph
10. cardiac

B. Matching

1. E
2. H
3. B
4. J
5. A
6. C
7. I
8. F
9. D
10. G

C. Concept Maps

I - Autonomic Nervous System

1. sympathetic
2. motor neurons
3. smooth muscle
4. first-order neurons
5. ganglia outside CNS
6. postganglionic

II - Sympathetic Division of ANS

7. thoracolumbar
8. spinal segments T1-L2
9. second-order neurons (postganglionic)
10. sympathetic chain of ganglia (paired)
11. visceral effectors
12. adrenal medulla (paired)
13. general circulation

III - Parasympathetic Divisions of ANS

14. craniosacral
15. brain stem
16. ciliary ganglion
17. CN VII
18. nasal, tear, salivary glands
19. otic ganglia
20. N X
21. segments S2-S4
22. intramural ganglia
23. lower-abdominopelvic cavity

D. Crossword Puzzle

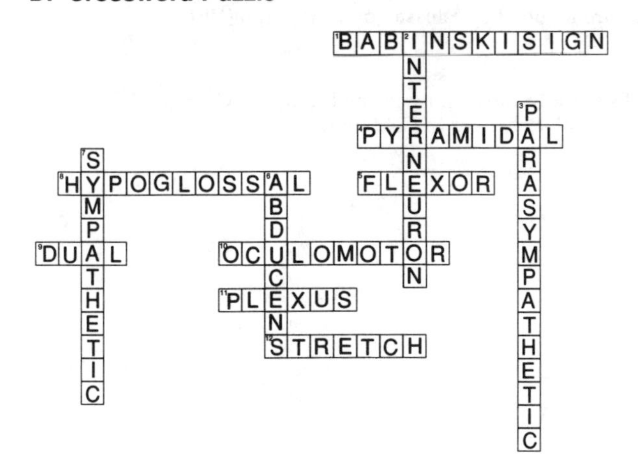

E. Short Answer Questions (Answers)

1. Posterior column, spinothalamic, and spinocerebellar pathways

2. Corticobulbar tracts, lateral corticospinal tract, and anteriorcorticospinal tract.

3. Rubrospinal tract, reticulospinal tract, vestibulospinal tract, and tectospinal tract.

4. The pyramidal system provides a rapid and direct mechanism for voluntary somatic motor control of skeletal muscles. The extrapyramidal system provides less precise control of motor functions, especially ones associated with overall body coordination and cerebellar function.

5. (a) Preganglionic (first-order) neurons located between segments T1 and L2 of the spinal cord.
   (b) Ganglionic (second-order) neurons located in ganglia near the vertebral column (sympathetic chain ganglia, collateral ganglia).
   (c) Specialized second-order neurons in the interior of the adrenal gland.

6. (a) preganglionic neurons in the brain stem and in sacral segments of the spinal cord;
   (b) ganglionic neurons in peripheral ganglia located within or adjacent to the target organs.

7. When dual innervation exists, the two divisions of the ANS often but not always have opposite effects. Sympathetic-parasympathetic opposition can be seen along the digestive tract, at the heart, in the lungs and elsewhere throughout the body.

8. Stimulation of the parasympathetic system leads to a general increase in the nutrient content of the blood. Cells throughout the body respond to this increase by absorbing nutrients and using them to support growth and other anabolic activities.

9. The sympathetic division stimulates tissue metabolism, increases alertness, and generally prepares the body to deal with emergencies.

10. The parasympathetic division conserves energy and promotes sedentary activities, such as digestion.

11. (a) All preganglionic fibers are cholinergic; they release acetylcholine (Ach) at their synaptic terminals. The effects are always excitatory.
    (b) Postganglionic fibers are also cholinergic, but the effects may be excitatory or inhibitory, depending on the nature of the receptor.
    (c) Most postganglionic sympathetic terminals are adrenergic; they release norepinephrine(NE). The effects are usually excitatory.

12. (a) A reduction in brain size and weight; (b) A decrease in blood flow to the brain; (c) Changes in synaptic organization of the brain; (d) Intracellular and extracellular changes in CNS neurons.

CHAPTER 10: *Sensory Functions*

Part I: Objective Based Questions

Objective 1

1. C
2. A
3. A
4. B

Objective 2

1. C
2. D
3. C
4. A
5. D
6. B
7. C
8. D

Objective 3

1. A
2. C
3. B
4. B
5. D
6. **Figure 10-1**
   A. olfactory tract
   B. cribiform plate
   C. olfactory bulb
   D. afferent nerve fiber
   E. basal cell
   F. olfactory gland
   G. olfactor receptor cell
   H. cilia
   I. mucus layer

Objective 4

1. D
2. A
3. C
4. B
5. **Figure 10-2**
   A. bitter taste
   B. sour taste
   C. sweet taste
   D. salty taste
   E. taste buds
   F. stratified squamous epithelium
   G. supporting cells
   H. gustatory cell
   I. cranial nerve
   J. microvilli

Objective 5

1. B
2. C
3. A
4. C
5. C
6. A
7. D
8. A
9. sclera
10. pupil
11. rods
12. cones
13. occipital
14. **Figure 10-3**
    A. ciliary body
    B. suspensory ligament
    C. iris
    D. aqueous humor
    E. lens
    F. cornea
    G. vitreous humor
    H. optic disc
    I. optic nerve
    J. fovea centralis
    K. sclera
    L. choroid coat
    M. retina
15. **Figure 10-4**
    A. superior oblique
    B. superior rectus
    C. lateral rectus
    D. inferior rectus
    E. inferior oblique

Objective 6

1. C
2. B
3. A
4. C
5. B
6. D
7. iris
8. myopia
9. macula lutea
10. accommodation
11. **Figure 10-5**
    A. pigment layer of retina
    B. rod
    C. cone
    D. amacrine cell
    E. horizontal cell
    F. bipolar cells
    G. ganglion cells

Objective 7

1. B
2. C
3. D

Objective 8

1. B
2. C
3. B
4. D
5. D
6. C

Objective 9

1. A
2. D
3. B
4. C
5. B
6. C
7. C
8. **Figure 10-6**
   A. outer ear
   B. middle ear
   C. inner ear
   D. pinna
   E. malleus
   F. incus
   G. stapes
   H. vestibular complex
   I. temporal bone
   J. cochlea
   K. vestibulocochlear nerve
   L. bony labyrinth
   M. auditory tube
   N. tympanum
   O. external auditory canal

Part II: Chapter Comprehensive Exercises

A. Word Elimination

1. touch
2. balance
3. stereoreceptors
4. cornea
5. choroid
6. propriopia
7. tympanum
8. otolith
9. auditory tube
10. cerumen

B. Matching

1. F
2. H
3. A
4. J
5. C
6. B
7. I
8. E
9. D
10. G

C. Concept Maps

Concept Map I - Special Senses

1. olfaction
2. smell
3. tongue
4. taste buds
5. ears
6. balance and hearing
7. audition
8. retina
9. rods and cones

Concept Map II - General Senses

1. thermoreceptors
2. pain
3. tactile
4. pressure
5. dendritic processes
6. Merkel's discs
7. Pacinian corpuscles
8. baroreceptors
9. aortic sinus
10. proprioception
11. muscle spindles

D. Crossword Puzzle

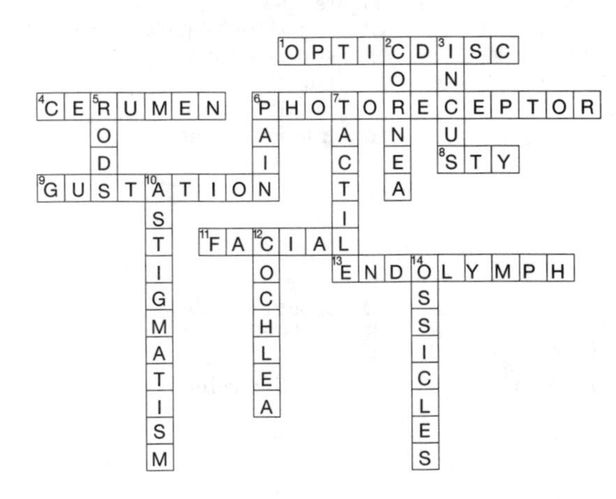

E. Short Answer Questions (Answers)

1. Sensations of temperature, pain, touch, pressure, vibration, and proprioception.

2. Smell (olfaction), taste (gustation), balance (equilibrium), hearing (audition), vision (sight).

3. (a) Nociceptors – variety of stimuli usually associated with tissue damage.
   (b) Thermoreceptors – changes in temperature.
   (c) Mechanoreceptors – stimulated or inhibited by physical distortion, contact, or pressure on their cell membranes.
   (d) Chemoreceptors – respond to presence of specific molecules.

4. Baroreceptors monitor changes in pressure. Proprioceptors monitor the position of joints, the tension in tendons and ligaments, and the state of muscular contraction.

5. Sensations leaving the olfactory bulb travel along the olfactory tract (N I) to reach the olfactory cortex, the hypothalamus, and portions of the limbic system.

6. Sweet, salty, sour, bitter.

7. Receptors in the saccule and utricle provide sensations of gravity and linear acceleration.

8. (a) Provides mechanical support and physical protection.
   (b) Serves as an attachment site for the extrinsic eye muscles.
   (c) Assists in the focusing process.

9. (a) Provides a route for blood vessels and lymphatics to the eye.
   (b) Secretes and reabsorbs aqueous humor.
   (c) Controls the shape of the lens; important in focusing process.

10. The pigment layer: (a) absorbs light after it passes through the retina; (b) biochemically interacts with the photoreceptor layer of the retina.

    The retina contains: (a) photoreceptors that respond to light; (b) supporting cells and neurons that perform preliminary processing and integration of visual information; (c) blood vessels supplying tissues lining the vitreous chamber.

11. During accommodation the lens becomes rounder to focus the image of a nearby object on the retina.

12. As aging proceeds, the lens becomes less elastic, takes on a yellowish hue, and eventually begins to lose its transparency. Visual clarity begins to fade, and when the lens turns completely opaque, the person becomes functionally blind despite the fact that the retinal receptors are alive and well.

CHAPTER 11: *The Endocrine System*

Part I: Objective Based Questions

Objective 1

1. A
2. C
3. B
4. C

Objective 2

1. D
2. B
3. C
4. A

Objective 3

1. D
2. B
3. D
4. C
5. A
6. C

Objective 4

1. B
2. A
3. C
4. B
5. B
6. D

Objective 5

1. posterior pituitary
2. oxytocin
3. ACTH
4. FSH
5. MSH
6. calcitonin
7. iodine
8. parathyroid
9. thymus
10. androgens
11. aldosterone
12. kidneys
13. heart
14. pancreas
15. glucagon
16. testes
17. estrogen
18. pineal
19. **Figure 11-1**
    A. hypothalamus
    B. pituitary
    C. thyroid
    D. thymus
    E. adrenals
    F. ovaries
    G. pineal
    H. parathyroids
    I. heart (atria)
    J. pancreas
    K. testes
20. **Figure 11-2**
21. A
22. H
23. E
24. C
25. F
26. G
27. D
28. B
29. L
30. M
31. K
32. I
33. J
34. N
35. Q
36. O
37. P

Objective 6

1. D
2. A
3. B
4. B

Objective 7

1. D
2. D
3. B

Objective 8

1. A
2. C
3. A
4. B
5. C

Objective 9

1. B
2. C
3. A
4. D

Objective 10

1. nervous
2. cardiovascular
3. digestive
4. urinary
5. muscular

Part II: Chapter Comprehensive Exercises

A. Word Elimination

1. prostate
2. keratin
3. PTH
4. elastin
5. aldosterone
6. hemoglobin
7. calcitonin
8. testes
9. melatonin
10. permissive

B. Matching

1. E
2. H
3. A
4. J
5. B
6. G
7. C
8. F
9. D
10. I

C. Concept Maps

Concept Map I - Endocrine Glands

1. hormones
2. epinephrine
3. peptide hormones
4. testosterone
5. pituitary
6. parathyroids
7. heart
8. male/female gonads
9. pineal
10. bloodstream

Concept Map II - Endocrine System Functions

1. cellular communication
2. homeostasis
3. target cells
4. contraction
5. ion channel opening
6. hormones

D. Crossword Puzzle

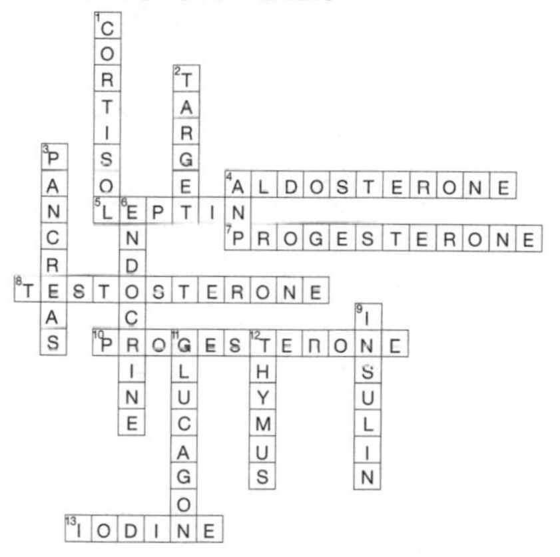

E. Short Answer Questions (Answers)

1. The hypothalamus: (a) Contains autonomic centers that exert direct neural control over the endocrine cells of the adrenal medulla. Sympathetic activation causes the adrenal medulla to release hormones into the bloodstream; (b) acts as an endocrine organ to release hormones into the circulation at the posterior pituitary; (c) secretes regulatory hormones that control activities of endocrine cells in the pituitary glands.

2. (a) Control by releasing hormones; (b) control by inhibiting hormones; (c) regulation by releasing and inhibiting hormones.

3. Thyroid hormones elevate oxygen consumption and rate of energy consumption in peripheral tissues, causing an increase in the metabolic rate. As a result, more heat is generated, replacing the heat lost to the environment.

4. Erythropoietin stimulates the production of red blood cells by the bone marrow. The increase in the number of RBCs elevates the blood volume, causing an increase in blood pressure.

5. The secretion of melatonin by the pineal gland is lowest during daylight hours and highest in the darkness of night. The cyclic nature of this activity parallels daily changes in physiological processes that follow a regular pattern.

6. (a) The two hormones may have opposing, or antagonistic, effects; (b) the two hormones may have an additive, or synergistic, effect; (c) one can have a permissive effect on another. In such cases the first hormone is needed for the second to produce its effect; (d) the hormones may have integrative effects, i.e., the hormones may produce different but complementary results in specific tissues and organs.

CHAPTER 12: *Blood*

Part I: Objective Based Questions

Objective 1

1. B
2. A
3. D
4. C
5. D

Objective 2

1. D
2. C
3. D
4. C
5. A

Objective 3

1. B
2. C
3. B
4. D
5. erythropoiesis
6. leukopoiesis
7. thrombopoiesis

Objective 4

1. A
2. D
3. D
4. C
5. A
6. A
7. D
8. C

Objective 5

1. B
2. D
3. C
4. A
5. C
6. A
7. D
8. **Figure 12-1**
   A. B Antigen
   B. A Antigen
9. **Figure 12-2**
   A. compatible
   B. incompatible
   C. incompatible
   D. compatible
   E. incompatible
   F. compatible
   G. incompatible
   H. compatible
   I. compatible
   J. compatible
   K. compatible
   L. compatible
   M. incompatible
   N. incompatible
   O. incompatible
   P. compatible

Objective 6

1. C
2. D
3. A
4. A
5. D
6. B
7. A
8. C
9. B
10. A
11. **Figure 12-3**
    A. neutrophil
    B. eosinophil
    C. monocyte
    D. basophil
    E. lymphocyte

Objective 7

1. B
2. D
3. C
4. B
5. D
6. A
7. C
8. B
9. D
10. B

Part II: Chapter Comprehensive Exercises

A. Word Elimination

1. analysis
2. lymph
3. serum
4. vitamin D
5. urinanalysis
6. Rh factor
7. platelets
8. RBC
9. platelet
10. transferrin

B. Matching

1. E
2. H
3. A
4. J
5. C
6. B
7. D
8. F
9. G
10. I

C. Concept Map I - Whole Blood

1. plasma
2. solutes
3. albumins
4. oxygen
5. leukocytes
6. neutrophils
7. monocytes

Concept Map II - Hemostasis

8. vascular spasm
9. platelet phase
10. forms plug
11. forms blood clots
12. clot retraction
13. clot dissolves
14. plasminogen
15. plasmin

D. Crossword Puzzle

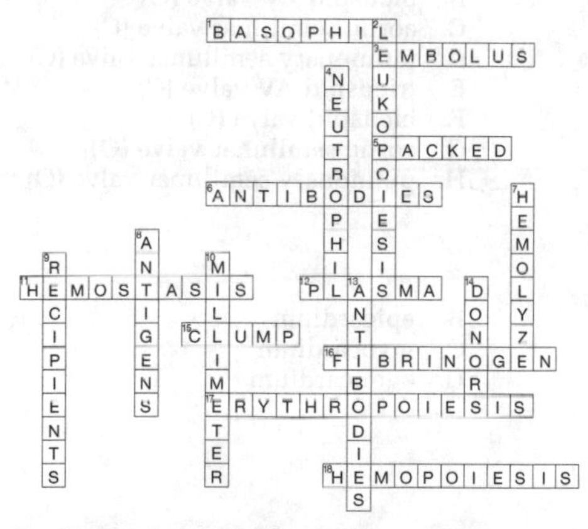

E. Short Answer Questions (Answers)

1. Blood: (a) "transports" dissolved gases; (b) "distributes" nutrients; (c) "transports" metabolic wastes; (d) "delivers" enzymes and hormones; (e) "regulates" the pH and electrolyte composition of interstitial fluid; (f) "restricts" fluid losses through damaged vessels; (g) "defends" the body against toxins and pathogens; (h) helps "regulate" body temperature by absorbing and redistributing heat.

2. (a) water; (b) electrolytes; (c) nutrients; (d) organic wastes; (e) proteins.

3. Granular leukocytes: neutrophils, eosinophils, basophils
   Agranular leukocytes: monocytes, lymphocytes

4. (a) transport of chemicals important to the clotting process
   (b) formation of a plug in the walls of damaged blood vessels
   (c) active contraction after clot formation has occurred

5. (a) vascular phase: spasm in damaged smooth muscle
   (b) platelet phase: platelet aggregation and adhesion
   (c) coagulation phase: activation of clotting system and clot formation
   (d) clot retraction: contraction of blood clot
   (e) clot destruction: enzymatic destruction of clot

6. embolus: a drifting blood clot
   thrombosis: a blood clot that sticks to the wall of an intact blood vessel

CHAPTER 13: The Heart

Part I: Objective Based Questions

Objective 1

1. D
2. C
3. A
4. A
5. B
6. D
7. A
8. B
9. D
10. atria
11. fibrous skeleton
12. intercalated discs
13. **Figure 13-1**
    A. right side
    B. coronary vessels
    C. aortic arch
    D. pulmonary trunk
    E. base of heart
    F. left side
    G. apex

Objective 2

1. A
2. C
3. B
4. D
5. C
6. A
7. B
8. C
9. D
10. C
11. A
12. B
13. C
14. A
15. **Figure 13-2**
    A. pulmonary semilunar valve
    B. superior vena cava
    C. right pulmonary arteries
    D. right atrium
    E. right pulmonary veins
    F. right pulmonary arteries
    G. tricuspid valve
    H. chordae tendinae
    I. right ventricle
    J. inferior vena cava
    K. aortic arch
    L. pulmonary trunk
    M. left atrium
    N. left pulmonary arteries
    O. left pulmonary veins
    P. bicuspid valve
    Q. aortic semilunar valve
    R. interventricular septum
    S. left ventricle
    T. myocardium
16. **Figure 13-3**
    A. tricuspid valve (O)
    B. bicuspid AV valve (O)
    C. aortic semilunar valve (C)
    D. pulmonary semilunar valve (C)
    E. tricuspid AV valve (C)
    F. bicuspid valve (C)
    G. aortic semilunar valve (O)
    H. pulmonary semilunar valve (O)

Objective 3

1. C
2. A
3. B
4. B
5. **Figure 13-4**
    A. intercalated disc
    B. epicardium
    C. myocardium
    D. endocardium

Objective 4

1. C
2. D
3. C
4. A

Objective 5

1. B
2. D
3. C
4. A
5. D
6. **Figure 13-5**
    A. sinoatrial (SA node)
    B. AV node
    C. AV bundle
    D. Bundle branches
    E. Purkinje fibers
7. **Figure 13-6**
    A. 1
    B. 2
8. P
9. QRS
10. T

Objective 6

1. C
2. C
3. D
4. B
5. A

Objective 7

1. B
2. A
3. B
4. B
5. C

Part II: Chapter Comprehensive Exercises

A. Word Elimination

1. L. pulmonary artery
2. semilunar
3. chordae tendinae
4. anastomoses
5. systole
6. tachycardia
7. auricles
8. AV valve
9. multinucleated
10. aortic arch

B. Matching

1. I
2. F
3. G
4. J
5. H
6. P
7. C
8. A
9. B
10. E

C. Concept Map I - The Heart

1. two atria
2. blood from atria
3. endocardium
4. tricuspid
5. oxygenated blood
6. two semilunar
7. aortic
8. deoxygenated blood
9. epicardium
10. pacemaker cells

Concept Map II - Path of Blood Flow through the Heart

1. superior vena cava
2. right atrium
3. tricuspid valve
4. right ventricle
5. pulmonary semilunar valve
6. pulmonary arteries
7. pulmonary veins
8. left atrium
9. bicuspid valve
10. left ventricle
11. aortic semilunar valve
12. L. common carotid artery
13. aorta
14. systemic arteries
15. systemic veins
16. inferior vena cava

D. Crossword Puzzle

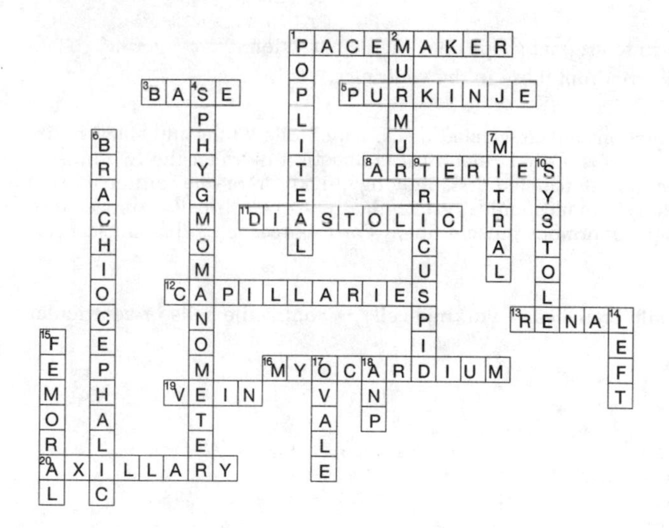

E. Short Answer Questions (Answers)

1. CO = SV x HR
   CO = 75 ml x 80 beats/min = 6000 ml/min
   6000 ml = 6.0 *l*/min

2. % increase = 5 *l*/min. = .50 = 50%
   10 *l*/min.

3. decreasing CO, decreasing SV, decreasing length of diastole, decreasing ventricular filling

4. SV = EDV - ESV
   SV = 140 ml – 60 m*l* = 80 m*l*

5. CO = SV x HR

Therefore, $\frac{CO}{HR} = \frac{SV \times \cancel{HR}}{\cancel{HR}}$

Therefore, $SV = \frac{CO}{HR}$

$SV = \frac{5\ l/min}{100\ B/min} = 0.05\ l/beat$

6. The visceral pericardium, or epicardium, covers the outer surface of the heart. The parietal pericardium lines the inner surface of the pericardial sac that surrounds the heart.

7. The chordae tendinae and papillary muscles are located in the right ventricle, the trabeculae carnae are found on the interior walls of both ventricles, and the pectinate muscles are found on the interior of both atria and a part of the right atrial wall.

8. The chordae tendinae are tendinous fibers that brace each cusp of the tricuspid valve and are connected to the papillary muscles. The trabeculae carneae of the ventricles contain a series of deep grooves and folds.

   The pectinate muscles are prominent muscular ridges that run along the surfaces of the atria and across the anterior atrial wall.

9. (a) epicardium; (b) myocardium; (c) endocardium
   (a) Stabilizes positions of muscle fibers and valves in heart.
   (b) Provides support for cardiac muscle fibers and blood vessels and nerves in the myocardium.
   (c) Helps distribute the forces of contraction.
   (d) Adds strength and prevents overexpansion of the heart.
   (e) Helps to maintain the shape of the heart.
   (f) Provides elasticity that helps the heart return to original shape after each contraction.
   (g) Physically isolates the muscle fibers of the atria from those in the ventricles.

10. Cardiac muscle fibers are connected by gap junctions at intercalated discs, which allow ions and small molecules to move from one cell to another. This creates a direct electrical connection between the two muscle fibers, and an action potential can travel across an intercalated disk, moving quickly from one cardiac muscle fiber to another. Because the cardiac muscle fibers are mechanically, chemically, and electrically connected to one another, the entire tissue resembles a single, enormous muscle fiber. For this reason cardiac muscle has been called a functional syncytium.

11. SA node → AV node → bundle of His → bundle branches → Purkinje cells → contractile cells of ventricular myocardium

12. bradycardia: heart rate slow than normal
    tachycardia: faster than normal heart rate

13. (a) Ion concentrations in extracellular fluid:
    (*Note*: EC = extracellular)
    decreasing EC K^+ → decreasing HR
    increasing EC Ca^{++} → increasing excitability and prolonged contraction
    (b) Changes in body temperature:
    decreasing temp → decreasing HR and decreasing strength of contractions
    increasing temp → increasing HR and increasing strength of contraction
    (c) Autonomic activity:
    parasympathetic stimulation →releases ACh → decreasing heart rate
    sympathetic stimulation → releases norepinephrine → increasing heart rate

CHAPTER 14: *Blood Vessels and Circulation*

Part I: Objective Based Questions

Objective 1

1. A
2. D
3. B
4. C
5. D
6. A
7. **Figure 14-1**
   A. artery
   B. arteriole
   C. capillaries
   D. venules
   E. veins

Objective 2

1. C
2. D
3. D
4. C
5. C

Objective 3

1. B
2. A
3. B
4. A
5. hydrostatic pressure
6. osmotic pressure

Objective 4

1. D
2. B
3. B
4. D
5. D
6. B

Objective 5

1. B
2. A
3. C
4. D
5. B

Objective 6

1. B
2. A
3. D
4. C
5. D
6. A

Objective 7

1. C
2. C
3. C
4. D
5. C

Objective 8

1. Circle of Willis
2. pulmonary veins
3. pulmonary arteries
4. internal jugular
5. axillary
6. brachial
7. radial, ulnar
8. basilar
9. diaphragm
10. common iliac
11. femoral, deep femoral
12. superior vena cava
13. brain
14. brachial
15. brachiocephalic
16. superior vena cava
17. femoral
18. inferior vena cava
19. hepatic portal
20. inferior vena cava
21. **Figure 14-2**
   A. R. common carotid
   B. vertebral
   C. R. subclavian
   D. brachiocephalic
   E. ascending aorta
   F. celiac
   G. brachial
   H. radial
   I. ulnar
   J. palmar arches
   K. external iliac
   L. popliteal
   M. posterior tibial
   N. anterior tibial
   O. peroneal
   P. plantar arch
   Q. L. common carotid
   R. aortic arch
   S. L. subclavian
   T. axillary
   U. descending aorta
   V. renal
   W. superior mesenteric
   X. gonadal
   Y. inferior mesenteric
   Z. common iliac
   AA. internal iliac
   BB. deep femoral
   CC. femoral
   DD. dorsalis pedis
22. **Figure 14-3**
   A. external jugular
   B. vertebral
   C. subclavian
   D. axillary
   E. cephalic
   F. brachial
   G. basilic
   H. hepatics
   I. median cubital
   J. cephalic
   K. median antebrachial
   L. ulnar
   M. palmar venous arches
   N. digital veins
   O. great saphenous
   P. popliteal
   Q. small saphenous
   R. peroneal
   S. dorsal venous arch
   T. plantar venous arch
   U. internal jugular
   V. brachiocephalic
   W. superior vena cava

Objective 8 (continued)

X. intercostals
Y. inferior vena cava
Z. renal
AA. gonadal
BB. lumbar
CC. common iliac
DD. external iliac
EE. internal iliac
FF. deep femoral
GG. femoral
HH. posterior tibial
II. anterior tibial

23. **Figure 14-4**
A. anterior cerebral
B. internal carotid
C. middle cerebral
D. basilar
E. vertebral
F. anterior communicating
G. anterior cerebral
H. posterior communicating
I. posterior cerebral
J. Circle of Willis

24. **Figure 14-5**
A. inferior vena cava
B. hepatic veins
C. liver
D. cystic vein
E. hepatic portal vein
F. superior mesenteric vein
G. colic veins
H. ascending colon
I. aorta
J. esophagus
K. stomach
L. L. gastric vein
M. gastroepiploic
N. spleen
O. splenic vein
P. pancreas
Q. L. colic vein
R. inferior mesenteric vein
S. descending colon
T. sigmoid branches
U. small intestine
V. superior rectal view

Objective 9

1. B
2. D
3. D
4. B

Objective 10

1. lymphatic
2. nervous
3. urinary
4. reproductive
5. skeletal

Part II: Chapter Comprehensive Exercises

A. Word Elimination

1. valves
2. tunica lumen
3. compression
4. baroreceptors
5. Ach
6. increased pH
7. pulmonary vein
8. carotid artery
9. phrenic
10. increased hematocrit

B. Matching

1. H
2. B
3. G
4. J
5. A
6. I
7. E
8. F
9. C
10. D

C. Concept Map I - The Cardiovascular System

1. pulmonary veins
2. arteries and arterioles
3. veins and venules
4. pulmonary arteries
5. systemic circuit

Concept Map II - Major Branches of the Aorta

1. ascending aorta
2. brachiocephalic artery
3. L. subclavian artery
4. thoracic artery
5. celiac trunk
6. superior mesenteric artery
7. R. gonadal artery
8. L. common iliac artery

Concept Map III - Major Veins

1. superior vena cava
2. azygous vein
3. L. hepatic veins
4. R. suprarenal vein
5. L. renal vein
6. L. common iliac vein

D. Crossword Puzzle

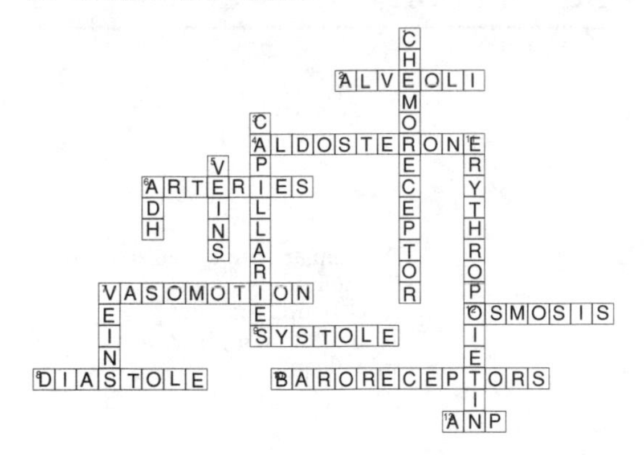

E. Short Answer Questions (Answers)

1. heart → arteries → arterioles → capillaries (gas exchange area) → venules → veins → heart

2. (a) Sinusoids are specialized fenestrated capillaries.
   (b) They are found in the liver, bone marrow, and the adrenal glands.
   (c) They form fattened, irregular passageways, so blood flows through the tissues slowly, maximizing time for absorption and secretion and molecular exchange.

3. In the pulmonary circuit, oxygen stores are replenished, carbon dioxide is excreted, and the "reoxygenated" blood is returned to the heart for distribution in the systemic circuit.

   The systemic circuit supplies the capillary beds in all parts of the body with oxygenated blood, and returns deoxygenated blood to the heart of the pulmonary circuit for removal of carbon dioxide.

4. (a) vascular resistance, viscosity, turbulence
   (b) Only vascular resistance can be adjusted by the nervous and endocrine systems.

5. $F = \frac{BP}{PR}$

   Flow is directly proportional to the blood pressure and inversely proportional to peripheral resistance; i.e. , increasing pressure, increasing flow; decreasing pressure, decreasing flow; increasing PR, decreasing flow; decreasing PR, increasing flow.

6. $\frac{120 \text{ mm Hg}}{80 \text{ mm Hg}}$ is a "normal" blood pressure reading.

   The top number, 120 mm Hg, is the systolic pressure, i.e., the peak blood pressure measured during ventricular systole.

   The bottom number, 80 mm Hg, is the diastolic pressure, i.e., the minimum blood pressure at the end of ventricular diastole.

7. MAP = 1/3 pulse pressure (p.p.) + diastolic pressure
   Therefore, p.p. = 110 mm Hg – 80 mm Hg = 30 mm Hg
   MAP = 1/3 (.30) = 10 + 80 mm Hg
   MAP = 90 mm Hg

8. Cardiac output, blood volume, peripheral resistance.

9. Aortic baroreceptors, carotid sinus baroreceptors, atrial baroreceptors.

10. Epinephrine and norepinephrine, ADH, angiotensin II, erythropoietin, and atrial natriuretic peptide.

11. Arteries lose their elasticity, the amount of smooth muscle they contain decreases, and they become stiff and relatively inflexible.

CHAPTER 15: *The Lymphatic System and Immunity*

Part I: Objective Based Questions

Objective 1

1. D
2. C
3. D
4. D
5. B
6. thoracic duct
7. lymph nodes
8. **Figure 15-1**
   A. R. lymphatic duct
   B. thymus
   C. thoracic duct
   D. lumbar lymph nodes
   E. inguinal lymph nodes
   F. L. lymphatic duct
   G. axillary lymph nodes
   H. spleen

Objective 2

1. A
2. B
3. A
4. C
5. A
6. B

Objective 3

1. C
2. D
3. C
4. B
5. A
6. **Figure 15-2**
   A. physical barriers
   B. phagocytes
   C. immunological surveillance
   D. complement system
   E. inflammatory response
   F. fever
   G. interferons

Objective 4

1. A
2. C
3. D
4. C
5. B
6. C
7. B

Objective 5

1. D
2. A
3. A
4. D
5. B

Objective 6

1. B
2. D
3. D
4. B

Objective 7

1. B
2. C
3. D
4. D

Objective 8

1. C
2. B
3. A
4. B
5. B
6. haptens
7. immunological competence

Objective 9

1. C
2. B
3. D
4. C

Objective 10

1. D
2. C
3. A
4. B

Objective 11

1. reproductive
2. A
3. respiratory
4. muscular
5. skeletal

Part II: Chapter Comprehensive Exercises

A. Word Elimination

1. pineal gland
2. antigens
3. salivary
4. nephron
5. antibodies
6. blood
7. complement
8. compatibility
9. IgB
10. enzymes

B. Matching

1. H
2. E
3. J
4. A
5. G
6. B
7. D
8. F
9. C
10. I

C. Concept Map I

1. nonspecific immunity
2. phagocytic cells
3. inflammation
4. specific immunity
5. innate
6. acquired
7. active
8. active immunization
9. transfer of antibodies via placenta
10. passive immunization

Concept Map II

1. viruses
2. macrophages
3. natural killer cells
4. helper T cells
5. B cells
6. antibodies
7. killer T cells
8. suppressor T cells
9. memory T and B cells

D. Crossword Puzzle

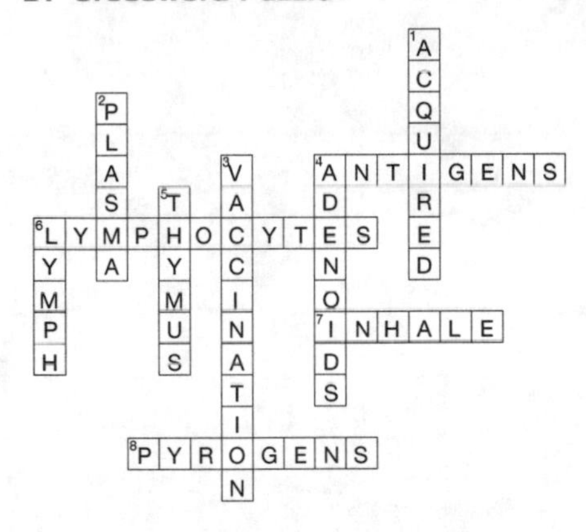

E. Short Answer Questions (Answers)

1. (a) lymphatic vessels; (b) lymph; (c) lymphatic organs

2. (a) production, maintenance, and distribution of lymphocytes
   (b) maintenance of normal blood volume
   (c) elimination of local variations in the composition of the interstitial fluid

3. (a) T cells (thymus); (b) B cells (bone-marrow); (c) NK cells - natural killer (bone-marrow)

4. (a) Cytotoxic T cells – cell-mediated immunity
   (b) helper T cells – release lymphokines that coordinate specific and nonspecific defenses
   (c) Suppressor T cells – depress responses of other T cells and B cells

5. Stimulated B cells differentiate into plasma cells that are responsible for production and secretion of antibodies. B cells are said to be responsible for humoral immunity.

6. (a) lymph nodes; (b) thymus; (c) spleen

7. NK (natural killer) cells are sensitive to the presence of abnormal cell membranes and respond immediately. When the NK cell makes contact with an abnormal cell it releases secretory vesicles that contain proteins called perforins. The perforins create a network of pores in the target cell membrane, releasing free passage of intracellular materials necessary for homeostasis, thus causing the cell to disintegrate.

   T cells and B cells provide defenses against specific threats, and their activation requires a relatively complex and time-consuming sequence of events.

8. (a) destruction of target cell membranes
   (b) stimulation of inflammation
   (c) attraction of phagocytes
   (d) enhancement of phagocytosis

9. (a) specificity
   (b) versatility
   (c) memory
   (d) tolerance

10. Active immunity appears following exposure to an antigen, as a consequence of the immune response.

    Passive immunity is produced by transfer of antibodies from another individual.

CHAPTER 16: *The Respiratory System*

Part I: Objective Based Questions

Objective 1

1. D
2. B
3. A
4. C
5. B
6. A

Objective 2

1. C
2. B
3. D
4. A
5. A
6. **Figure 16-1**

A. internal nares
B. nasopharynx
C. pharyngeal tonsil
D. auditory tube
E. soft palate
F. palatine tonsil
G. oropharynx
H. epiglottis
I. glottis
J. laryngopharynx
K. vocal cord
L. esophagus
M. frontal sinus
N. nasal conchae
O. nasal vestibule
P. external nares
Q. hard palate
R. oral cavity
S. tongue
T. mandible
U. hyoid bone
V. thyroid cartilage
W. cricoid cartilage
X. trachea

Objective 3

1. D
2. C
3. B
4. A
5. C
6. B
7. A
8. A
9. C
10. external nares
11. vestibule
12. oropharynx
13. pharynx
14. nasopharynx
15. larynx
16. trachea
17. lobes of the lungs
18. primary bronchi
19. alveoli
20. **Figure 16-2**

A. sphenoidal sinus
B. pharynx
C. epiglottis
D. vocal cords (folds)
E. esophagus
F. right lungs
G. frontal sinus
H. nasal conchae
I. tongue
J. hyoid
K. thyroid cartilage
L. cricoid cartilage
M. larynx
N. tracheal cartilage
O. left bronchus
P. left lung
Q. diaphragm

Objective 4

1. A
2. A
3. C
4. B
5. D
6. B
7. C
8. B

Objective 5

1. A
2. A
3. C
4. B

Objective 6

1. B
2. B
3. C
4. B
5. C
6. A
7. **Figure 16-3**

A. 23%
B. 7%
C. 93%
D. 70%

Objective 7

1. D
2. A
3. D
4. D
5. C
6. A
7. lungs
8. CO_2

Objective 8

1. A
2. D
3. A
4. A
5. A

Objective 9

1. A
2. B
3. C
4. A

Objective 10

1. D
2. urinary
3. endocrine
4. nervous
5. skeletal

Part II: Chapter Comprehensive Exercises

A. Word Elimination

1. alveoli
2. mandibular
3. glottis
4. middle lobe
5. primary bronchi
6. internal intercostals
7. P_{CO_2}
8. NH_4
9. chemoreceptor
10. surfactant

B. Matching

1. F
2. I
3. J
4. B
5. H
6. E
7. A
8. C
9. G
10. D

C. Concept Map I

1. upper respiratory tract
2. pharynx and larynx
3. paranasal sinuses
4. speech production
5. lungs
6. ciliated mucous membrane
7. alveoli
8. blood
9. O_2 from alveoli into blood

D. Crossword Puzzle

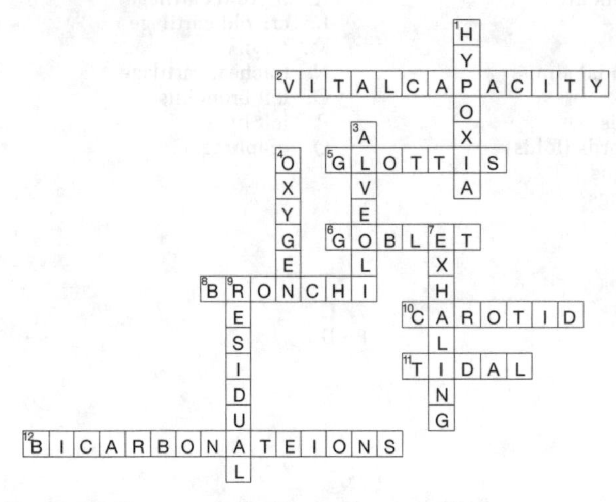

E. Short-Answer Questions (Answers)

1. (a) Provides an area for gas exchange between air and blood.
   (b) To move air to and from exchange surfaces.
   (c) Protects respiratory surfaces from abnormal changes or variations.
   (d) Defends the respiratory system and other tissues from pathogenic invasion.
   (e) Permits communication via production of sound.
   (f) Participates in the regulation of blood volume, pressure, and body fluid pH.

2. The air is warmed to within 1 degree of body temperature before it enters the pharynx. This results from the warmth of rapidly flowing blood in an extensive vasculature of the nasal mucosa. For humidifying the air, the nasal mucosa is supplied with small mucus glands that secrete a mucoid fluid inside the nose. The warm and humid air prevents drying of the pharynx, trachea, and lungs, and facilitates the flow of air through the respiratory passageways without affecting environmental changes.

3. Surfactant cells are scattered among the simple squamous epithelial cells of the respiratory membranes. They produce an oily secretion containing a mixture of phospholipids that coat the alveolar epithelium and keep the alveoli from collapsing like bursted bubbles.

4. External respiration includes the diffusion of gases between the alveoli and the circulating blood.
   Internal respiration is the exchange of dissolved gases between the blood and the interstitial fluids in peripheral tissues.
   Cellular respiration is the absorption and utilization of oxygen by living cells via biochemical pathways that generate carbon dioxide.

5. The vital capacity is the maximum amount of air that can be moved into and out of the respiratory system in a single respiratory cycle.

   Vital capacity = inspiratory reserve + expiratory reserve + tidal volume.

6. Alveolar ventilation, or V_e, is the amount of air reaching the alveoli each minute. It is calculated by subtracting the anatomic dead space, V_d, from the tidal volume (V_t) using the formula:

   $$V_e = F_x (V_t - V_d)$$

7. (a) CO_2 may be dissolved in the plasma (7 percent).
   (b) CO_2 may be bound to the hemoglobin of RBC (23 percent).
   (c) CO_2 may be converted to a molecule of carbonic acid (70 percent).

8. (a) mechanoreceptor reflexes
   (b) chemoreceptor reflexes
   (c) protective reflexes

9. A rise in arterial Pco_2 immediately elevates cerebrospinal fluid CO_2 levels and stimulates the chemoreceptor neurons of the medulla. These receptors stimulate the respiratory contor causing an increase in the rate and depth of respiration, or hyperventilation.

CHAPTER 17: *The Digestive System*

Part I: Objective Based Questions

Objective 1

1. C
2. A
3. D
4. B
5. **Figure 17-1**
   A. oral cavity, teeth, tongue
   B. liver
   C. gallbladder
   D. large intestine
   E. salivary glands
   F. pharynx
   G. esophagus
   H. stomach
   I. pancreas
   J. small intestine

Objective 2

1. D
2. A
3. secretion
4. absorption
5. excretion

Objective 3

1. A
2. D
3. B
4. C
5. A
6. D
7. C
8. A
9. **Figure 17-2**
   A. mesenteric artery and vein
   B. mesentery
   C. visceral peritoneum
   D. plica
   E. mucosa
   F. submucosa
   G. lumen
   H. circular muscular layer
   I. muscularis externa
   J. mucosal gland
   K. visceral peritoneum (serosa)

Objective 4

1. B
2. A
3. A
4. A
5. B
6. C
7. C

Objective 5

1. D
2. D
3. D
4. C
5. C
6. A
7. C
8. B
9. C
10. incisors
11. incisor
12. molars
13. parotid
14. uvula
15. **Figure 17-3**
   A. cuspid
   B. incisors
   C. molars
   D. bicuspid
16. **Figure 17-4**
   A. crown
   B. neck
   C. root
   D. pulp cavity
   E. enamel
   F. dentin
   G. gingival sulcus
   H. cementum
   I. periodontal ligament
   J. root canal
   K. alveolar bone
   L. blood vessels and nerve

Objective 6

1. C
2. D
3. B
4. HCl
5. pepsinogen
6. cardia
7. fundus
8. pylorus
9. rugae
10. proteins
11. **Figure 17-5**
   A. esophagus
   B. body
   C. lesser curvature
   D. lesser omentum
   E. pylorus
   F. diaphragm
   G. fundus
   H. cardia
   I. greater curvature
   J. rugae
   K. greater omentum

Objective 7

1. B
2. D
3. B
4. C
5. A
6. C
7. B
8. D
9. B
10. C

Objective 8

1. A
2. gallbladder
3. lobule
4. hepatocyte
5. D
6. C
7. D
8. A

Objective 9

1. C
2. A
3. A
4. B
5. A
6. haustrae
7. taenia coli
8. cecum
9. appendix
10. **Figure 17-6**
    A. hepatic portal vein
    B. inferior vena cava
    C. transverse colon
    D. ascending colon
    E. ileocecal valve
    F. cecum
    G. ileum
    H. vermiform appendix
    I. aorta
    J. splenic vein
    K. splenic flexure
    L. greater omentum
    M. descending colon
    N. haustra
    O. taenia coli
    P. sigmoid colon
    Q. rectum

Objective 10

1. C
2. D
3. B
4. A
5. D

Objective 11

1. B
2. C
3. A
4. D

Objective 12

1. urinary
2. cardiovascular
3. endocrine
4. nervous
5. integumentary

Part II: Chapter Comprehensive Exercises

A. Word Elimination

1. circulation
2. mesentery
3. excretion
4. tongue
5. dentin
6. chyme
7. intrinsic factor
8. HCl
9. amylase
10. haustra

B. Matching

1. G
2. E
3. A
4. I
5. C
6. B
7. J
8. F
9. D
10. H

C. Concept Map I - Digestive System

1. amylase
2. pancreas
3. bile
4. hormones
5. digestive tract movements
6. stomach
7. large intestine
8. hydrochloric acid
9. intestinal mucosa

Concept Map II - Chemical Events in Digestion

10. esophagus
11. small intestine
12. polypeptides
13. amino acids
14. complex sugars and starches
15. disaccharides and trisaccharides
16. simple sugars
17. monoglycerides, fatty acids in micelles
18. triglycerides
19. lacteal

D. Crossword Puzzle

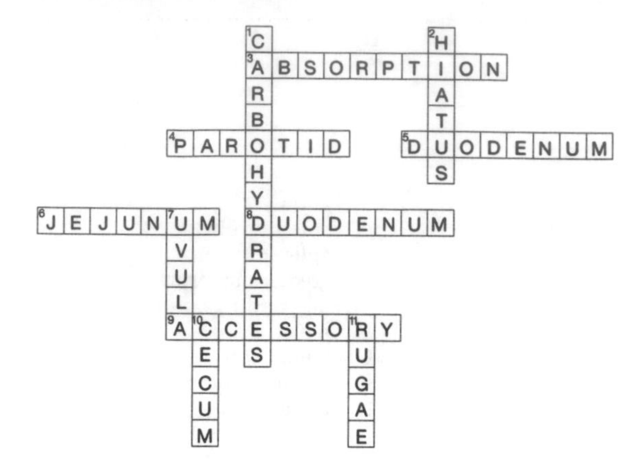

E. Short Answer Questions (Answers)

1. (a) ingestion
   (b) mechanical processing
   (c) digestion
   (d) secretion
   (e) absorption
   (f) compaction
   (g) excretion (defecation)

2. (a) parotid glands
   (b) sublingual glands
   (c) submandibular glands

3. (a) incisors – clipping or cutting
   (b) cuspids (canines) – tearing or slashing
   (c) bicuspids (premolars) – crushing, mashing, and grinding
   (d) molars – crushing, mashing, and grinding

4. Parietal cells and chief cells are types of secretory cells found in the wall of the stomach. Parietal cells secrete intrinsic factors and hydrochloric acid. Chief cells secrete an inactive proenzyme, pepsinogen.

5. (a) cephalic, gastric, intestinal
   (b) CNS regulation; release of gastrin into circulation; enterogastric reflexes, secretion of cholecystokinin (CCK) and secretin

6. Secretin, cholecystokinin, and glucose-dependent insulinotropic peptide (GIP)

7. (a) resorption of water and compaction of feces
   (b) the absorption of important vitamins liberated by bacterial action
   (c) the storing of fecal material prior to defecation

8. (a) metabolic regulation
   (b) hematological regulation
   (c) bile production

9. (a) Endocrine function – pancreatic islets secrete insulin and glucagon into the bloodstream.
   (b) Exocrine function – secrete a mixture of water, ions, and digestive enzymes into the small intestine.

CHAPTER 18: Nutrition and Metabolism

Part I: Objective Based Questions

Objective 1

1. D
2. C
3. B
4. C

Objective 2

1. A
2. B
3. C
4. B
5. A
6. C

Objective 3

1. C
2. D
3. D
4. Acetyl-CoA
5. Triglycerides
6. LDLs - low density lipoproteins
7. HDLs - high density lipoproteins

Objective 4

1. B
2. C
3. A
4. D
5. D

Objective 5

1. B
2. C
3. B
4. B

Objective 6

1. D
2. C
3. A
4. B
5. C
6. B
7. D
8. **Figure 18-1**
   A. fats, oils, and sweets
   B. milk, yogurt, cheese
   C. vegetable group
   D. bread, cereal, rice
   E. meat, poultry, fish, nuts, beans
   F. bread, cereal, rice

Objective 7

1. chloride ion
2. calcium
3. phophate ion
4. iron
5. copper
6. vitamin A
7. vitamin D
8. vitamin K
9. riboflavin
10. niacin
11. vitamin B_6
12. folacin

Objective 8

1. Calorie
2. C
3. A
4. B

Objective 9

1. D
2. D
3. D
4. A

Objective 10

1. A
2. B
3. C
4. thermoregulation
5. pyrexia

Part II: Chapter Comprehensive Exercises

A. Word Elimination

1. glycolysis
2. anaerobic
3. NAD
4. ATP
5. EAA
6. linoleic
7. minerals
8. zinc
9. vitamin C
10. thermoregulation

B. Matching

1. G
2. D
3. A
4. J
5. B
6. H
7. E
8. C
9. F
10. I

C. Concept Map I - Food Intake

1. vegetables
2. meat
3. carbohydrates
4. 9 cal/gram
5. proteins
6. tissue growth and repair
7. vitamins
8. metabolic regulators

Concept Map II - Anabolism and Catabolism

1. lipolysis
2. lipogenesis
3. glycolysis
4. beta oxidation
5. gluconeogenesis
6. amino acids

D. Crossword Puzzle

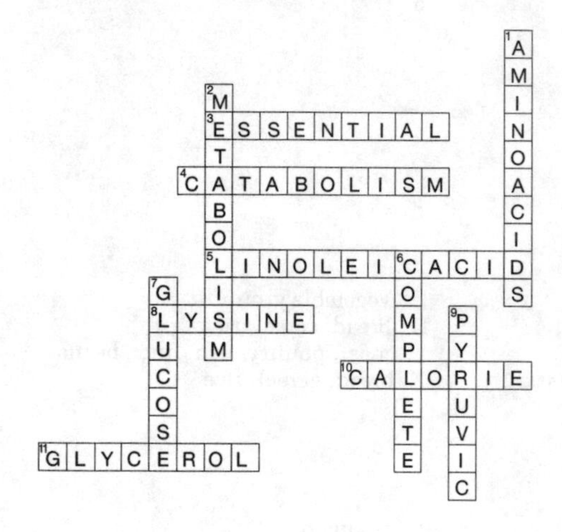

E. Short Answer Questions (Answers)

1. (a) proteins are difficult to break apart; (b) their energy yield is less than that of lipids; (c) the byproduct, ammonia, is a toxin that can damage cells; (d) proteins form the most important structural and functional components of any cell; extensive protein catabolism threatens homeostasis at the cellular and system levels.

2. When nucleic acids are broken down, only the sugar and pyrimidine bases (cytosine, thymine, uracil) provide energy. Purine bases (adenine, guanine) cannot be catabolized; instead, they are deaminated (the amine is removed) and excreted as uric acid, a nitrogenous waste.

3. (a) liver; (b) adipose tissue; (c) skeletal muscle; (d) neural tissue; (e) other peripheral tissue.

4. (a) milk, yogurt, cheese; (b) meat, poultry, fish, nuts, beans; (c) vegetable group; (d) fruit group; (e) bread cereal rice; (f) fats, oils, and sweets.

5. minerals (i.e., inorganic ions) are important because they: (a) determine the osmolarities of fluids; (b) play major roles in important physiological processes; (c) are essential co-factors in a variety of enzymes.

6. (a) fat soluble: ADEK; (b) water-soluble: B complex and C (ascorbic acid)

CHAPTER 19: *The Urinary System*

Part I: Objective Based Questions

Objective 1

1. B
2. C
3. A
4. D
5. D
6. C
7. **Figure 19-1**
   A. kidneys
   B. ureters
   C. urinary bladder
   D. urethra

Objective 2

1. B
2. D
3. C
4. D
5. C
6. B
7. **Figure 19-2**
   A. minor calyx
   B. renal pelvis
   C. ureter
   D. renal column
   E. renal pyramid
   F. renal vein
   G. major calyx
   H. renal capsule
   I. cortex

Objective 3

1. A
2. C
3. B
4. D
5. C
6. C
7. **Figure 19-3**
   A. efferent arteriole
   B. afferent arteriole
   C. renal corpuscle
   D. glomerulus
   E. proximal convoluted tubule
   F. loop of Henle
   G. distal convoluted tubule
   H. collecting duct
   I. papillary duct

Objective 4

1. C
2. B
3. A
4. A
5. D
6. B

Objective 5

1. D
2. C
3. B
4. A
5. D
6. C

Objective 6

1. B
2. D
3. C
4. B
5. A
6. C

Objective 7

1. B
2. C
3. C
4. internal sphincter
5. rugae
6. neck
7. **Figure 19-4**
   A. ureter
   B. urethral openings
   C. internal sphincter
   D. external sphincter
   E. detrusor muscle
   F. trigone
   G. prostate gland
   H. urethra

Objective 8

1. C
2. C
3. A
4. D
5. B

Objcctive 9

1. integumentary
2. respiratory
3. digestive

Objective 10

1. C
2. D
3. B
4. A
5. C

Objective 11

1. B
2. D
3. A
4. D
5. B
6. ADH
7. aldosterone
8. kidneys
9. aldosterone
10. calcitriol

Objective 12

1. B
2. A
3. B
4. D
5. A
6. C
7. C
8. A

Objective 13

1. D
2. B
3. A
4. C

Part II: Chapter Comprehensive Exercises

A. Word Elimination

1. prostate
2. nephron
3. calyces
4. vasa recta
5. urine
6. glucose
7. ADH
8. adrenalin
9. urine
10. ureter

B. Matching

1. F
2. H
3. I
4. C
5. J
6. D
7. A
8. G
9. B
10. E

C. Concept Map I

1. ureters
2. urinary bladder
3. nephrons
4. glomerulus
5. PCT
6. collecting tubules
7. medulla
8. renal sinus
9. minor calyces

Concept Map II

10. renal artery
11. arcuate artery
12. afferent artery
13. efferent artery
14. interlobular vein
15. interlobar vein

D. Crossword Puzzle

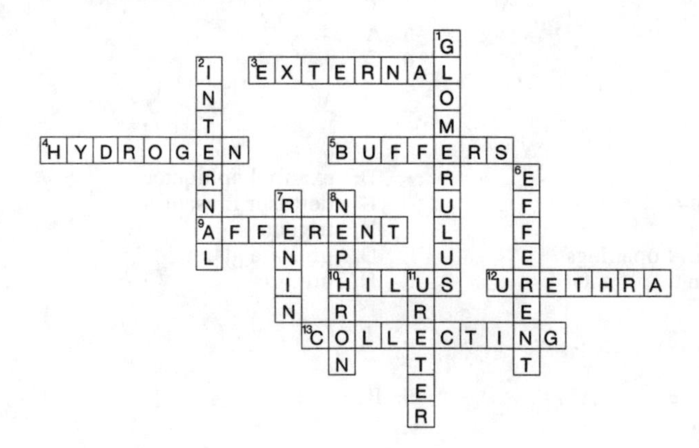

E. Short Answer Questions (Answers)

1. (a) regulates plasma concentrations of ions; (b) regulates blood volume and blood pressure; (c) contributes to stabilization of blood pH; (d) conserves valuable nutrients; (e) eliminates organic wastes; (f) assists liver in detoxification and deamination.

2. (a) produces a powerful vasoconstriction of the afferent arteriole, thereby decreasing the glomerular filtration rate and slowing the production of filtrate; (b) stimulation of renin release; (c) direct stimulation of water and sodium ion reabsorption.

3. (a) autoregulation; (b) hormonal regulation; (c) autonomic regulation

4. (a) ADH – decreased urine volume; (b) renin – causes angiotensin II production; stimulates aldosterone production; (c) aldosterone – increased sodium ion reabsorption; decreased urine concentration and volume; (d) atrial natriuretic peptide (ANP) – inhibits ADH production; results in increased urine production.

5. (a) stimulates water conservation at the kidney, reducing urinary water losses; (b) stimulates the thirst center to promote the drinking of fluids; the combination of decreased water loss and increased water intake gradually restores normal plasma osmolarity.

6. (a) secrete and/or absorb hydrogen ions; (b) control excretion of acids and bases; (c) generate additional buffers when necessary.

CHAPTER 20: *The Reproductive System*

Part I: Objective Based Questions

Objective 1

1. D
2. B
3. D
4. C
5. D
6. B
7. D

Objective 2

1. A
2. C
3. B
4. D
5. C
6. A
7. **Figure 20-1**
   A. pubic symphysis
   B. penile urethra
   C. penis
   D. urethral meatus
   E. scrotum
   F. urinary bladder
   G. ureter
   H. rectum
   I. seminal vesicle
   J. prostate gland
   K. ejaculatory duct
   L. bulbourethral gland
   M. ductus deferens
   N. epididymis
   O. testis

Objective 3

1. B
2. A
3. C
4. B
5. D
6. A

Objective 4

1. C
2. D
3. A
4. B
5. B
6. C
7. A
8. penis
9. prepuce
10. corpus spongiosum
11. corpus cavernosa

Objective 5

1. D
2. B
3. C
4. A
5. B

Objective 6

1. B
2. C
3. D
4. B
5. D
6. cervix
7. lobules
8. vestibule
9. labia majora
10. areola
11. **Figure 20-2**
    A. ovarian follicle
    B. ovary
    C. uterine tube
    D. urinary bladder
    E. pubic symphysis
    F. urethra
    G. greater vestibular gland
    H. clitoris
    I. labium minus
    J. labium majus
    K. uterus
    L. endometrium
    M. sigmoid colon
    N. fornix
    O. cervix
    P. vagina
    Q. anus

Objective 7

1. C
2. B
3. A
4. A
5. B
6. B

Objective 8

1. A
2. C
3. B
4. D
5. A
6. D
7. FSH
8. LH
9. progesterone
10. progesterone
11. FSH

Objective 9

1. B
2. A
3. C
4. A
5. C
6. B

Objective 10

1. menopause
2. climacteric
3. B
4. C

Objective 11

1. reproductive
2. cardiovascular
3. endocrine
4. nervous

Part II: Chapter Comprehensive Exercises

A. Word Elimination

1. cremaster muscle
2. scrotum
3. mitosis
4. epididymis
5. penis
6. corpus luteum
7. areola
8. vestibular glands
9. androgens
10. coitus

B. Matching

1. G
2. E
3. A
4. J
5. C
6. B
7. I
8. D
9. F
10. H

C. Concept Map I - Male Reproductive Tract

1. ductus deferens
2. penis
3. seminiferous tubules
4. produce testosterone
5. FSH
6. seminal vesicles
7. bulbouretheral glands
8. urethra

Concept Map II - Female Reproductive Tract

1. uterine tubes
2. follicles
3. granulosa and thecal cells
4. endometrium
5. supports fetal development
6. vagina
7. vulva
8. labia majora and minora
9. clitoris
10. nutrients

D. Crossword Puzzle

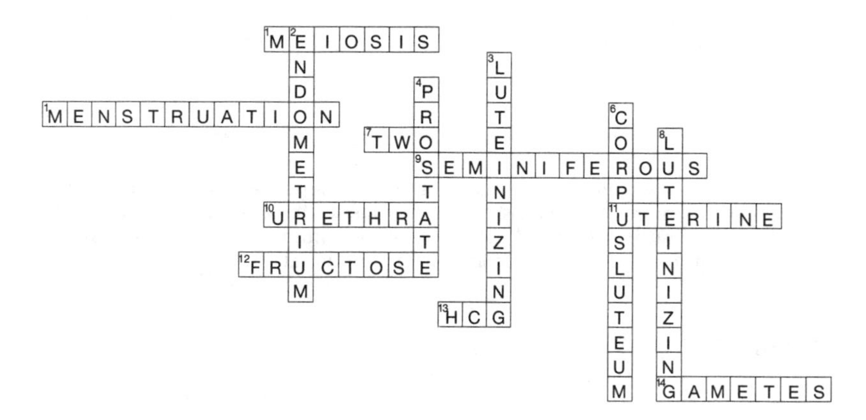

E. Short Answer Questions (Answers)

1. (a) seminal vesicles, prostate gland, bulbourethral glands
   (b) activates the sperm, provides nutrients for sperm motility; provides sperm motility, produces buffers to counteract acid conditions

2. Seminal fluid is the fluid component of semen. Semen consists of seminal fluid, sperm, and enzymes.

3. *Emission* involves peristaltic contractions of the ampulla, pushing fluid and spermatozoa into the prostatic urethra. Contractions of the seminal vesicles and prostate gland move the seminal mixture into the membranous and penile walls of the prostate gland.

   Ejaculation occurs as powerful, rhythmic contractions of the ischiocavernosus and bulbocavernosus muscles push semen toward the external urethral orifice.

4. (a) Promotes the functional maturation of spermatozoa.
   (b) Maintains accessory organs of male reproductive tract.
   (c) Responsible for male secondary sexual characteristics.
   (d) Stimulates sexual behaviors and sexual drive.

5. (a) Serves as a passageway for the elimination of menstrual fluids.
   (b) Receives penis during coitus; holds sperm prior to passage into uterus.
   (c) In childbirth it forms the lower portion of the birth canal.

6. (a) *Arousal* – parasympathetic activation leads to an engorgement of the erectile tissues of the clitoris and increased secretion of the greater vestibular glands.
   (b) *Coitus* – rhythmic contact with the clitoris and vaginal walls provides stimulation that eventually leads to orgasm.
   (c) *Orgasm* – accompanied by peristaltic contractions of the uterine and vaginal walls and rhythmic contractions of the bulbocavernosus and ischiocavernosus muscles giving rise to pleasurable sensations.

7. Step 1: Formation of primary follicles
   Step 2: Formation of secondary follicle
   Step 3: Formation of a tertiary follicle
   Step 4: Ovulation
   Step 5: Formation and degeneration of the corpus luteum

8. (a) menses
   (b) proliferative phase
   (c) secretory phase

9. (a) human chorionic gonadotrophin (HCG)
   (b) relaxin
   (c) human placental lactogen (HPL)
   (d) estrogens and progestins

10. By the end of the sixth month of pregnancy the mammary glands are fully developed, and the glands begin to produce colostrum. This contains relatively more proteins and far less fat than milk, and it will be provided to the infant during the first two or three days of life. Many of the proteins are immunoglobulins that may help the infant ward off infections until its own immune system becomes fully functional.

F. Formation of Gametes: Gametogenesis

1. Spermatogenesis
   (A) spermatogonia
   (B) primary spermatocytes
   (C) secondary spermatocytes
   (D) spermatids
   (E) sperm

2. Oogenesis
   (A) Oogonia
   (B) primary oocyte
   (C) secondary oocyte
   (D) ovum

CHAPTER 21: *Development and Inheritance*

Part I: Objective Based Questions

Objective 1

1. C
2. B
3. D
4. capacitation
5. amphimixis
6. fertilization

Objective 2

1. C
2. A
3. B
4. C
5. C
6. A
7. D
8. A
9. cleavage
10. implantation
11. blastomeres
12. morula
13. ectoderm
14. mesoderm
15. endoderm

Objective 3

1. C
2. B
3. D
4. A
5. B

Objective 4

1. C
2. D
3. C
4. A
5. B
6. A

Objective 5

1. A
2. A
3. C
4. D

Objective 6

1. C
2. D
3. A
4. maturity
5. senescence

Objective 7

1. C
2. A
3. B
4. D
5. A
6. B
7. C
8. A
9. meiosis
10. gametogenesis
11. autosomal
12. homozygous
13. heterozygous

14. Color blindness is an X-linked trait. The Punnett square shows that sons produced by a normal father and a heterozygous mother will have a 50 percent chance of being color blind, while the daughters will all have normal color vision.

Paternal alleles \ Maternal alleles	X^C	X^c
X^C	X^CX^C	X^CX^c
Y	X^CY	X^cY (color blind)

15. The Punnett square reveals that 50 percent of their offspring have the possibility of inheriting albinism.

Paternal alleles \ Maternal alleles	a	a
A	Aa	Aa
a	aa (albino)	aa (albino)

16. Both the mother and father are heterozygous-dominant.
T–tongue roller t–non-tongue roller
The Punnett square reveals that there is a 25 percent chance of having children who are not tongue rollers and a 75 percent chance of having children with the ability to roll the tongue.

Paternal alleles \ Maternal alleles	T	t
T	TT (yes)	Tt (yes)
t	Tt (yes)	tt (no)

Part II: Chapter Comprehensive Exercises

A. Word Elimination

1. inheritance
2. embryology
3. gestation
4. placenta
5. colostrum
6. fetus
7. Marfan's syndrome
8. curly hair
9. gastrulation
10. amnion

B. Matching

1. E
2. H
3. A
4. J
5. B
6. I
7. C
8. F
9. D
10. G

C. Concept Map I - Fertilization and Development

1. zygote
2. germ layer
3. endoderm
4. muscle
5. yolk sac
6. allantois
7. progesterone
8. relaxin

Concept Map II - Cleavage and Blastocyst Formation

1. secondary oocyte
2. fertilization
3. zygote
4. 2-cell stage
5. 8-cell stage
6. morula
7. early blastocyst
8. implantation

D. Crossword Puzzle

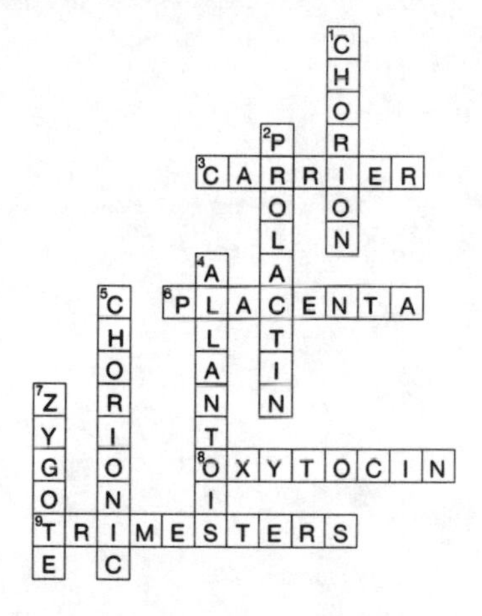

E. Short Answer Questions (Answers)

1. (a)
   - yolk sac
   - amnion
   - allantois
   - chorion

   (b)
   - endoderm and mesoderm
   - ectoderm and mesoderm
   - endoderm and mesoderm
   - mesoderm and trophoblast

2. (a) the respiratory rate goes up and the tidal volume increases; (b) the maternal blood volume increases; (c) the maternal requirements for nutrients increase; (d) the glomerular filtration rate increases; (e) the uterus increases in size.

3. (a) secretion of relaxin by the placenta - softens symphysis pubis; (b) weight of the fetus - deforms cervical orifice; (c) rising estrogen levels; (d) both b and c promote release of oxytocin.

4. (a) hypothalamus - increasing production of GnRH; (b) increasing circulatory levels of FSH and LH (ICSH) by the anterior pituitary; (c) FSH and LH initiate gametogenesis and the production of male or female sex hormones that stimulate the appearance of secondary sexual characteristics and behaviors.

5. (a) some cell populations grow smaller throughout life; (b) the ability to replace other cell populations decreases; (c) genetic activity changes over time; (d) mutations occur and accumulate.

6. (a) dilation stage (b) expulsion stage (c) placental stage

7. infancy, childhood, adolescence, maturity, senescence